## CONTRIBUTORS TO THIS VOLUME

K. L. Chopra

C. Ghosh

G. Hass

J. B. Heaney

W. R. Hunter

R. C. Kainthla

S. M. Ojha

D. K. Pandya

A. P. Thakoor

# Physics of Thin Films

*Advances in Research and Development*

*Edited by*

### GEORG HASS

*U.S. Army Electronics Research and Development Command
Night Vision and Electro-Optics Laboratory
Fort Belvoir, Virginia*

### MAURICE H. FRANCOMBE

*Research and Development Center
Westinghouse Electric Corporation
Pittsburgh, Pennsylvania*

### JOHN L. VOSSEN

*RCA Laboratories
David Sarnoff Research Center
Princeton, New Jersey*

VOLUME 12

1982

ACADEMIC PRESS
A Subsidiary of Harcourt Brace Jovanovich, Publishers

New York   London
Paris   San Diego   San Francisco   São Paulo   Sydney   Tokyo   Toronto

COPYRIGHT © 1982, BY ACADEMIC PRESS, INC.
ALL RIGHTS RESERVED.
NO PART OF THIS PUBLICATION MAY BE REPRODUCED OR
TRANSMITTED IN ANY FORM OR BY ANY MEANS, ELECTRONIC
OR MECHANICAL, INCLUDING PHOTOCOPY, RECORDING, OR ANY
INFORMATION STORAGE AND RETRIEVAL SYSTEM, WITHOUT
PERMISSION IN WRITING FROM THE PUBLISHER.

ACADEMIC PRESS, INC.
111 Fifth Avenue, New York, New York 10003

*United Kingdom Edition published by*
ACADEMIC PRESS, INC. (LONDON) LTD.
24/28 Oval Road, London NW1 7DX

LIBRARY OF CONGRESS CATALOG CARD NUMBER: 63–16561

ISBN 0-12-533012-X

PRINTED IN THE UNITED STATES OF AMERICA

82 83 84 85    9 8 7 6 5 4 3 2 1

# Contents

| | |
|---|---|
| Contributors to Volume 12 | vii |
| Preface | ix |
| Articles Planned for Future Volumes | xi |
| Contents of Previous Volumes | xiii |

### Reflectance and Preparation of Front Surface Mirrors for Use at Various Angles of Incidence from the Ultraviolet to the Far Infrared

#### *G. Hass, J. B. Heaney, and W. R. Hunter*

| | |
|---|---|
| I. Introduction | 2 |
| II. Reflectance Measurements and Preparation of Mirror Coatings | 3 |
| III. Reflectance of Metallic Front Surface Mirrors without Overcoating | 16 |
| IV. Mirror Coatings with Protective Layers and Reflectance-Enhancing Surface Films | 22 |
| V. Metal–Dielectric Mirrors for Use as Reflection-Type Filters | 42 |
| VI. Water Absorption in Evaporated Dielectric Films | 46 |
| References | 49 |

### Photoemissive Materials

#### *C. Ghosh*

| | |
|---|---|
| I. Introduction | 54 |
| II. The Mechanism of Photoemission | 54 |
| III. The Ag–O–Cs (S-1) Photocathode | 70 |
| IV. The Alkali Antimonides | 84 |
| V. Negative Electron Affinity Materials | 112 |
| VI. Applications of Photoemissive Materials | 140 |
| VII. Conclusions | 157 |
| References | 158 |

### Chemical Solution Deposition of Inorganic Films

#### *K. L. Chopra, R. C. Kainthla, D. K. Pandya, and A. P. Thakoor*

| | |
|---|---|
| I. Introduction | 168 |
| II. Spray Pyrolytic Process | 169 |
| III. Characteristic Features of the Spray Pyrolytic Process | 178 |

|      |                                                    |     |
|------|----------------------------------------------------|-----|
| IV.  | Multicomponent Doping and Alloying Effects         | 181 |
| V.   | Structural Properties                              | 187 |
| VI.  | Electrical and Optical Properties                  | 192 |
| VII. | Solution Growth Process                            | 201 |
| VIII.| Impurity and Dopant Effects                        | 211 |
| IX.  | Multicomponent Films                               | 212 |
| X.   | Oxide Films                                        | 213 |
| XI.  | Structure                                          | 214 |
| XII. | Transport Properties                               | 217 |
| XIII.| Some Large-Area Application                        | 223 |
| XIV. | Concluding Remarks                                 | 230 |
|      | References                                         | 232 |

## Plasma-Enhanced Chemical Vapor Deposition of Thin Films

### S. M. Ojha

|       |                                    |     |
|-------|------------------------------------|-----|
| I.    | Introduction                       | 237 |
| II.   | Deposition Techniques and Systems  | 238 |
| III.  | Preparation and Properties of Films| 246 |
| IV.   | Plasma Oxidation                   | 275 |
| V.    | Plasma Carburizing                 | 283 |
| VI.   | Glow-Discharge Nitriding           | 286 |
| VII.  | Conclusions                        | 289 |
|       | References                         | 289 |

| | |
|---|---|
| AUTHOR INDEX | 297 |
| SUBJECT INDEX | 312 |

## Contributors to Volume 12

Numbers in parentheses indicate the pages on which the authors' contributions begin.

K. L. CHOPRA (*167*), Thin Film Laboratory, Indian Institute of Technology, Delhi, New Delhi-110016, India

C. GHOSH (*53*), Bhabha Atomic Research Center, Optoelectronics Section, Bombay 400085, India

G. HASS* (*1*), U.S. Army Electronics Research and Development Command, Night Vision and Electro-Optics Laboratory, Fort Belvoir, Virginia 22060

J. B. HEANEY (*1*), National Aeronautics and Space Administration, Goddard Space Flight Center, Greenbelt, Maryland 20771

W. R. HUNTER (*1*), U.S. Naval Research Laboratory, Washington, D. C. 20375

R. C. KAINTHLA (*167*), Thin Film Laboratory, Indian Institute of Technology, Delhi, New Delhi-110016, India

S. M. OJHA (*237*), Standard Telecommunications Laboratories Limited, Harlow, Essex, England CM17 9NA

D. K. PANDYA (*167*), Thin Film Laboratory, Indian Institute of Technology, Delhi, New Delhi-110016, India

A. P. THAKOOR (*167*), Thin Film Laboratory, Indian Institute of Technology, Delhi, New Delhi-110016, India

*Present address: 7728 Lee Avenue, Alexandria, Virginia 22308.

# Preface

This twelfth volume of *Physics of Thin Films* contains four articles which, although emphasizing primarily optical topics, cover a wide range of preparative approaches, physics phenomena, and applications.

The first article by G. Hass, J. B. Heaney, and W. R. Hunter reviews recent progress in the development of metal coatings and protective layers for front surface mirrors used from the ultraviolet to the far infrared. New thin-film materials and deposition conditions are described, suitable for minimizing reflectance changes with angle of incidence. Novel oxide protective coatings offering enhanced chemical stability and mechanical durability have been developed.

In the second article, C. Ghosh presents a comprehensive treatment of the important technological field of photoemissive materials. After giving a rather detailed review of the physics of photoemission, the main classes of thin-film photoemitters, including Ag–O–Cs, alkali antimonides, and negative-electron affinity photocathodes are discussed. A description of field-assisted cathodes, such as transferred-electron structures and field-emission arrays, potentially suitable for wavelengths beyond 1.1 $\mu$m is also given.

The growing need for low-cost thin-film production processes has prompted increased interest in nonvacuum deposition approaches capable of high rate and large area. Spray pyrolysis is a promising method and in the third article on chemical solution deposition of inorganic films, K. L. Chopra and his co-workers discuss this technique and the solution growth technique and review their recent applications to the deposition of a wide range of semiconductor and insulator compounds. The techniques are suitable for II–VI and IV–VI sulfides and selenides, as well as many oxides, nitrides, and carbides. In particular, good success is reported for CdS-based compositions suitable for high-efficiency thin-film solar cells.

The fourth and final article by S. M. Ojha reviews recent developments in the field of plasma-enhanced chemical vapor deposition. The use of both plasma etching and plasma deposition in the processing of semiconductor devices and circuits is becoming especially attractive because both are compatible, low-temperature, "dry-processing" techniques. This article describes the experimental conditions required for a range of element and compound materials and points to some of the unusual film properties and structures achieved by this approach.

G. Hass
M. H. Francombe
J. L. Vossen

# Articles Planned for Future Volumes

The Activated Reactive Evaporation Process
  *R. F. Bunshah*

Ionized Cluster Beam Deposition of Thin Films
  *T. Takagi*

Ion Plating
  *D. G. Teer*

Contacts to III–V Semiconductors
  *J. M. Woodall and T. Jackson*

Laser Coatings
  *H. E. Bennett and J. M. Bennett*

Superconducting Thin Films
  *M. Ashkin, J. A. Gavaler, M. A. Jonocka, and J. H. Parker*

Ferroelectric Films
  *M. H. Francombe, S. Y. Wu, and W. J. Takei*

# Contents of Previous Volumes

## Volume 1

**Ultra-High Vacuum Evaporators and Residual Gas Analysis**
*Hollis L. Caswell*

**Theory and Calculations of Optical Thin Films**
*Peter H. Berning*

**Preparation and Measurement of Reflecting Coatings for the Vacuum Ultraviolet**
*Robert P. Madden*

**Structure of Thin Films**
*Rudolf E. Thun*

**Low Temperature Films**
*William B. Ittner, III*

**Magnetic Films of Nickel-Iron**
*Emerson W. Pugh*

AUTHOR INDEX · SUBJECT INDEX

## Volume 2

**Structural Disorder Phenomena in Thin Metal Films**
*C. A. Neugebauer*

**Interaction of Electron Beams with Thin Films**
*C. J. Calbick*

**The Insulated-Gate Thin-Film Transistor**
*Paul K. Weimer*

**Measurement of Optical Constants of Thin Films**
*O. S. Heavens*

**Antireflection Coatings for Optical and Infrared Optical Materials**
*J. Thomas Cox and Georg Hass*

**Solar Absorptance and Thermal Emittance of Evaporated Coatings**
*Louis F. Drummeter, Jr. and Georg Hass*

**Thin Film Components and Circuits**
*N. Schwartz and R. W. Berry*

AUTHOR INDEX · SUBJECT INDEX

## Volume 3

**Film-Thickness and Deposition-Rate Monitoring Devices and Techniques for Producing Films of Uniform Thickness**
*Klaus H. Behrndt*

**The Deposition of Thin Films by Cathode Sputtering**
*Leon I. Maissel*

**Gas-Phase Deposition of Insulating Films**
*L. V. Gregor*

**Methods of Activating and Recrystallizing Thin Films of II–VI Compounds**
*A. Vecht*

**The Mechanical Properties of Thin Condensed Films**
*R. W. Hoffman*

**Lead Salt Detectors**
*D. E. Bode*

AUTHOR INDEX · SUBJECT INDEX

## Volume 4

**Precision Measurements in Thin Film Optics**
*H. E. Bennett and Jean M. Bennett*

**Nucleation Processes in Thin Film Formation**
*J. P. Hirth and K. L. Moazed*

**Evaporated Single-Crystal Films**
*J. W. Matthews*

**The Growth and Structure of Electrodeposits**
*Kenneth R. Lawless*

**Thin Glass Films**
*W. A. Pliskin, D. R. Kerr, and J. A. Perri*

**Hot-Electron Transport and Electron Tunneling in Thin Film Structures**
*C. R. Crowell and S. M. Sze*

AUTHOR INDEX · SUBJECT INDEX

## Volume 5

**Interference Photocathodes**
*D. Kossel, K. Deutscher, and K. Hirschberg*

Design of Multilayer Interference Filters
*Alfred Thelen*

Oxide Layers Deposited from Organic Solutions
*H. Schroeder*

The Preparation and Properties of Semiconductor Films
*M. H. Francombe and J. E. Johnson*

The Preparation of Films by Chemical Vapor Deposition
*W. M. Feist, S. R. Steele, and D. W. Readey*

AUTHOR INDEX · SUBJECT INDEX

## Volume 6

Anodic Oxide Films
*C. J. Dell'Oca, D. L. Pulfrey, and L. Young*

Size-Dependent Electrical Conduction in Thin Metal Films and Wires
*D. C. Larson*

Optical Properties of Metallic Films
*F. Abelès*

Interactions in Multilayer Magnetic Films
*Arthur Yelon*

Diffusion in Metallic Films
*C. Weaver*

AUTHOR INDEX · SUBJECT INDEX

## Volume 7

Electron Diffraction Analysis of the Local Atomic Order in Amorphous Films
*D. B. Dove*

The Preparation and Use of Unbacked Metal Films as Filters in the Extreme Ultraviolet
*W. R. Hunter*

Properties and Applications of III–V Compound Films Deposited by Liquid Phase Epitaxy
*H. Kressel and H. Nelson*

Electromigration in Thin Films
*F. M. d'Heurle and R. Rosenberg*

Built-Up Molecular Films and Their Applications
*V. K. Srivastava*

AUTHOR INDEX · SUBJECT INDEX

## Volume 8

Dielectric Film Materials for Optical Applications
*Elmar Ritter*

Inhomogeneous and Coevaporated Homogeneous Films for Optical Applications
*R. Jacobsson*

Discontinuous and Cermet Films
*Z. H. Meiksin*

Electrical Conduction in Disordered Nonmetallic Films
*A. K. Jonscher and R. M. Hill*

Topologically Structured Thin Films in Semiconductor Device Operation
*H. C. Nathanson and J. Guldberg*

SUBJECT INDEX

## Volume 9

Transparent Conducting Films
*J. L. Vossen*

Metal-Dielectric Interference Filters

Surface Plasma Oscillations and Their Applications
*H. Raether*

Magnetic Bubble Films
*P. Chaudhari, J. J. Cuomo, R. J. Gambino, and E. A. Giess*

AUTHOR INDEX · SUBJECT INDEX

## Volume 10

Spectrally Selective Surfaces for Photothermal Solar Energy Conversion
*R. E. Hahn and B. O. Seraphin*

The Use of Evaporated Films for Space Applications—Extreme Ultraviolet Astronomy and Temperature Control of Satellites
*G. Hass and W. R. Hunter*

**Scattering by All-Dielectric Multilayer Bandpass Filters and Mirrors for Lasers**
  *Jay M. Eastman*

**Thin Films for Integrated Optics**
  *D. B. Ostrowsky and C. Vanneste*

**Correction of Optical Elements by the Addition of Evaporated Films**
  *J. R. Kurdock and R. R. Austin*

AUTHOR INDEX · SUBJECT INDEX

**Volume 11**

**Preparation and Testing of Reflectance Coatings for Diffraction Gratings in the Extreme Ultraviolet**
  *W. R. Hunter and G. Hass*

**Progress, Problems, and Applications of Molecular-Beam Epitaxy**
  *Colin E. C. Wood*

**Thin-Film IV–VI Semiconductor Photodiodes**
  *H. Holloway*

**The Universal Dielectric Response: A Review of Data and Their New Interpretation**
  *A. K. Jonscher*

AUTHOR INDEX · SUBJECT INDEX

# Physics of Thin Films

*Advances in Research and Development*

VOLUME 12

# Reflectance and Preparation of Front Surface Mirrors for Use at Various Angles of Incidence from the Ultraviolet to the Far Infrared

G. HASS

*U. S. Army Electronics Research and Development Command*
*Night Vision and Electro-Optics Laboratory*
*Fort Belvoir, Virginia*\*

J. B. HEANEY

*National Aeronautics and Space Administration*
*Goddard Space Flight Center*
*Greenbelt, Maryland*

W. R. HUNTER

*U.S. Naval Research Laboratory*
*Washington, D.C.*

| | |
|---|---|
| I. Introduction | 2 |
| II. Reflectance Measurements and Preparation of Mirror Coatings | 3 |
|     1. Reflectance Measurements | 3 |
|     2. Preparation Techniques | 8 |
| III. Reflectance of Metallic Front Surface Mirrors without Overcoatings | 16 |
| IV. Mirror Coatings with Protective Layers and Reflectance-Enhancing Surface Films | 22 |
|     1. General Remarks | 22 |
|     2. Evaporated Al Mirrors with Overcoatings | 25 |
|     3. Reflection-Type Polarizers for the 10.6-$\mu$m $CO_2$ Laser Emission Line Using $Al_2O_3$-Coated Al Mirrors | 32 |
|     4. Protected Al Mirrors with High Reflectance in the 8–12 $\mu$m Region from Normal to High Angles of Incidence | 35 |
|     5. Adherent Ag Mirrors with Protective Coatings | 36 |
|     6. Reflectance-Enhancing Coatings for Rh and Other Metallic Front Surface Mirrors | 39 |
| V. Metal–Dielectric Mirrors for Use as Reflection-Type Filters | 42 |
| VI. Water Absorption in Evaporated Dielectric Films | 46 |
| References | 49 |

\* Present address: 7728 Lee Avenue, Alexandria, Virginia 22308.

## I. Introduction

Today, reflecting optics are extensively employed in optical devices and equipment designed to perform studies and measurements in the ultraviolet (UV), visible, and infrared (IR). Their increased use, inspired in large part by the novel requirements of space, laser, and IR applications, has stimulated a great amount of research into the properties of coatings for front surface mirrors. Important progress toward the development of reflecting coatings with increased reflectance, improved mechanical and chemical durability, and better adherence to various substrates has been made in recent decades, and research in pursuit of these goals continues. The main factors responsible for the improvements made in the preparation of mirror coatings are the following:

(1) The development of vacuum systems that make it possible to deposit films at very low pressures, or in the presence of oxygen and other gases of well-controlled pressures for reactive evaporation

(2) New deposition techniques such as electron gun evaporation, which allows the deposition of many new optical coatings such as high-melting-point metals and oxides

(3) Improved devices for monitoring film thicknesses during film deposition

(4) The availability of purer and well-outgassed starting materials

The preparation and reflectance of coatings for vacuum ultraviolet (VUV) spectroscopy and space astronomy have been reported in recent summary articles ($1$–$3$). This article presents data on the UV, visible, and IR reflectances of the most frequently used mirror coatings, Al, Ag, Au, and Rh, and discusses their reflectance behavior at various angles of incidence. General rules for predicting the effect of incidence angle on the reflectance and polarization of the reflected beam are given for metal coatings.

For many mirror applications utilizing highly reflecting metal films, hard, transparent, and adherent single and multilayer overcoatings must be applied to improve the chemical and mechanical durability of the mirror coatings. The most frequently used protective layers are films of silicon oxides: SiO, $SiO_x$, $SiO_2$ ($4$–$7$), and $Al_2O_3$ ($8$). For most purposes, layers of these materials 1000–2000 Å thick are sufficient to give adequate protection against abrasion. It will be shown, however, that such silicon oxide- and $Al_2O_3$-protected mirrors have essentially the same high reflectance as unprotected surfaces in the 8–12 $\mu$m region of the IR at close to normal incidence, but greatly decreased reflectance at angles greater than 40° ($9$, $10$). Therefore, mirrors of this type are unsuitable for use in devices at 45° angle of incidence, or in scanning optics, in this wavelength region. The 8–12 μm wavelength band is emphasized because it is one of the atmospheric windows. The drastic IR reflectance decrease at higher angles of incidence is caused by the fact that the optical constants $n$ and $k$ of the above oxide films have values of less than 1 in this region. Since only the $R_p$ component is responsible for the reflectance decrease, film combinations of this type can be used to produce highly efficient reflection-type polarizers for the IR, e.g., for the $CO_2$ laser emission line at 10.6 μm (11). Protected mirror coatings that have high reflectance in the 8–12 μm region from normal to high angles of incidence will also be described (12). Since many dielectric overcoatings used on front surface mirrors absorb water when exposed to air, the effect of this water absorption on the reflectance of overcoated mirrors is discussed (13, 14). The ability to employ metal–oxide combinations in the production of front surface mirrors with spectrally selective reflectances for use as reflection-type filters in optical devices is also described (15).

## II. Reflectance Measurements and Preparation of Mirror Coatings

### 1. Reflectance Measurements

The reflectance of front surface mirrors should be as high as possible for the wavelength region and angles of incidence for which they are to be used. To ensure that high reflectances have been achieved, accurate reflectance measurements of the mirrors must be made. These measurements enable one to predict the efficiency of devices and equipment using reflecting optics.

When a beam of radiant energy is incident on an opaque mirror coating, part of it is reflected and part is absorbed. Reflectance is defined as the ratio of the reflected radiant energy $I$ to the incident radiant energy $I_0$. It is usually denoted by $R = I/I_0$. The reflected beam can consist of two parts, the specularly reflected component, which is an extension of the incident beam as defined by the geometrical laws of reflection, and the diffusely reflected component, which is the part of the radiant energy scattered from the surface due to roughness. Since we are dealing in this article only with rather smooth reflecting coatings deposited onto well polished substrates, all discussions are limited to the treatment of the specular reflectance of front surface mirrors.

In addition to the term reflectance, defined above, another term frequently employed is reflectivity. We suggest that this term should be used to define the ideal reflectance of a material; for example, the reflectivity of Ag at

$\lambda = 5000$ Å is 97.9%. The term reflection represents a process and should not be used in connection with numbers.

Methods for measuring $R$ can be divided into two groups: (1) methods that measure the reflectance without a comparison standard; devices of this type are called absolute reflectometers; and (2) techniques that determine $R$ of a sample by comparison with a standard mirror of known reflectance.

There are various techniques for determining $R$ without the use of a reflectance standard. A combination goniometer–reflectometer is the most straightforward device for determining $R$ of mirror samples at close to normal and over a range of other angles of incidence. Figure 1 (16) is a schematic diagram of such a reflectometer. Because no reference sample is used, the detector must measure the radiant energy incident on the mirror as well as that reflected from the mirror. $P'$ and $P$ represent the detector at two different positions—$P'$ for the incident measurement and $P$ for the reflected measurement. $M$ represents the mirror in position for the reflected measurement and $M'$ shows the mirror moved out of the optical path to enable the incident measurement to be made. In this figure, the detector and mirror rotate about a common axis so that $P$ is always equidistant from the center of $M$ and from the exit slit of the monochromator. Such an arrangement is convenient for measuring the angle of incidence at which the reflectance is determined. It is also necessary if the radiation beam is not parallel because of the inherent nonuniformity of detectors. If the slit–detector distances for $P'$ and $P$ are different, the change of illumination of the detector due to nonparallelism of the incident radiation could give rise

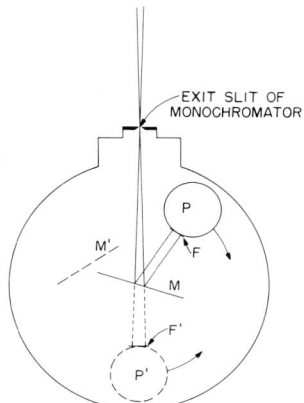

Fig. 1. Schematic diagram of a reflectometer for absolute reflectance measurements over a wide range of incidence angles; $P$ ($P'$) and $M$ ($M'$) represent the radiation detector and mirror, respectively.

to an error in $I/I_0$. The cause of errors in measuring $R$ with this type of reflectometer has been diagnosed by Hunter (17).

The measurement of $R$ at angles other than normal incidence is of interest if the mirrors are used at various angles, and also for the determination of the optical constants from the reflectance data obtained at many different angles of incidence from 10° to 80° (18–20). The goniometer–reflectometer can easily be adapted for use in a vacuum system so that $R$ measurements can be made *in situ*. Madden and Canfield (21) used such equipment to measure $R$ of freshly deposited metal films before and after exposure to air, and also used it to determine the VUV optical properties of various metals. By employing different light sources, monochromators, and detectors, the use of this instrument can be extended to $R$ measurements from the UV to the IR. Measurements before and after air inlet furnish important information on the effect of oxidation and water absorption on the reflectance of metal mirrors with and without overcoatings.

The Strong-type reflectometer (22) shown in Fig. 2 is also capable of giving absolute reflectance measurements. In the left panel of the figure, radiation is incident on a comparison mirror before being reflected to the photometer. Another identical comparison mirror is provided, so that when the test mirror is in place the path length is the same, as is the angle of incidence on the test and comparison mirrors. One thus obtains a measurement of $R^2$ of the test mirror at a fixed angle of incidence, usually near normal incidence. In practice only one comparison mirror is used, in order to avoid the problems inherent in making two *identical* comparison mirrors. The

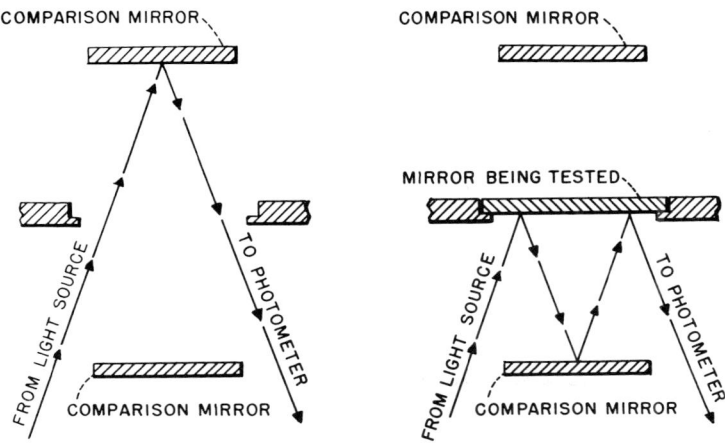

FIG. 2. A Strong-type reflectometer, with which the absolute value of $R^2$ can be measured. This instrument is usually capable of measuring reflectances at only one angle of incidence.

comparison-test mirror assembly is then rotated so that one position is equivalent to the situation shown in the left panel and the second position to that in the right panel.

A precision Strong-type reflectometer suitable for measuring the absolute $R$ values of mirrors with high accuracy at close to normal incidence has been described by Bennett and Koehler (23) and Bennett and Bennett (24). With this instrument they measured $R$ of highly reflecting samples in the region 0.3–32 $\mu$m with an absolute accuracy of 0.1% and an average deviation of 0.04%. The precision of this instrument has been improved to $\pm 0.01\%$ in reflectance (25). This technique can be used to provide data for reflectance standards at normal incidence. In a review article (24), Bennett and Bennett describe the design of many other devices suitable for measuring the reflectance of mirror coatings.

Multiple-pass reflectometers have become one of the most important tools for measuring the absolute reflectance of highly reflecting mirrors with great accuracy. Gates et al. (26) used a varying number of multiple reflections from identical parallel specularly reflecting mirrors to determine the absolute infrared reflectances of various metals at incidence angles between 20° and 60° with an estimated accuracy of 0.2%. Figure 3 is a schematic diagram of their apparatus. The source is directed and collimated by $M_1$, $M_2$, and $M_3$; $M_4$ and $M_5$ are the two identical and parallel test mirrors. After multiple reflections between the test mirrors, $M_6$ and $M_7$ direct the radiation beam to an IR spectrometer.

Harris and Fowler (27), using a similar technique, measured the reflectance

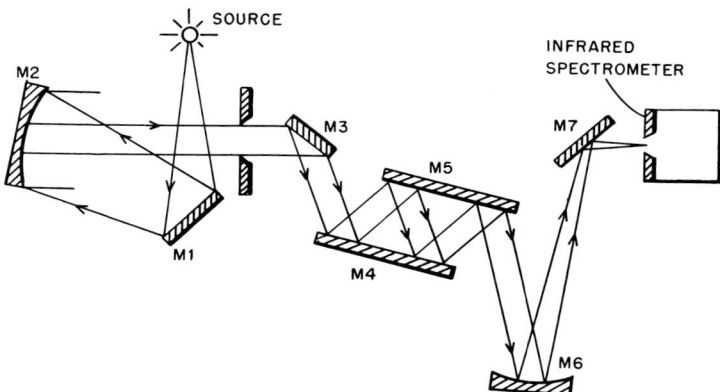

FIG. 3. A multiple-pass reflectometer designed for the measurement of absolute reflectance in the infrared. M4 and M5 have identical mirror surfaces to be measured and are always parallel. By varying the spacing between M4 and M5 the number of reflections can be changed, and by varying their orientation the angle of incidence can be changed.

of Au in the region 8.5–84 μm. Herriott and Schulte (28) used a large number of multiple reflections to measure with very high precision the extremely high reflectance of all-dielectric mirrors at normal incidence. Perry (29) used the Herriott–Schulte reflectometer to determine a reflectance of 99.8% for numerous alternating layers of $ZnS/ThF_4$. The measurements were made using 100 passes. Arnon and Baumeister (30) have described a versatile high-precision multiple-pass reflectometer. In this instrument, the mirrors can be either plane or curved with a few diopters of power, and the angle of incidence can vary from 5° to 70°. Multilayer stacks consisting of 33 layers of $Ta_2O_5/SiO_2$ showed a reflectance of 99.96% ± 0.02% at $\lambda = 520$ nm. Arnon and Baumeister also give many references to other articles that describe various types of multiple-pass reflectometers.

Since reflectance values determined with absolute reflectometers are usually made wavelength-by-wavelength, the time required to obtain reflectance curves over an extended wavelength region can be quite long. Therefore, most reflectance versus wavelength measurements of samples are made by comparison with calibrated standards in recording double-beam spectrophotometers that are commercially available, such as those produced by Perkin-Elmer and Beckman. Figure 4 is a schematic diagram of such a double-beam spectrophotometer. A common source S illuminates two identical optical paths. In one path the sample and reference mirror can be interchanged. After the reflections, the radiation beams enter a spectrophotometer. After dispersion, the ratio of the two beam intensities is obtained and usually presented on a strip chart or $x$–$y$ recorder. Sometimes the beam from the source is dispersed and then split before traversing the optical path containing the mirrors.

Since the reflectance data obtained with these instruments are relative to that of a calibrated standard, it is necessary to have reliable standards on hand. Mirror coatings of Rh, freshly deposited Al and Au, measured with

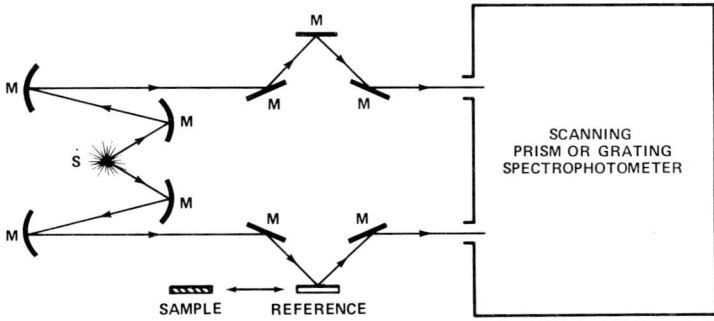

FIG. 4. Schematic diagram of a double-beam spectrophotometer.

a high-precision reflectometer, are very suitable standards. With this method, highest accuracy is obtained by leaving the mirror in one beam untouched while using the second beam to record the $R$ curves of the standard and samples to be measured. In this way the accuracy of relative $R$ measurements can be better than 0.5%, and an average deviation of less than 0.1% can be obtained by the use of an expanded $R$ scale. However, there is a difficulty associated with these devices. The reflectance is usually determined for a particular angle of incidence close to normal. It is only recently that additional accessories have been developed to measure $R$ at greater angles of incidence.

## 2. Preparation Techniques

By far the most widely used technique for depositing reflecting coatings is evaporation in high vacuum. With no other method can films of highest reflectance and of any desired thickness be prepared with such complete control. By the use of suitable shutters and rotating sector wheels, and by movement of the mirror substrates during the evaporation, films of extremely uniform thickness over large areas may be obtained. This article is therefore confined to reflecting coatings prepared by high-vacuum evaporation performed at various deposition rates and pressures onto substrates of various temperatures.

Aluminum is the most frequently used metal for depositing reflecting coatings on front surface mirrors because it has high reflectance from the vacuum UV to the far IR, adheres well to glass and other substrates, and does not tarnish in normal air. It is also easy to evaporate from helical tungsten coils or by electron bombardment. Obviously, Al coatings are especially important for astronomical mirrors and reflection gratings for which high reflectance in the UV is required. In order to produce highly reflecting films of materials, such as Al, that readily adsorb oxygen and other gases, extreme care must be taken to insure that the evaporated films are not contaminated by residual gases present during the deposition. It is well known that Al films of highest reflectance can be produced in conventional evaporators only using extremely high deposition rates and the purest grade of Al (99.999% pure) as the starting material (*16, 31, 32*). Highly reflecting Al films can also be produced by slower evaporation in ultrahigh-vacuum systems (*33, 34*). Both of these methods eliminate the absorption of impurities during the deposition. Hutcheson *et al.* (*35*) have made a comparison of the reflectances of Al films evaporated in conventional and ultrahigh-vacuum systems. In both of their vacuum systems, the evaporation sources consisted of six multistrand helical tungsten coils, which were heated simultaneously. The coils were outgassed and freed from impurities before being charged

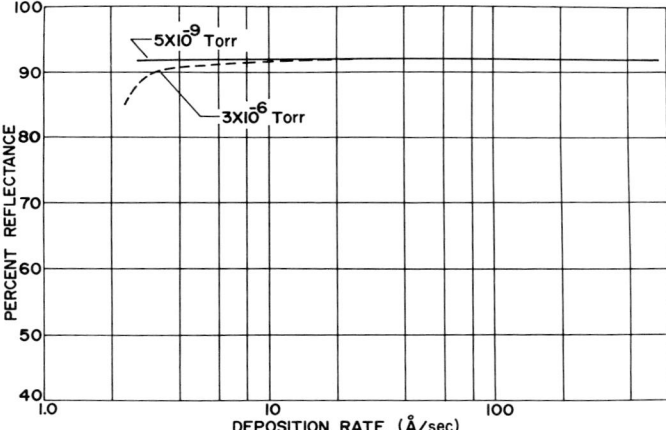

FIG. 5. The reflectance of Al as a function of deposition rate in high- and ultrahigh-vacuum conditions, $3 \times 10^{-6}$ and $5 \times 10^{-9}$ Torr, respectively. Measurements were made at a wavelength of 400 nm. All films aged 24 hr in air.

with 99.999% pure Al. This was followed by a second heating to melt the Al and to remove its adsorbed gases. With this arrangement, Al can be deposited at rates up to 1000 Å sec$^{-1}$ at a distance of about 50 cm. The effect of deposition rate on the reflectance at $\lambda = 400$ and 200 nm of Al films deposited at $3 \times 10^{-6}$ and $5 \times 10^{-9}$ Torr is shown in Fig. 5 and Fig. 6, respectively.

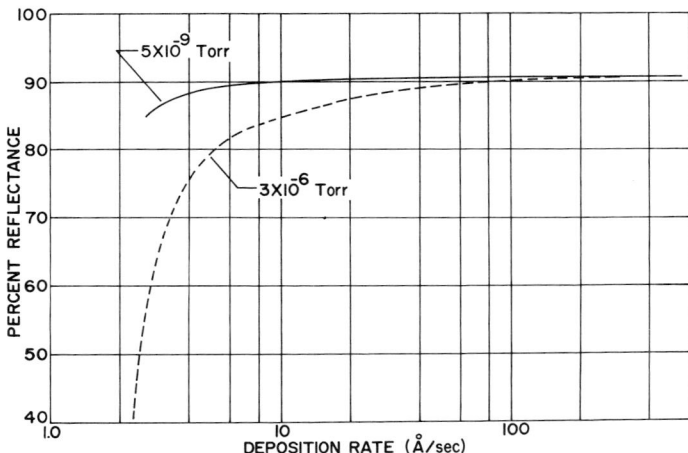

FIG. 6. The reflectance of Al as a function of deposition rate in high- and ultrahigh-vacuum conditions, $3 \times 10^{-6}$ and $5 \times 10^{-9}$ Torr, respectively. Measurements were made at a wavelength of 200 nm. All films aged 24 hr in air.

For Al films deposited at rates of 300 Å sec$^{-1}$ and greater there is no measurable difference in reflectance of the two types of films. However, if lower deposition rates are used, the reflectance of Al films deposited at $5 \times 10^{-9}$ Torr is much less dependent on the deposition rate than that of films condensed at $3 \times 10^{-6}$ Torr. At $\lambda = 400$ nm the reflectance difference is still quite small and is noticeable only at rates lower than 20–30 Å sec$^{-1}$. At $\lambda = 200$ nm the reflectance difference of the two types of films becomes very pronounced, and can be as high as 30% if deposition rates of about 3 Å sec$^{-1}$ are used. At such low rates even films condensed at $5 \times 10^{-9}$ Torr show a considerable decrease in reflectance. Since for most metals high deposition rates result in denser films with smoother surfaces (36), rapid evaporation is recommended for almost all mirror coatings.

The fact that evaporated Al does not tarnish in normal atmosphere is the result of the good protective qualities of its natural oxide film, which grows to an ultimate thickness of about 40 Å (37) in air. This thin oxide film results in a negligible loss of the Al reflectance for wavelengths longer than 1.0 μm and a loss of only 0.3% at 500 nm. However, this is not true for the VUV region, in which the oxide film causes a drastic decrease in reflectance (38).

A detailed description of a 2-m evaporator suitable for coating large mirrors with Al and Al plus overcoatings has been published by Bradford *et al.* (39). A schematic diagram of the evaporator with some of its features is shown in Fig. 7. The coating chamber is 2 m in diameter and about 2 m high. It is constructed of 304 stainless steel and consists of a vertical cylinder

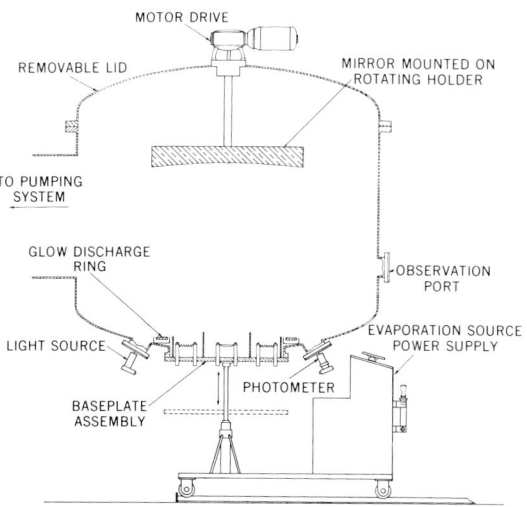

FIG. 7. Schematic diagram of a 2-m-diameter evaporator suitable for coating large mirrors.

with a curved bottom and a flanged removable lid. A flanged opening at the center of the bottom is fitted with a 75-cm-diameter baseplate for supporting the evaporation sources. The sources are connected to copper leads large enough to carry several thousand amperes without excessive heating. This permits the use of heavy heating elements and makes it possible to deposit Al films from helical tungsten coils at very high rates using a low-voltage 15-kW power supply. For evaporating platinum and iridium films a 10-kW electron gun has been used.

A mechanical feedthrough shaft with a motor drive support, located in the center of the lid, allows mirror forms to be rotated during the film deposition in order to achieve coatings of uniform thickness over large areas. Three viewing ports at eye level, shielded from the evaporants with solenoid-operated Venetian-blind-type shutters, allow the observation of the interior of the chamber during the evaporation process. Two quartz-glass windows adjacent to the baseplate opening allow film thicknesses to be monitored during their deposition by reflectance measurements with monochromatic light. A visible or UV light source can be used. The light source and photomultiplier housing of the reflectance monitoring system are indicated in Fig. 7. Three additional openings in the bottom of the chamber accommodate feedthroughs for vacuum gauges, electrical and mechanical controls, and gas inlets and needle valves. A circular aluminum glow-discharge cathode consisting of a 10-cm-wide aluminum ring is placed, insulated from the grounded tank, at the bottom of the chamber. It is connected to a dc high-voltage power supply capable of furnishing 15 kV and 500 mA. A glow discharge produced with 6 kV and 500 mA was found to be sufficient to clean the substrates before the coatings are applied.

Shutters are placed over the evaporation sources to avoid contamination of the mirror substrates while the filaments and boats with the evaporants are being outgassed and brought to evaporation temperature. The chamber is lined with interchangeable stainless-steel shields to facilitate removal of accumulated coating material.

Methods for producing Al mirrors and overcoated Al mirrors with diameters greater than 50 cm must necessarily start with a consideration of the thickness uniformity of the aluminum coating and its influence on the mirror contour. Using a circular array of filaments about 55 cm in diameter to deposit an Al film 1000 Å thick at the center of a 1-m-diameter mirror located 1 m above the vapor sources results in a thickness distribution that is nearly symmetrical about the center of the mirror and decreases to about 600 Å at the edge. This thickness is still sufficient to ensure high reflectance for all wavelengths longer than 1000 Å, but significantly alters the contour of a mirror figured to 1/50 of a wavelength at $\lambda = 5000$ Å. For accurately figured mirrors, therefore, a technique for correcting the aluminum thickness

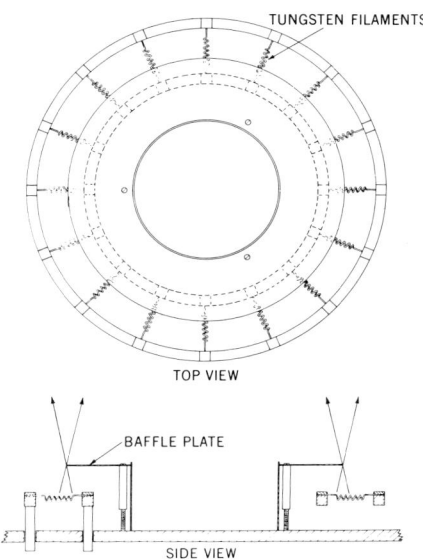

FIG. 8. Arrangement of filaments and baffle plate used in the 2-m evaporator of Fig. 7 for producing a uniform thickness of Al over a diameter of 1 m.

distribution must be employed. The method selected for correcting the aluminum thickness consists of placing a circular baffle above the filaments at a height that partially masks the center of the mirror, as illustrated in Fig. 8 (*39*). The correct diameter and height were estimated by geometrical construction and refined by experiment. Typical results for the thickness distributions, obtained with and without the baffle plate, are shown in Fig. 9 (*39*). Thicknesses for the thick films, as determined interferometrically using the Tolansky method, have an average value of 720 Å with a root-mean-square error of ±60 Å. The distribution for the thinner films, as determined from transmittance measurements corrected for oxidation effects, yielded an average of 175 Å with a variation of approximately ±15 Å. Data for the thinner films were included because aluminum films in the 100–200 Å thickness range may be used for producing large-diameter beam splitters or UV filters.

The arrangement used for depositing Al and dielectric films of uniform thickness on larger mirrors and for monitoring the thickness of the dielectric films during deposition is illustrated in Fig. 10 (*39*). The dielectric is evaporated from a centrally located shallow tungsten boat and the array of filaments described above is employed for the deposition of Al. Behrndt (*40*)

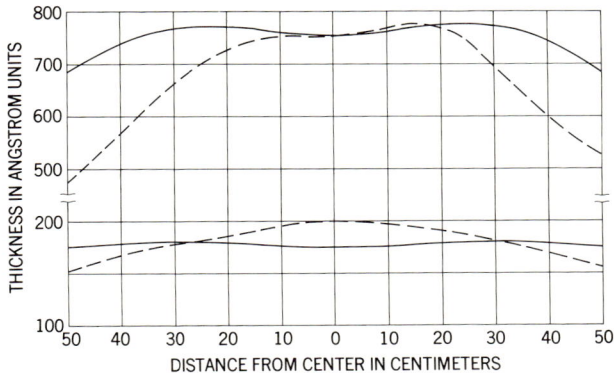

FIG. 9. Typical thickness distributions over a 1-m diameter, obtained using the baffle–filament arrangement of Fig. 8. The dashed lines show the distribution without, and the solid lines with, the baffle plate.

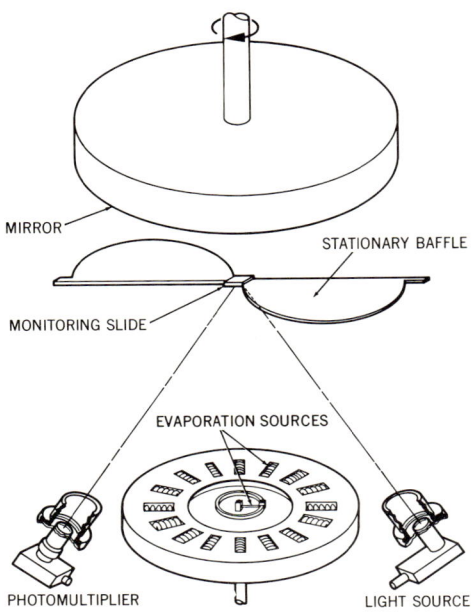

FIG. 10. Arrangement of evaporator sources and baffle used in the 2-m evaporator of Fig. 7 to achieve uniform Al + dielectric coatings. The mirror to be coated is rotated behind a stationary baffle.

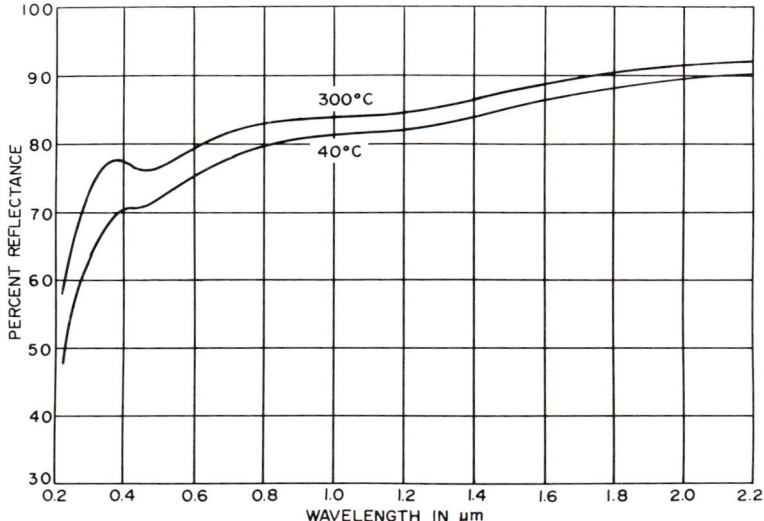

FIG. 12. Reflectance of opaque Rh films deposited on glass at 40° and 300°C in the wavelength region 0.2–2.2 μm.

glass and fused $SiO_2$ can be overcome by evaporating a binding layer of nichrome or chromium onto the substrate before the Ag or Au mirror coating is deposited. The smoothness of all evaporated metallic mirror films is also affected by the vapor angle of incidence during evaporation. Holland (46) showed that diffusely reflecting surfaces form easily as the vapor incidence angle is increased. The preparation and properties of dielectric overcoatings are discussed in connection with the metallic reflecting films on which they are used.

## III. Reflectance of Metallic Front Surface Mirrors without Overcoatings

Table I lists the UV, visible, and IR normal-incidence reflectance of freshly deposited, opaque metal films of Ag, Al, Au, Cu, Rh, and Pt, as reported by Hass (47). The only film material that has high reflectance in all these regions is Al. Its high reflectance of about 90% extends even into the vacuum UV down to about 1000 Å if its surface is completely free of

## TABLE I

Percent Reflectance at Normal Incidence of Freshly Evaporated Mirror Coatings of Aluminum, Silver, Gold, Copper, Rhodium, and Platinum from the Ultraviolet to the Infrared[a]

| $\lambda(\mu m)$ | Al | Ag | Au | Cu | Rh | Pt |
|---|---|---|---|---|---|---|
| 0.220 | 91.5 | 28.0 | 27.5 | 40.4 | 57.8 | 40.5 |
| 0.240 | 91.9 | 29.5 | 31.6 | 39.0 | 63.2 | 46.9 |
| 0.260 | 92.2 | 29.2 | 33.6 | 35.5 | 67.7 | 51.5 |
| 0.280 | 92.3 | 25.2 | 37.8 | 33.0 | 70.7 | 54.9 |
| 0.300 | 92.3 | 17.6 | 37.7 | 33.6 | 73.4 | 57.6 |
| 0.315 | 92.4 | 5.5 | 37.3 | 35.5 | 75.0 | 59.4 |
| 0.320 | 92.4 | 8.9 | 37.1 | 36.3 | 75.5 | 60.0 |
| 0.340 | 92.5 | 72.9 | 36.1 | 38.5 | 76.9 | 62.0 |
| 0.360 | 92.5 | 88.2 | 36.3 | 41.5 | 78.0 | 63.4 |
| 0.380 | 92.5 | 92.8 | 37.8 | 44.5 | 78.1 | 64.9 |
| 0.400 | 92.4 | 95.6 | 38.7 | 47.5 | 77.4 | 66.3 |
| 0.450 | 92.2 | 97.1 | 38.7 | 55.2 | 76.0 | 69.1 |
| 0.500 | 91.8 | 97.9 | 47.7 | 60.0 | 76.6 | 71.4 |
| 0.550 | 91.5 | 98.3 | 81.7 | 66.9 | 78.2 | 73.4 |
| 0.600 | 91.1 | 98.6 | 91.9 | 93.3 | 79.7 | 75.2 |
| 0.650 | 90.5 | 98.8 | 95.5 | 96.6 | 81.1 | 76.4 |
| 0.700 | 89.7 | 98.9 | 97.0 | 97.5 | 82.0 | 77.2 |
| 0.750 | 88.6 | 99.1 | 97.4 | 97.9 | 82.6 | 77.9 |
| 0.800 | 86.7 | 99.2 | 98.0 | 98.1 | 83.1 | 78.5 |
| 0.850 | 86.7 | 99.2 | 98.2 | 98.3 | 83.4 | 79.5 |
| 0.900 | 89.1 | 99.3 | 98.4 | 98.4 | 83.6 | 80.5 |
| 0.950 | 92.4 | 99.3 | 98.5 | 98.4 | 83.9 | 80.6 |
| 1.0 | 94.0 | 99.4 | 98.6 | 98.5 | 84.2 | 80.7 |
| 1.5 | 97.4 | 99.4 | 99.0 | 98.5 | 87.7 | 81.8 |
| 2.0 | 97.8 | 99.4 | 99.1 | 98.6 | 91.4 | 81.8 |
| 3.0 | 98.0 | 99.4 | 99.3 | 98.6 | 95.0 | 90.6 |
| 4.0 | 98.2 | 99.4 | 99.4 | 98.7 | 95.8 | 93.7 |
| 5.0 | 98.4 | 99.5 | 99.4 | 98.7 | 96.4 | 94.9 |
| 6.0 | 98.5 | 99.5 | 99.4 | 98.7 | 96.8 | 95.6 |
| 7.0 | 98.6 | 99.5 | 99.4 | 98.7 | 97.0 | 95.9 |
| 8.0 | 98.7 | 99.5 | 99.4 | 98.8 | 97.2 | 96.0 |
| 9.0 | 98.7 | 99.5 | 99.4 | 98.8 | 97.4 | 96.1 |
| 10.0 | 98.7 | 99.5 | 99.4 | 98.9 | 97.6 | 96.2 |
| 15.0 | 98.9 | 99.6 | 99.4 | 99.0 | 98.1 | 96.5 |
| 20.0 | 99.0 | 99.6 | 99.4 | | | |
| 30.0 | 99.2 | 99.6 | 99.4 | | | |

[a] Hass (47).

oxide (2, 3). The reflectance of the other metal films drops rapidly in the visible or UV. Aluminum is, therefore, the only film material for mirrors and gratings that require high reflectance in the UV. Evaporated Ag films have the highest reflectance of any mirror coating from the short-wavelength region of the visible to the IR. A great amount of research has therefore been performed and is continuing on techniques for producing adherent and well protected front surface Ag mirrors. In the IR at $\lambda = 10$ $\mu$m, Al, Au, Cu, and Ag reflect almost equally well, having reflectances between 98.7% and 99.5%. At this wavelength even Rh and Pt films have reflectances of 97.6 and 96.2%, respectively.

The optical properties of metals are usually characterized by two parameters, the index of refraction $n$ and the extinction coefficient $k$. Knowledge of these two parameters, which are called the optical constants, or the complex index $N = n - ik$ is required to calculate the reflectance of a metal surface at various angles of incidence, the effect of surface films on the reflectance, and the phase change on reflection. A rather complete list of the optical constants of the metals discussed in this article with reference to their origin is presented in Table II (47).

The reflectances $R_s$ and $R_p$ of opaque metal coatings characterized by optical constants $n$ and $k$ and their dependence on angle of incidence $\theta$ are given by Eqs. (1)–(3). The subscripts s and p refer to light polarized perpendicular and parallel, respectively, to the plane of incidence.

Normal incidence:

$$R_s = R_p = \frac{(n-1)^2 + k^2}{(n+1)^2 + k^2} \tag{1}$$

Reflectances as a function of angle of incidence $\theta$:

$$R_s = \frac{a^2 + b^2 - 2a\cos\theta + \cos^2\theta}{a^2 + b^2 + 2a\cos\theta + \cos^2\theta} \tag{2}$$

$$R_p = R_s \frac{a^2 + b^2 - 2a\sin\theta\tan\theta + \sin^2\theta\tan^2\theta}{a^2 + b^2 + 2a\sin\theta\tan\theta + \sin^2\theta\tan^2\theta} \tag{3}$$

where

$$2a^2 = [(n^2 - k^2 - \sin^2\theta)^2 + 4n^2k^2]^{1/2} + [n^2 - k^2 - \sin^2\theta]$$
$$2b^2 = [(n^2 - k^2 - \sin^2\theta)^2 + 4n^2k^2]^{1/2} - [n^2 - k^2 - \sin^2\theta]$$

In the case of unpolarized light with equal amplitudes of perpendicular and parallel components, the reflectance is $R = \frac{1}{2}(R_s + R_p)$.

The calculated dependence of the reflectances of Ag and Al on the angle of

## TABLE II

Optical Constants and Calculated Normal Incidence Reflectance of the Most Important Mirror Coatings[a]

| Metal | λ (μm) | n | k | R (% calc.) | Metal | λ (μm) | n | k | R (% calc.) |
|---|---|---|---|---|---|---|---|---|---|
| Al | 0.220 | 0.14 | 2.35 | 91.8 | Ag | 0.400 | 0.075 | 1.93 | 93.9 |
|  | 0.260 | 0.19 | 2.85 | 92.0 |  | 0.500 | 0.050 | 2.87 | 97.9 |
|  | 0.300 | 0.25 | 3.33 | 92.1 |  | 0.600 | 0.060 | 3.75 | 98.4 |
|  | 0.340 | 0.31 | 3.80 | 92.3 |  | 0.700 | 0.075 | 4.62 | 98.7 |
|  | 0.380 | 0.37 | 4.25 | 92.6 |  | 0.800 | 0.090 | 5.45 | 98.8 |
|  | 0.436 | 0.47 | 4.84 | 92.7 |  | 0.950 | 0.110 | 6.56 | 98.9 |
|  | 0.492 | 0.64 | 5.50 | 92.2 |  | 2.0 | 0.48 | 14.4 | 99.1 |
|  | 0.546 | 0.82 | 5.99 | 91.6 |  | 4.0 | 1.89 | 28.7 | 99.1 |
|  | 0.650 | 1.30 | 7.11 | 90.7 |  | 6.0 | 4.15 | 42.6 | 99.1 |
|  | 0.700 | 1.55 | 7.00 | 88.8 |  | 8.0 | 7.14 | 56.1 | 99.1 |
|  | 0.800 | 1.99 | 7.05 | 86.4 |  | 10.0 | 10.69 | 69.0 | 99.1 |
|  | 0.950 | 1.75 | 8.50 | 91.2 | Au | 0.55 | 0.33 | 2.32 | 81.5 |
|  | 2.0 | 2.30 | 16.5 | 96.8 |  | 0.60 | 0.20 | 2.90 | 91.9 |
|  | 4.0 | 5.97 | 30.3 | 97.5 |  | 0.70 | 0.13 | 3.84 | 96.7 |
|  | 6.0 | 11.0 | 42.4 | 97.7 |  | 0.80 | 0.15 | 4.65 | 97.4 |
|  | 8.0 | 17.0 | 55.0 | 98.0 |  | 0.90 | 0.17 | 5.34 | 97.8 |
|  | 10.0 | 25.4 | 67.3 | 98.0 |  | 1.0 | 0.18 | 6.04 | 98.1 |
| Cu | 0.55 | 0.76 | 2.46 | 66.9 |  | 2.0 | 0.54 | 11.2 | 98.3 |
|  | 0.60 | 0.19 | 2.98 | 92.8 |  | 4.0 | 1.49 | 22.2 | 98.8 |
|  | 0.80 | 0.17 | 4.84 | 97.3 |  | 6.0 | 3.01 | 33.0 | 98.9 |
|  | 1.0 | 0.20 | 6.27 | 98.1 |  | 8.0 | 5.05 | 43.5 | 99.0 |
|  | 3.0 | 1.22 | 17.1 | 98.4 |  | 10.0 | 7.41 | 53.4 | 99.0 |
|  | 7.0 | 5.25 | 40.7 | 98.8 | Rh | 0.546 | 1.62 | 4.63 | 77.1 |
|  | 10.8 | 12.6 | 64.3 | 98.8 |  |  |  |  |  |

[a] The sources from which these optical constants are quoted can be found in Ref. 47.

incidence is shown in Figs. 13 and 14. The calculations for each metal were made for two wavelengths, one in the visible and one in the IR, using Eqs. (2) and (3) and the optical constants listed in Table II. The reflected intensities follow a certain pattern in all cases: $R_s$, the perpendicular component, increases steadily from the normal incidence value up to 100% at grazing incidence, while $R_p$, the parallel component, first decreases to a minimum and then rises rather rapidly to 100% at grazing incidence. The essential features of the reflectance curves are given by three quantities:

(1) normal incidence reflectance, which is the same for both components, and which is given by Eq. (1)
(2) the angle at which $R_p$ reaches a minimum
(3) the reflectance value of $R_p$ at the minimum

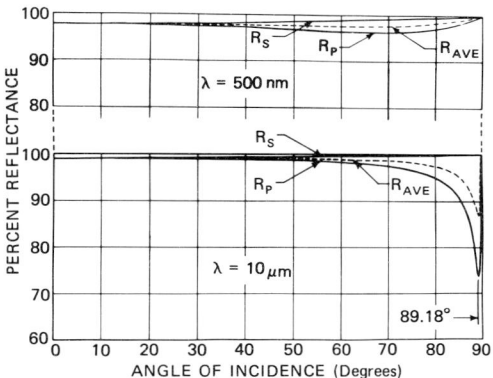

FIG. 13. Calculated reflectance of evaporated Ag as a function of incidence angle for $\lambda = 500$ nm and $\lambda = 10$ μm.

In all cases in which the values of the optical constants are such that $(n^2 + k^2) \gg 1$, which is true for most metals in the visible and IR, the angle $\phi$ at which $R_p$ reaches a minimum can be obtained to a very good approximation from

$$\phi = \cos^{-1}\left(\frac{\{1 + [4/(n^2 + k^2)]\}^{1/2} - 1}{\{1 + [4/(n^2 + k^2)]\}^{1/2} + 1}\right)^{1/2} \qquad (4)$$

With the above assumption, the minimum value of $R_p/R_s = (\tan \Psi)^2$ can be derived from

$$\tan \Psi = \frac{k/n}{1 + [1 + (k^2/n^2)]^{1/2}} \qquad (5)$$

For large values of $(n^2 + k^2)$, the minimum of $R_p$ occurs very nearly at the angle at which the relative phase change between the reflected s and p components is 90°, which is called the principal angle of incidence.

Since for most highly reflecting metals $R_s$ is nearly 100% at the principal angle of incidence, the minimum value of $R_p$ can be derived from

$$R_{p(min)} = \tan^2 \Psi \qquad (6)$$

For most metallic reflecting coatings used from the visible to the IR it is possible to make an assessment of their reflectance behavior as a function of incidence angle rather simply without resorting to extensive calculations. The ratio $k/n$ and the value of $(n^2 + k^2)$ are the dominant factors for predicting the reflectance behavior as a function of incidence angle. They can be used here to explain the difference between the reflectance curves of Ag and

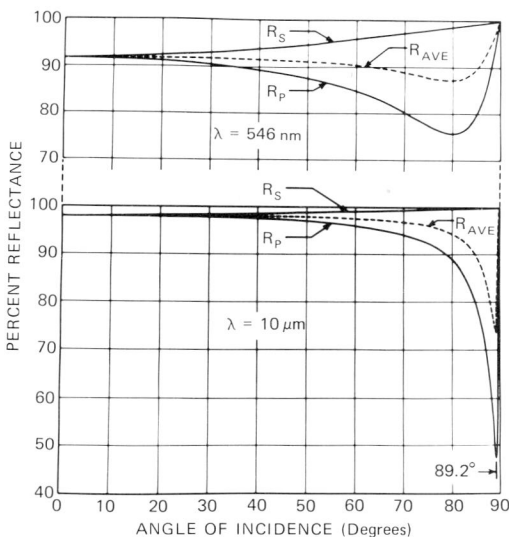

FIG. 14. Calculated reflectance of evaporated Al as a function of incidence angle for $\lambda = 546$ nm and $\lambda = 10$ μm.

Al shown in Figs. 13 and 14. The angle of incidence at which $R_p$ reaches a minimum moves closer to 90° as the value of $(n^2 + k^2)$ increases. This is demonstrated by the following tabulated values calculated from Eqs. (4)–(6):

|  | Ag ($\lambda = 500$ nm) | Al ($\lambda = 546$ nm) | Ag ($\lambda = 10$ μm) | Al ($\lambda = 10$ μm) |
|---|---|---|---|---|
| $(n^2 + k^2)$ | 8.2 | 36 | 4900 | 5200 |
| $\phi$ at $R_{p(min)}$ | 70° | 80° | 89.18° | 89.20° |

The ratio $k/n$ has a pronounced effect on the decrease in the magnitude of $R_p$ with angle of incidence. As $k/n$ gets larger, there is a corresponding increase in the minimum value of $R_p$, which occurs at or close to the principal angle of incidence. Silver at $\lambda = 500$ nm has an extremely high $k/n$ value of 57.4. At this wavelength, therefore, $R_p$ drops only 1.6% between normal incidence and the principal angle of incidence. For Al in the visible with $k/n = 7.2$, $R_p$ drops considerably lower. The great difference in the minimum reflectance of $R_p$ for Al and Ag at 10 μm can be explained by the pronounced difference in their $k/n$ values at this wavelength. Al, with a $k/n$ value of 2.6, has an $R_{p(min)}$ of 47%, while $R_{p(min)}$ of Ag, with $k/n = 6.5$, drops only to 73%. In the far IR, where for all metals $n$ and $k$ become very large and $k/n$ approaches unity, $R_p$ drops to a minimum value of 17% close to grazing incidence.

## IV. Mirror Coatings with Protective Layers and Reflectance-Enhancing Surface Films

### 1. General Remarks

All of the metal mirrors may be overcoated with protective layers which in some cases serve to increase or decrease the reflectance of the underlying metal in certain wavelength regions. At normal incidence the reflectance of a metal characterized by optical constants $n$ and $k$ and coated with a nonabsorbing surface film of refractive index $n_1$ and thickness $t_1$ is given by

$$R = \frac{r_1^2 + r_2^2 - 2r_1 r_2 \cos(2\phi - \delta)}{1 + r_1^2 r_2^2 - 2r_1 r_2 \cos(2\phi - \delta)}$$

where

$$r_1 = \frac{n_1 - 1}{n_1 + 1}, \qquad r_2 = \left[\frac{(n - n_1)^2 + k^2}{(n + n_1)^2 + k^2}\right]^{1/2}$$

$$\phi = \frac{360° n_1 t_1}{\lambda_0}, \qquad \tan \delta = \frac{2n_1 k}{n_1^2 - n^2 - k^2}$$

and $\delta$ is the absolute phase change at the dielectric–metal boundary. Overcoatings such as these inevitably alter the intrinsic reflectance and polarization properties of various metals differently, because the underlying metals have different optical constants. This is demonstrated in Fig. 15 (48), which

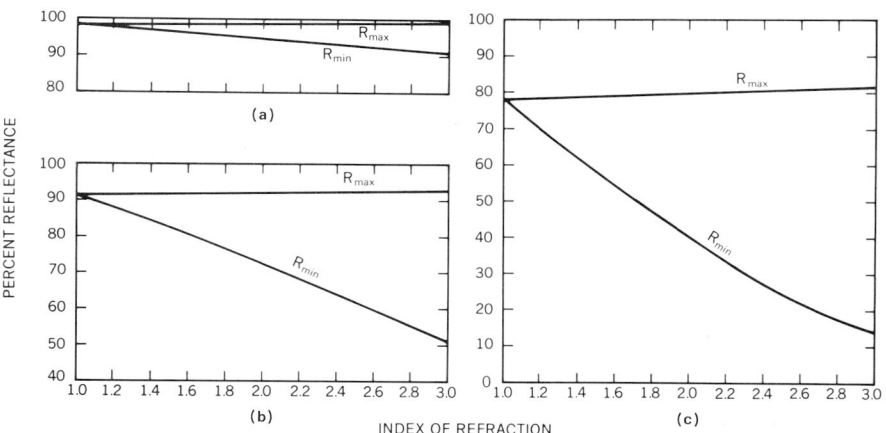

FIG. 15. Calculated reflectance of (a) Ag, (b) Al, and (c) Rh coated with effective $\lambda/4$ or odd multiple of $\lambda/4$ ($R_{min}$) and $\lambda/2$ or even multiple of $\lambda/4$ ($R_{max}$) thicknesses of nonabsorbing surface films with $n$ values ranging from 1.0 to 3.0 at $\lambda = 546$ nm.

shows the effect of single-layer overcoatings on the normal incidence reflectance of Ag, Al, and Rh at $\lambda = 546$ nm. The nonabsorbing surface films have $n$ values from 1 to 3 and were assumed to be effectively one-quarter ($R_{min}$) and one-half wavelength ($R_{max}$) thick. The optical constants of the three metals presented in Table II were used for the calculations. A typical protective coatings, such as $Al_2O_3$, with an $n$ of about 1.6, deposited to a thickness of $\lambda/4$ or an odd multiple of $\lambda/4$, will reduce the reflectance of Ag from 98.3 to 96.5%, of Al from 91.5 to 80%, and of Rh from 78.2 to 55%. This shows clearly that the intrinsic high reflectance of Ag is least affected by a nonabsorbing surface layer. At the $\lambda/2$ positions, the reflectance of the three metals are slightly increased by a nonabsorbing surface coating, and this is more pronounced for Rh than it is for Ag and Al. The reflectance increase caused by $\lambda/2$-thick surface films also increases with increasing $n$ values of the overcoatings.

In Fig. 16 (48), the calculated reflectance of Ag as a function of incidence angle is presented for the perpendicular ($R_s$), the parallel ($R_p$), and the average ($R_{av}$) components of reflectance at $\lambda = 550$ nm. The calculations were performed for bare Ag and for Ag overcoated with effectively $\lambda/4$- and $2\lambda/4$-thick single-layer films of $n = 1.6$, and a typical high–low index reflectance-enhancing film pair. It is evident that an evaporated Ag mirror without or with a nonabsorbing overcoating exhibits only a very small change in reflectance at $\lambda = 550$ nm for angles of incidence ranging from 0 to 90°. For all angles of incidence, the polarization $R_p/R_s$ remains close to unity. For a Ag mirror coated with a reflectance-enhancing double layer of

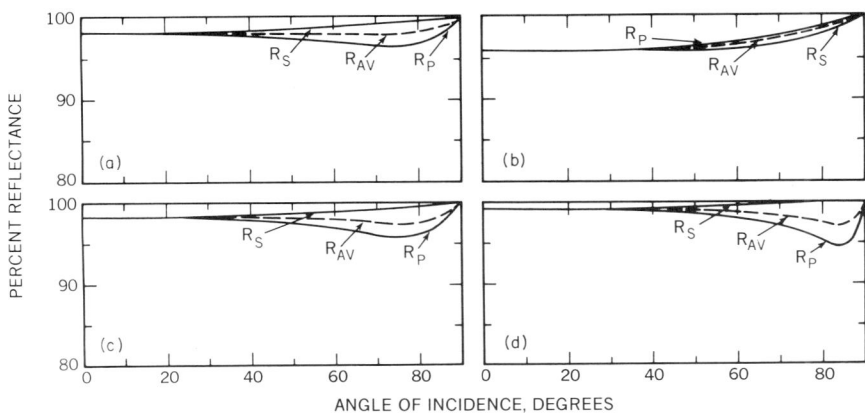

FIG. 16. Calculated reflectance of Ag and Ag plus three typical nonabsorbing surface films as a function of incidence angle at $\lambda = 550$ nm: (a) Ag; (b) Ag + $\lambda/4$, $n = 1.6$; (c) Ag + $2\lambda/4$, $n = 1.6$; (d) Ag + $\lambda/4$, $n_1 = 1.6$, + $\lambda/4$, $n_2 = 2.3$.

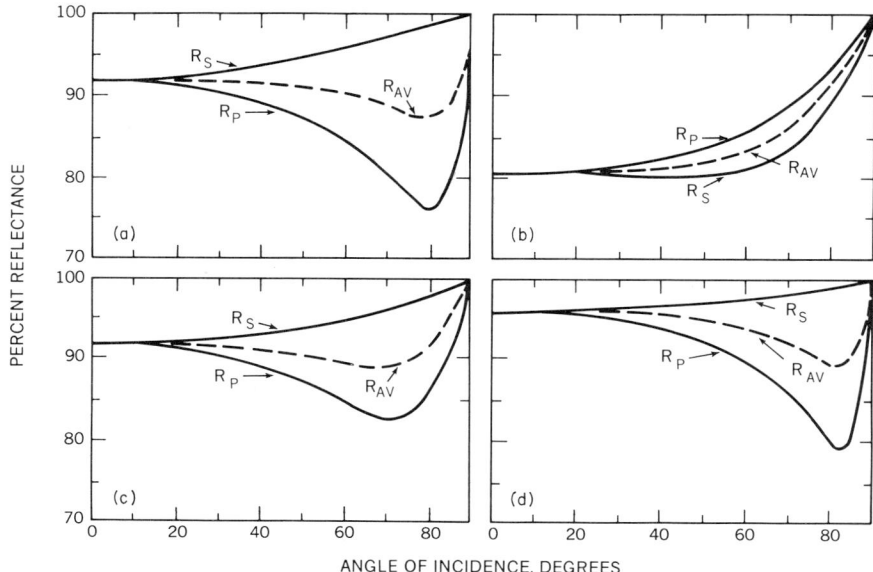

Fig. 17. Calculated reflectance of Al and Al plus three typical nonabsorbing surface films as a function of incidence angle at $\lambda = 546$ nm: (a) Al; (b) Al + $\lambda/4$, $n = 1.6$; (c) Al + $2\lambda/4$, $n = 1.6$; (d) Al + $\lambda/4$, $n_1 = 1.6$, + $\lambda/4$, $n_2 = 2.3$.

$n_1 = 1.6$ ($Al_2O_3$) and $n_2 = 2.3$ ($CeO_2$) (49), the normal incidence reflectance increases from 98.3 to 99.3%, and remains virtually unchanged up to angles of incidence of 50°.

Figures 17 and 18 (48) contain similarly calculated angular reflectance behavior for Al and Rh, respectively. Both metals show the same general trend as Ag without and with surface films, but in each case the average reflectances are considerably lower than they are for Ag, and there is a much greater difference between the $R_s$ and $R_p$ components with increasing angle of incidence. This means that the reflected beams may be highly polarized at high incidence angles. The figures also show that the normal incidence reflectances of Al and Rh mirrors coated with reflectance-enhancing film pairs of $n_1 = 1.6$ and $n_2 = 2.3$ are increased from 91.5 to 95.0% and from 78.2 to 89.0%, respectively. A more complete discussion of reflectance-increasing film stacks will be presented in a later part of this article.

The preceding series of figures clearly shows that the evaporated Ag films have the highest reflectance in the visible and that they also introduce less polarization into an optical system than Al and Rh mirrors when used at angles of incidence greater than 40°.

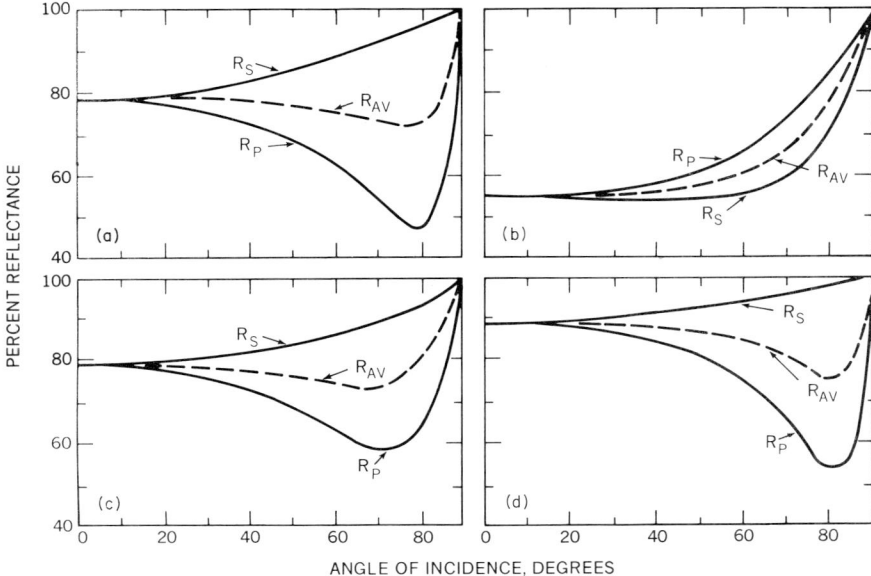

FIG. 18. Calculated reflectance of Rh and Rh plus three typical nonabsorbing surface films as a function of incidence angle at $\lambda = 546$ nm: (a) Rh; (b) Rh $+ \lambda/4$, $n = 1.6$; (c) Rh $+ 2\lambda/4$, $n = 1.6$; (d) Rh $+ \lambda/4$, $n_1 = 1.6$, $+ \lambda/4$, $n_2 = 2.3$.

## 2. Evaporated Al Mirrors with Overcoatings

Evaporated Al is undoubtedly the most frequently used coating for front surface mirrors. However, for many mirror applications the natural oxide film on Al is too thin to furnish sufficient protection, especially if the mirrors require frequent cleaning. Therefore, various methods for overcoating Al with hard and transparent protective overlayers have been developed.

Thin films produced by evaporation of silicon monoxide (SiO), coatings of $SiO_2$ or $Al_2O_3$ deposited by electron gun evaporation (8), and anodically produced $Al_2O_3$ surface films (50) are the most commonly used protective layers for evaporated Al mirrors.

SiO is usually evaporated from directly heated tantalum or molybdenum boats and containers filled with pieces of SiO about 3 mm thick. The containers and their charges have to be well outgassed before being used to deposit optical coatings. The composition and optical properties of SiO layers and the reflectance characteristics of Al coated with such protective layers, however, depend greatly upon the conditions under which SiO is evaporated. Films deposited at high deposition rates of at least 20 Å sec$^{-1}$

and low pressures of less than $1 \times 10^{-5}$ Torr consist of true SiO. Such films show rather strong absorptance in the UV and the short-wavelength region of the visible. They also have a very high index of refraction, which reaches about 2.0 in the visible and 1.9 in the near IR (*51*). They are, therefore, not suitable as protective layers on Al if high reflectances in the visible and UV are required. However, they are excellent antireflection coatings for silicon and germanium in the near IR (*52*). To produce films with negligible absorptance in the visible and the near IR, low deposition rates at rather high pressures of oxygen must be used (rates of 4 Å sec$^{-1}$ at $p = 8-10 \times 10^{-5}$ Torr of oxygen). Films produced under such a strongly oxidizing evaporation condition, which is called reactive evaporation, frequently consist of $Si_2O_3$ and have *n* values of about 1.55 in the visible. Some films may have more or less oxygen than $Si_2O_3$, depending on the oxygen pressure and deposition rate. Therefore, films produced by reactive evaporation are usually called $SiO_x$ coatings. Figure 19 shows the effect of SiO evaporation conditions on the visible and UV reflectance of silicon oxide-protected Al mirrors. All three protective coatings are chosen to be effectively one-half wavelength thick at $\lambda = 550$ nm in order to produce the highest reflectance in the visible. The bottom curve shows the reflectance of Al coated with SiO at low pressure and a high deposition rate of 20 Å sec$^{-1}$. This true SiO film produces high absorptance for all wavelengths shorter than 500 nm and some absorptance even at longer wavelengths. A mirror coated in this way has a yellow appearance. The middle curve, which still exhibits high absorptance below 400 nm, was obtained by evaporating SiO onto Al at a much lower

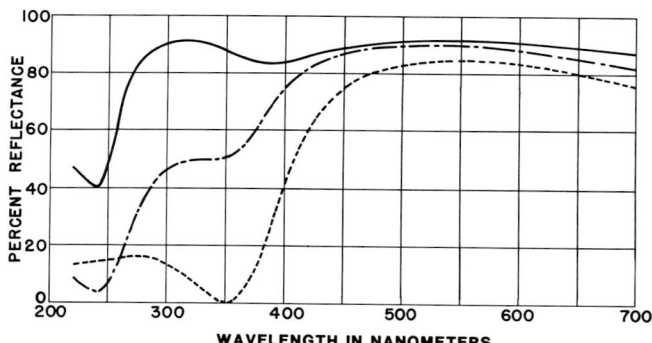

FIG. 19. Effect of SiO evaporation conditions on the visible and UV reflectance of silicon oxide-protected Al mirrors: (—) $P = 8 \times 10^{-5}$ Torr of oxygen deposition rate, 3 Å/sec; (–--) $P = 1 \times 10^{-5}$ Torr, deposition rate 4 Å/sec; (---) $P = 8 \times 10^{-6}$ Torr, deposition rate 20 Å/sec. Protective coatings effectively $\lambda/2$ thick at $\lambda = 550$ nm.

deposition rate and at pressures of $1 \times 10^{-5}$ Torr and without the introduction of oxygen. The top curve represents the reflectance of Al coated with a strongly oxidized silicon oxide film ($SiO_x$) prepared with a low deposition rate of 3 Å sec$^{-1}$ and an oxygen pressure of $8 \times 10^{-5}$ Torr. This $SiO_x$ protective layer gives the Al mirror high reflectance down to 300 nm. Reactively evaporated $SiO_x$ is, therefore, the preferred coating for Al mirrors that are to be used in the visible and the near UV. These oxidized $SiO_x$ films are less dense than those of true SiO because they are deposited at lower rates and rather high pressures. However, they make excellent protective coatings since they adhere strongly to Al and harden rapidly by further surface oxidation when exposed to air. Below 300 nm, $SiO_x$ films still show rather high absorptance, which increases with decreasing wavelength. This had limited the usefulness of $SiO_x$-coated Al mirrors until Bradford and Hass (5, 7) discovered that UV irradiation eliminates completely the UV absorption in these $SiO_x$ films, making it possible to produce well protected Al mirrors with 91% reflectance down to 200 nm.

Figure 20 shows the effect of UV irradiation on the visible and UV reflectance of evaporated Al protected with true SiO and strongly oxidized $SiO_x$ films. The UV irradiations were performed with a 435-W Hanovia quartz mercury burner at a distance of about 20 cm. Both films are again effectively one-half wavelength thick at $\sim \lambda = 550$ nm. It can be seen that 5 hr of UV treatment completely eliminates the initial high UV absorptance of the $SiO_x$ film deposited slowly at a high pressure of oxygen, while the same UV irradiation has very little effect on the coating produced at much lower

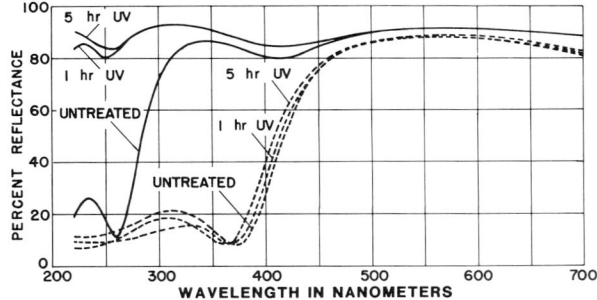

FIG. 20. Effect of UV irradiation with a 435-W quartz mercury burner on the reflectance of Al protected with unoxidized and strongly oxidized silicon oxide coatings: (—) $P = 9 \times 10^{-5}$ Torr of oxygen deposition rate, 4.5 Å/sec; (---) $P = 4 \times 10^{-6}$ Torr, deposition rate 5.5 Å/sec. Protective coatings effectively $\lambda/2$ thick at 550 to 600 nm. Irradiations performed at a distance of 20 cm.

pressure and without oxygen. Additional experiments with much thicker reactively deposited $SiO_x$ ($t > 1$ µm) films showed that their UV absorptance could also be completely removed by UV irradiation.

Two completely different effects were found to be responsible for the optical changes in reactively deposited $SiO_x$ caused by UV treatment (7):

(1) ultraviolet irradiation rearranges the oxygen atoms and molecules gathered during the evaporation process and forms well defined silicon oxide molecules that are highly transparent in the UV;

(2) exposure to air or oxygen increases the oxygen content in the deposited films, which removes their UV absorptance and decreases their index of refraction.

An alternate method of increasing the oxidation state of reactively evaporated SiO films and to eliminate their UV absorptance without UV treatment has been described by Heitmann (53). He performed his evaporations in ionized oxygen. The ionized gas is produced in a gas discharge tube of high current density which is located inside the vacuum system. The ionized gas emerges from a nozzle in the wall of the discharge tube directly into the vacuum area, where the evaporations are performed. Films produced in this way are free of absorption down to 190 nm and have refractive indices identical to those of the fused silica substrate. A complete arrangement of Heitmann's vacuum system suitable for performing reactive evaporations in ionized gases is described in reference (53).

$SiO_2$ and $Al_2O_3$ films properly deposited by electron beam evaporation also form hard, durable coatings on Al that are nonabsorbing in the visible and UV down to below 200 nm (8). When deposited onto evaporated Al, they show only the expected maxima and minima in their reflectance curves. Films of $Al_2O_3$ deposited by electron beam evaporation to thicknesses of $30\lambda/4$ ($\lambda = 550$ nm) for spacecraft temperature-control applications were found to be highly stable and adherent with no evident internal stress in the coating (54).

For Al front surface mirrors that are to be used in the IR, it is important to study the effect of absorptances in the protective layers on IR reflectance. All silicon oxides—SiO, $SiO_x$, and $SiO_2$—have strong absorption bands in the important 8–12 µm atmospheric window region. In practice, silicon oxide layers 1000 to 2000 Å thick have been found to be thick enough to give adequate protection against abrasion and humidity. Numerous measurements (47, 51, 55) have established that the near-normal IR reflectance of Al mirrors protected with such thin coatings is nearly the same as that of uncoated Al in the 8–12 µm region. To explain this effect, one has to remember that light reflected from highly reflecting metal surfaces produces a standing wave pattern with a node, or condition of zero vibration, at the metal surface. Under these circumstances, absorbing films that are thin compared to the wavelength produce very little absorptance. If thicker protective layers of silicon oxides are applied, significant reflectance decreases can be observed. An Al mirror coated with an SiO layer approximately 1 $\mu$m thick exhibits a near-normal reflectance of almost zero at 10 $\mu$m. The wavelengths at which thick silicon oxide films produce the greatest reflectance decreases depend upon the composition or ratio of Si to O. For SiO the greatest reflectance decrease occurs between 10.0 and 10.2 $\mu$m; for $Si_2O_3$, between 9.7 and 11.5 $\mu$m; and for $SiO_2$, between 9.4 and 12.5 $\mu$m. Al mirrors coated with thick silicon oxide layers are therefore unsuitable as IR reflectance coatings for the 8–12 $\mu$m region. However, their IR absorption properties have been exploited to produce surface coatings with low solar absorptivity and controllable high thermal emissivity for controlling the temperature of many spacecraft (56–58).

All the IR reflectance data for Al coated with thin protective layers of silicon oxides that were reported above showed that the normal-incidence reflectances of such protected mirrors have nearly the same high reflectance as bare Al in the 8–12 $\mu$m region, provided that the oxide layers are kept thin enough ($t = 1000-2000$ Å). However, this situation changes drastically if such mirrors are used at angles of incidence greater than 40°. At high angles of incidence the reflectances of silicon oxide-coated Al mirrors are far below their normal incidence values and the reflectance of bare Al at the same angle of incidence (9). This is demonstrated in Fig. 21, which shows the calculated reflectance of Al coated with 1500 Å of $SiO_2$ at 60° angle of incidence. The optical constants of $SiO_2$ published by Boeckner (59) and of Al listed in Table II were used for the calculations. The figure shows that the Al–$SiO_2$ film combination has its lowest reflectance at about 8.1 $\mu$m for 60° angle of incidence. At this angle $R_{av}$ is about 51% and the $R_s$

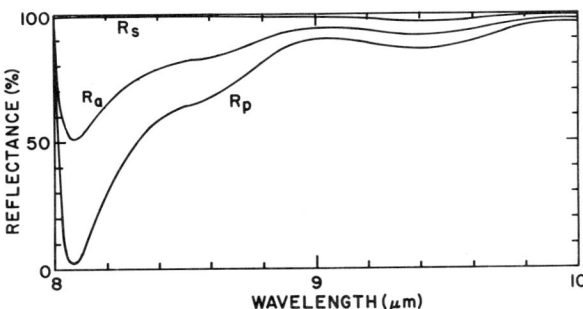

FIG. 21. Calculated reflectance of Al coated with 1500 Å of $SiO_2$ at 60° angle of incidence in the wavelength region of 8–10 $\mu$m.

and $R_p$ values are 99.0 and 2.5%, respectively. Therefore, at this wavelength and incidence angle the reflected beam is almost completely polarized, and the low $R_{av}$ is caused solely by the reflectance drop of the $R_p$ component. At 45° angle of incidence, $R_{av}$, $R_s$, and $R_p$ are 60, 98.5, and 25%, respectively. The region of high and moderate reflectance decrease extends from 8.05 to about 9.6 μm for mirrors used at 60° angle of incidence. Al + $SiO_x$ and Al + SiO show a similar reflectance behavior. Their reflectance minima at 45 and 60° occur at longer wavelengths and are not as low as those caused by an $SiO_2$ surface layer. This angle of incidence effect was found to be most severe when the optical constants of the thin protective layers have values of $n$ less than 1 and $k$ values between 0.2 and 0.8 ($N = n - ik$). Lettington and Ball (59a) report that this reflectance decrease will occur if $\cos \phi < (n_1^2 + k_1^2)^{-1/2}$ where $\phi$ is the angle of incidence and $n_1$ and $k_1$ are the optical constants of the protective overlayer. The optical constants of dielectric materials satisfy this condition at those wavelengths slightly shorter than their "reststrahlen" high-reflectance bands. If the $n$ of the surface layer is larger than 1, there is no great reduction in reflectance at 45° angle of incidence, no matter what $k$ is. Of course, this is true only if the protective layers are thin and the wavelength region of interest is 8–12 μm. The effect of reflectance reduction caused by thin surface layers with $n$ and $k$ both less than 1 is not peculiar to Al but is essentially the same for any highly reflecting metal and depends almost entirely on the optical constants of the protective layers. Table III shows the visible and IR reflectance and polarization characteristics of uncoated Ag and of Ag overcoated with 300 Å of $Al_2O_3$ and 1500 Å of $SiO_2$ at 0 and 45° angles of incidence (48). The thin $Al_2O_3$ layer is used to assure good adherence between the Ag and $SiO_2$ film. In the

TABLE III

CALCULATED VISIBLE VERSUS IR REFLECTANCE AND POLARIZATION CHARACTERISTICS FOR UNCOATED Ag AND Ag + 300 Å $Al_2O_3$ + 1500 Å $SiO_2$

|  | $\theta = 0°$ | $\theta = 45°$ | | | |
|---|---|---|---|---|---|
|  | $R$ | $R_s$ | $R_p$ | $R_{av}$ | $R_p/R_s$ |
| $\lambda = 0.55$ μm |  |  |  |  |  |
| Uncoated Ag | 98.2 | 98.7 | 97.5 | 98.1 | 0.988 |
| Ag + 300 Å $Al_2O_3$ + 1500 Å $SiO_2$ | 97.8 | 98.9 | 97.6 | 98.2 | 0.987 |
| $\lambda = 8.10$ μm |  |  |  |  |  |
| Uncoated Ag | 98.6 | 99.0 | 98.0 | 98.5 | 0.990 |
| Ag + 300 Å $Al_2O_3$ + 1500 Å $SiO_2$ | 98.5 | 98.9 | 30.6 | 64.7 | 0.309 |

visible at $\lambda = 550$ nm, where the surface films are nonabsorbing, the uncoated and overcoated Ag films have practically the same reflectance at 0 and 45° angles of incidence and the $R_p/R_s$ values are close to 1. In the IR at $\lambda = 8.1$ μm, the uncoated and overcoated mirrors have the same reflectance at normal incidence. At 45° angle of incidence, however, the reflectance of the protected Ag film drops to $R_p = 30.6\%$ and $R_{av} = 64.7\%$ and decreases the $R_p/R_s$ value to 0.31, while the bare Ag film shows practically no change in reflectance between 0 and 45° angles of incidence. Table III shows that the drop of $R_{av}$ is again caused entirely by the decrease of the $R_p$ component, and that $R_s$ remains high for all angles of incidence.

In search of a protective film that would not reduce the reflectance of Al in the 8–12 μm region at higher angles of incidence, electron beam evaporated $Al_2O_3$ appeared promising. It forms hard and adherent layers on Al (8) and is also one of the few protective coatings that adhere well to Ag (48, 60). The optical constants of a thin $Al_2O_3$ film, published by Harris (61), in the 8–12 μm region were reported to be $n = 1.4–1.7$ and $k = 0.0–0.7$. Protective layers with such optical constants should not cause any considerable reflectance decrease at higher angles of incidence in this region. However, other experimental measurements cast doubt on the validity of these optical constants of $Al_2O_3$ in the wavelength region 8–12 μm. In one of the experiments, the reflectance of single-crystal $Al_2O_3$ (sapphire) plates was measured in the region 7–15 μm. The measured reflectance at 9.5 μm was almost zero. This can occur only if the value of $n$ is close to 1 and $k$ is very small. Therefore at $\lambda = 10$ μm, $n$ should decrease to values of less than 1 since sapphire has the well known broad reststrahlen reflectance at 14–15 μm. Evaporated $Al_2O_3$ is amorphous and much less dense, and has $n$ values much lower than those of sapphire. The region over which Al coated with evaporated $Al_2O_3$ has low reflectance at high angles of incidence may therefore be quite large. To decide the accuracy of this prediction, Al was coated with $Al_2O_3$ layers of three different thicknesses, 1000, 1750, and 2300 Å, by electron gun evaporation in high vacuum. The reflectances of the samples were measured with unpolarized radiation in the 7–15 μm region at close to 0, 40, and 60° angles of incidence. The results are shown in Fig. 22, which also includes the reflectance values of unprotected Al in the same wavelength region and at the same angles of incidence (10). The reflectance curves show that the reflectance of uncoated Al remains high throughout the entire region from 7 to 15 μm and at all three angles of incidence, while $Al_2O_3$-protected Al mirrors exhibit considerable reflectance decreases over an extended wavelength region. This reflectance decrease gets progressively worse with increasing film thickness and angle of incidence. The greatest reflectance decrease of all $Al_2O_3$-protected Al mirrors occurs in the 10.6–10.7 μm region.

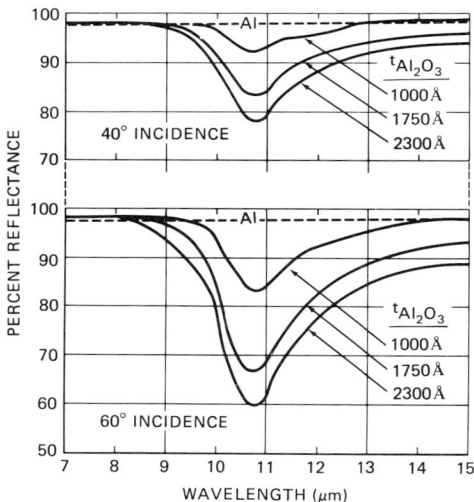

FIG. 22. Calculated reflectance of Al and measured reflectance of Al + Al$_2$O$_3$ films of three different thicknesses at 40° and 60° incidence angles in the region 7–15 μm.

For the previously reported mirror coatings protected with silicon oxides, the minimum reflectance at higher angles of incidence occurred at shorter wavelengths, because their reststrahlen reflectances are located at shorter wavelengths. Another important difference is that the low-reflectance region of Al$_2$O$_3$-protected Al is much broader than that of Al coated with silicon oxide layers. For example, Al + 1500 Å of SiO$_2$ used at 60° angle of incidence has an average reflectance of less than 90% from about 8 to 8.8 μm, while a coating of Al + 1750 Å of Al$_2$O$_3$ reflects less than 90% at the same angle of incidence in the region 9.8–13.3 μm. It should be mentioned again that in the regions where Al$_2$O$_3$- and silicon oxide-protected Al mirrors show low reflectance at high angles of incidence, the reflected IR radiation is highly polarized, since the reflectance decreases are entirely caused by the decrease of the $R_p$ component, while $R_s$ increases with increasing angle of incidence from about 99.0 to 100% at grazing incidence. This effect makes it possible to prepare efficient reflection-type polarizers for various regions in the far IR.

### 3. Reflection-Type Polarizers for the 10.6-μm CO$_2$ Laser Emission Line Using Al$_2$O$_3$-Coated Al Mirrors

As previously mentioned, the measured average reflectance of Al coated with Al$_2$O$_3$ has a minimum at about 10.6 μm at high angles of incidence.

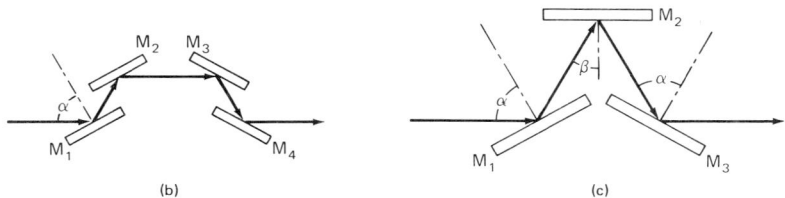

FIG. 23. Reflection-type polarizers for 10.6-$\mu$m $CO_2$ laser radiation using four and three $Al_2O_3$-coated Al mirrors: (a) Reflectance at 10.6 $\mu$m of Al coated with three thicknesses of $Al_2O_3$ at angles of incidence 40° and 60°. (b) Four-mirror polarizer; all angles of incidence are 60°. (c) Three-mirror polarizer; angles of incidence are given by $2\alpha - \beta = 90°$.

The reflectance of the $R_p$ component must be considerably lower, since $R_s$ can be assumed to be about 99%. It can be calculated using the relation $R_p = 2R_{av} - R_s$. In Fig. 23, $R_{av}$ and $R_p$ at 10.6 $\mu$m for Al coated with various thicknesses of $Al_2O_3$ at 40 and 60° angles of incidence are listed, and schematic diagrams of four- and three-mirror polarizers are shown (11). The lowest value of $R_p$ (and therefore the highest degree of polarization) is obtained with the thickest $Al_2O_3$ film used and at the highest angle of incidence, 60°, for which $R_{av}$ was measured. That $R_p$ becomes low while $R_s$ stays high makes it possible to design efficient reflection-type polarizers for 10.6-$\mu$m $CO_2$ laser radiation. The optimum angle for highest polarization depends on the optical constants $n$ and $k$ of the $Al_2O_3$ surface layer. In order to determine this optimum angle, the optical constants of $Al_2O_3$ were derived from a simultaneous fit of the measured data of the reflectance of a sapphire plate at normal incidence and the reflectance of Al + $Al_2O_3$ as a function of incidence angle. This analysis yielded values of $n \simeq 0.65-0.50$ and $k \simeq 0.3-0.5$ for the wavelength region 10.6–10.8 $\mu$m. The optimum angle of incidence, for which $R_p$ is a minimum for these optical constants, is about 75°. However, as shown below, very useful polarizers can be made which use angles of incidence less than 75°.

The four-mirror polarizer shown in Fig. 23 uses four mirrors coated with Al + 2300 Å of $Al_2O_3$ all at 60° angle of incidence. This results in a polarizer with a ratio $R_s/R_p$ of about 500 at $\lambda = 10.6$ $\mu$m. Consequently, an efficient polarizer for 10.6-$\mu$m radiation can be made using four identical mirrors in a simple arrangement in which the outgoing beam is in the same direction as the incident one. In addition, since $R_s = 99\%$, the throughput for the s component is very high (96%). Accurate alignment of the mirrors can easily be achieved by the use of a visible laser.

Another arrangement, which requires only three Al mirrors coated with $Al_2O_3$ films, is shown in the same figure. Here the beam direction can also be maintained without deviation, but the output beam will be inverted with respect to the input one. For most applications this may not be important. For this configuration, in contrast to the four-mirror design, the angles are not equal, but must obey the relationship $2\alpha - \beta = 90°$ in order to maintain an undeviated beam direction. If $\alpha$ is chosen to be 60°, $\beta$ must be 30°. However, for this choice of angles the calculated ratio of $R_s/R_p$ for the same $Al_2O_3$-coated mirrors would be only 30, as compared to the value of 500 for the four-mirror design. If, on the other hand, $\alpha$ is chosen to be 70°, for which $\beta$ would be 50°, the ratio of $R_s/R_p$ would be nearly 400, or almost as high as that of the four-mirror polarizer. The throughput for the s component in the three-mirror design would be about 97%.

The results for the above mirror designs are based on measurements of Al protected with $Al_2O_3$. As noted previously, the phenomenon of reduced reflectance at higher angles of incidence depends on the properties of the protective layers, not those of the underlying metal. It is most pronounced when the optical constants $n$ and $k$ of the protecting layer are both very much less than 1. This is typical for ionic solids at the short-wavelength side of their reststrahlen region. Here $n$ decreases usually to values far below 1 as the reststrahlen reflectance region is approached, while $k$ remains small. There are a large number of dielectric materials available that have reststrahlen bands over a very wide wavelength region from about 9 $\mu$m to beyond 50 $\mu$m. This offers the possibility of choosing almost any wavelength in this range for which a reflection-type polarizer can be made to operate efficiently. The width of any particular polarizing region will of course depend upon the optical properties and the thickness of the dielectric overcoating. Some will be rather narrow (e.g., $SiO_2$), and others will be much broader (e.g., $Al_2O_3$ and MgO).

It can therefore be concluded that rather simple and inexpensive reflection-type polarizers can be built for the IR which have $R_s/R_p$ ratios in excess of 500, s-component throughputs greater than 96%, and large apertures. Operation in many bands in the wavelength region from 9 $\mu$m to beyond 50 $\mu$m can easily be achieved by proper selection of the dielectric overcoating material, using a frame in which mirrors with various surface layers can be easily replaced and correctly mounted.

## 4. Protected Al Mirrors with High Reflectance in the 8–12 μm Region from Normal to High Angles of Incidence

In the preceding paragraphs it has been pointed out that Al mirrors protected with rather thin (1000–2000 Å) silicon oxide and $Al_2O_3$ films have greatly reduced reflectance in the 8–12 μm region when used at angles of incidence greater than 40°. The analysis of this effect also showed that the reduction in reflectance at high angles of incidence occurs in the wavelength region where $n$ is less than 1 and $k$ is very small. A survey of possible alternative dielectric materials that might lead to durable mirrors with high reflectance in the 8–12 μm region at higher angles of incidence led to a study of yttria ($Y_2O_3$) and hafnium dioxide ($HfO_2$) (12). Heitmann (62) has shown that reactively evaporated yttria and scandia ($Sc_2O_3$) films have low absorptance in the 8–12 μm region, with excellent mechanical stability and chemical durability. Therefore, Al mirrors overcoated with evaporated $Y_2O_3$ and $HfO_2$ were prepared for IR reflectance measurements. The evaporation of $HfO_2$ was made from a copper hearth with an electron gun, and a tungsten boat was used for the $Y_2O_3$ evaporation. The $HfO_2$ and $Y_2O_3$ layers had optical thicknesses of about one-half wavelength at 550 nm.

The resulting $HfO_2$- and $Y_2O_3$-protected Al mirror coatings are very hard and adherent. The IR reflectance at near-normal incidence is essentially the same as that of uncoated Al in the 8–12 μm region. Measured reflectance curves at 40 and 60° angles of incidence are shown in Fig. 24 along with values for uncoated Al in the wavelength region of 7–14 μm. Al mirrors

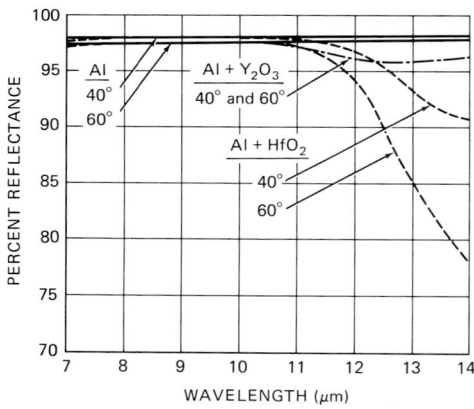

Fig. 24. Calculated reflectance of Al and measured reflectances of Al + $Y_2O_3$ and Al + $HfO_2$ at 40° and 60° incidence angles in the region 7–14 μm. Protective coatings $\sim \lambda/2$ thick at $\lambda$ = 550 nm.

coated with $Y_2O_3$ have reflectances nearly as high as uncoated Al throughout the entire 7–14 μm region at both 40 and 60°. The reflectance of Al mirrors coated with $HfO_2$ is also high up to a wavelength of about 12 μm but begins to fall off at longer wavelengths. In addition, the two protective layers described here, $Y_2O_3$ and $HfO_2$, are nonabsorbing down to the UV when properly deposited (62, 63). $Y_2O_3$ and $HfO_2$ are just two of several materials that can be used to form durable protective coatings for Al mirrors that have high reflectance in the 8–12 μm region for all angles of incidence from 0° to greater than 60°. Lettington and Ball (59a), for example, have shown that an IR transmitting diamond-like carbon coating can be used on metal mirrors in the 8–12-μm region with no appreciable reflectance loss at incidence angles of 45° and 60°.

## 5. Adherent Ag Mirrors with Protective Coatings

The superior reflectance and polarization properties of evaporated Ag compared to other mirror coatings such as evaporated Al and Rh has been clearly demonstrated in Figs. 12–17. However, the use of evaporated Ag as a front surface mirror coating has been severely restricted by its poor adhesion to mirror substrates and its susceptibility to sulfide tarnishing. Bennett et al. (64) and Burge et al. (65) have studied the growth of natural and induced sulfide tarnishing of evaporated Ag films. Their work showed that a thin ($t \sim 100$ Å) sulfide tarnish film will significantly decrease the near-normal reflectance of unprotected Ag mirrors at wavelengths below about 1 μm but will not alter the reflectance at longer wavelengths. Coatings selected for use between the Ag layer and its substrate and as protective films over Ag must, therefore, resolve the adherence and tarnishing problems without significantly decreasing the intrinsic high reflectance of evaporated Ag. New demands imposed by multimirror optical devices, such as those flown aboard earth-viewing spacecraft with scanning mirror systems (66), that require exceedingly high-reflecting, low-polarizing mirror elements, have stimulated research to overcome the inherent limitations of evaporated Ag coatings. Attempts to use evaporated silicon oxide and magnesium fluoride as protective films have been generally unsuccessful due to their poor adhesion to Ag. A large variety of other coating combinations have been used on front surface Ag mirrors with varying degrees of success. They share in common the need for an intermediate adherent layer between the evaporated Ag film and the mirror substrate and one or more dielectric outer layers to inhibit sulfide tarnishing and to protect against moisture attack.

One such combination that has been studied in detail (48) uses electron beam evaporated $Al_2O_3$ both as an adherent layer between the Ag film and its substrate and also to promote good adhesion between the Ag and a protecting outer layer of reactively deposited silicon oxide. Evaporated nichrome (80% Ni, 20% Cr) was found to work equally well as an adherent layer between the substrate and the Ag film. The optimum thickness of the silicon oxide outer layer necessary to confer protection with minimal loss of near-normal reflectance due to IR absorption was found to be between 1000 and 2000 Å. The near-normal IR reflectance of Ag and Ag + 300 Å $Al_2O_3$ + 1500 Å $SiO_x$ is presented in Fig. 25 (48). The reflectance of the protected Ag surface remains above 95% from 0.45 to 20 μm. At wavelengths shorter than 0.45 μm, the reflectance decreases due to a combination of effects: a roughness-related surface plasmon excitation in the Ag film coupled with a thickness-related interference minimum. Water absorbed by the $SiO_x$ film is responsible for the band at ~3 μm, whereas the 9.6 μm band is due to the intrinsic absorption of the 1500 Å-thick $SiO_x$ layer. Water absorption in evaporated dielectric thin films is discussed in a later section of this article.

The effects of intense $H_2S$ exposure on a Ag film protected with 300 Å of $Al_2O_3$ alone as an outer layer and an equivalently thin $Al_2O_3$ film between

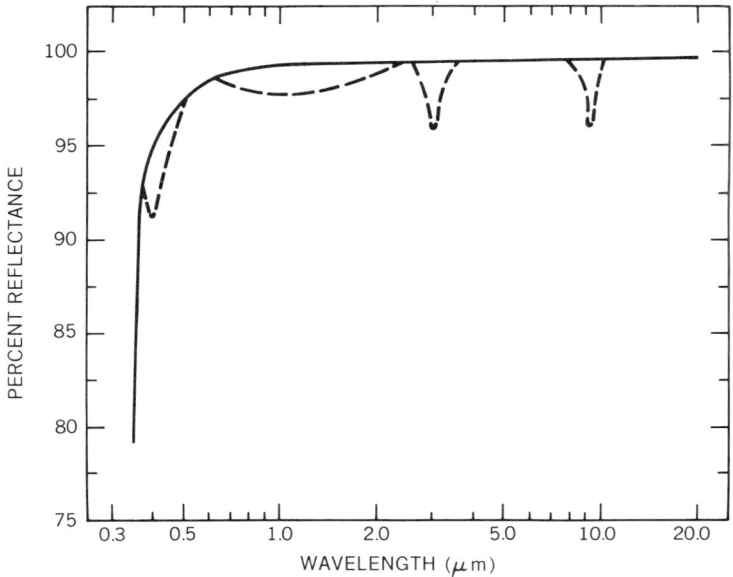

FIG. 25. Measured reflectance of freshly deposited Ag (—) (Ref. 67) and Ag + 300 Å $Al_2O_3$ + 1500 Å $SiO_x$ (---) in the region 0.36–20 μm.

the Ag and its substrate are compared to the reflectance of an unprotected Ag surface in Fig. 26 (*48*). Accelerated sulfide tarnishing was accomplished by exposing the test mirrors to humidified $H_2S$ gas mixed with air in a ratio of 1:9 by means of an apparatus and procedure described by Bennett *et al.* (*64*). The relatively rapid deterioration of the unprotected Ag film is an indication of the severity of the environment. The protected Ag film experienced a drop in reflectance of less than 1%, confined to the region of the surface plasmon effect. This indicates that a very thin, well adherent single layer of $Al_2O_3$ can inhibit sulfide tarnishing on Ag.

An additional outer layer of $SiO_x$, with a minimum thickness of about 1000 Å, was found to be necessary to prevent localized delamination of the Ag film caused by combined high humidity (95% r.h.) and temperature (50°C) exposure (*48*). This $Ag + Al_2O_3 + SiO_x$ combination was found to survive exposure to intense $H_2S$ and high-humidity environments with virtually no decrease in near-normal reflectance. However, as discussed in Section IV,2, the IR reflectance in the 8–12 $\mu$m region drops significantly for

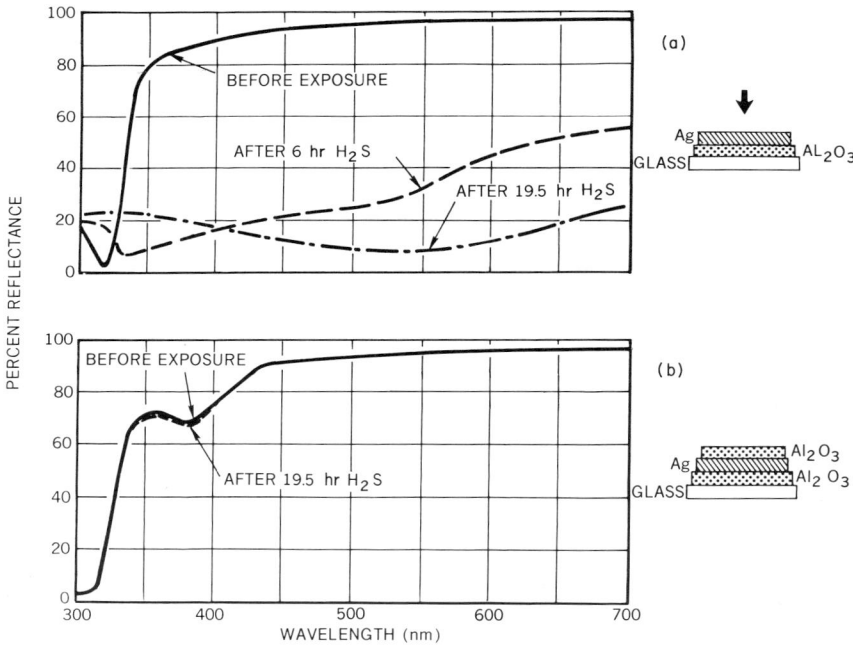

FIG. 26. Measured reflectance of unprotected (a) Ag and (b) Ag protected with 300 Å of $Al_2O_3$ before and after $H_2S$ exposure.

incidence angles above 40°. This is clearly demonstrated by the data of Table III. The search for an ideal protecting film combination for evaporated Ag mirrors that will resolve the adhesion, tarnishing, and IR incidence angle and reflectance loss problems is the subject of continuing research. Attention is currently focused on the same dielectric materials previously adopted for use on Al mirrors in the 7–14 μm region, i.e., $Y_2O_3$ and $HfO_2$, as discussed in Section IV,4 of this article and exemplified by the work of Lubezky et al. (68).

### 6. Reflectance-Enhancing Coatings for Rh and Other Metallic Front Surface Mirrors

As mentioned previously, evaporated Rh films are excellently suited as front surface mirrors because they are extremely hard and chemically very durable. However, their adherence to glass and fused silica substrates is often poor. Even films deposited in oil-free, ion titanium pumped vacuum systems on substrates of 300°C may frequently be lifted with scotch tape. This poor adherence of Rh coatings can be easily overcome by evaporating nichrome films about 25 Å thick onto the glass or fused silica substrates before the Rh mirror coating is deposited. This thin adherence-increasing nichrome inner layer was found to have no effect on the mechanical and chemical durability or on the reflectance of evaporated Rh mirrors. Its thickness can be monitored by transmittance measurements during deposition using monochromatic light of $\lambda = 546$ nm. Nichrome should be deposited until the transmittance of the substrate decreases from 92% to about 80%.

For some applications, the visible reflectance of plain Rh mirrors is too low. By applying pairs of dielectric coatings with alternately low and high indices of refraction, their reflectance can be enhanced over a rather broad region. The low-index film ($n_L$) adjacent to the metal must be effectively $\lambda/4$ and all other films optically $\lambda/4$ thick to obtain maximum reflectance at normal incidence.

The optical thickness of the effectively $\lambda/4$-thick film on the metal surface can be determined from $n_1 t_1 = (\lambda/4)(\delta/180°)$, where $\delta$ is the absolute phase change at the dielectric–metal boundary, as described in Section IV,1.

For normal incidence, the maximum reflectance of a metal with optical constants $n$ and $k$ when coated with low-index ($n_L$)–high-index ($n_H$) film pairs is given by

$$R = |(1 - Y^{2x}Z)/(1 + Y^{2x}Z)|^2$$

where $Y = n_H/n_L$, $Z = n_L|(1 + r_3)/(1 - r_3)|$, and $x$ is the number of film pairs, and

$$r_3 = \left(\frac{(n_L - n)^2 + k^2}{(n_L + n)^2 + k^2}\right)^{1/2}$$

Figure 27 shows the visible reflectance of evaporated Rh deposited at 300°C with and without a reflectance-enhancing film pair of $SiO_2$ and $TiO_2$. The reflectance curve of an evaporated Al mirror is shown on the same graph for comparison. The $SiO_2$ film was produced by electron beam evaporation and the $TiO_2$ layer was prepared by depositing metallic Ti and heat oxidizing it to $TiO_2$ at 420°C. $TiO_2$ films produced in this way consist of rutile and have the highest index of refraction ($n \simeq 2.7$) and best durability (69). To obtain the mirror coating with the reflectance shown in Fig. 27, $SiO_2$ is first evaporated onto freshly deposited Rh until the reflectance decreases to a minimum at about 530 nm. Then Ti is deposited until the transmittance of an uncoated monitoring glass slide decreases from 92 to 6% at the same wavelength. After the heat treatment, this results in a $TiO_2$ layer one-quarter wavelength thick at about 530 nm. The maximum reflectance of such $Rh-SiO_2-TiO_2$ mirror coatings is 93.1% and is even higher than that of plain Al over most of the visible region (44).

Rhodium mirrors with and without reflectance-enhancing surface films of $SiO_2 + TiO_2$ were boiled for 1 hr in 5% salt water and placed for 10 hr in 10% NaOH and 10% HCl. None of the mirrors showed any damage or change in reflectance after these treatments. The heat resistance of an $SiO_2$–

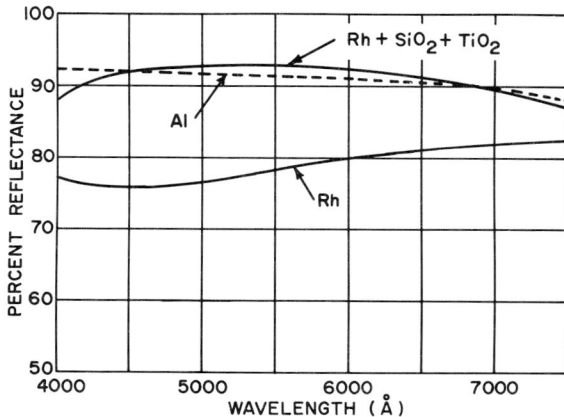

FIG. 27. Visible reflectance of Rh with and without a reflectance-enhancing film pair of $SiO_2$ and $TiO_2$. The reflectance of Al is included for comparison.

TiO$_2$ overcoated Rh mirror was found to be better than that of bare Rh. During a 16-hr heat treatment in air at 400°C, the reflectance of plain Rh decreased at $\lambda = 550$ nm from 78.2 to 68%, whereas the SiO$_2$-TiO$_2$-overcoated Rh mirrors showed no reflectance loss during the same heat treatment and during exposure to even higher temperatures (450–500°C). The formation of Rh$_2$O$_3$ on the surface of bare Rh was found to be responsible for its reflectance decrease. This oxidation of Rh occurs only at high temperatures.

Many $(n_L + n_H)$ $\lambda/4$ film pairs are available to increase the reflectance of metals from the UV to the IR. Films of SiO$_2$ + HfO$_2$ are suitable for increasing the reflectance of metals in the UV (63). MgF$_2$ + CeO$_2$, Al$_2$O$_3$ + TiO$_2$, and reactively deposited SiO$_x$ + TiO$_2$ are the most frequently used reflectance-enhancing film combinations for the visible. Figure 28 shows the visible reflectance of Al with and without a reflectance-enhancing film pair of MgF$_2$ + CeO$_2$ (47). The overcoated Al film has at normal incidence a maximum reflectance of 97% at $\lambda = 550$ nm. If two film pairs of MgF$_2$ and CeO$_2$ are applied, the maximum reflectance increases to 99.0%. A rather complete discussion of the preparation and properties of dielectric films suitable as protective and reflectance-enhancing coatings on metals has been published by Ritter (13).

Multilayer quarter-wave stacks consisting of alternating films of equal optical thickness but different refractive indices play an important role in the preparation of laser mirrors. Most laser mirrors are deposited onto well polished dielectric substrates and should have high reflectance with extremely low absorption. The film materials used for laser mirrors should have a

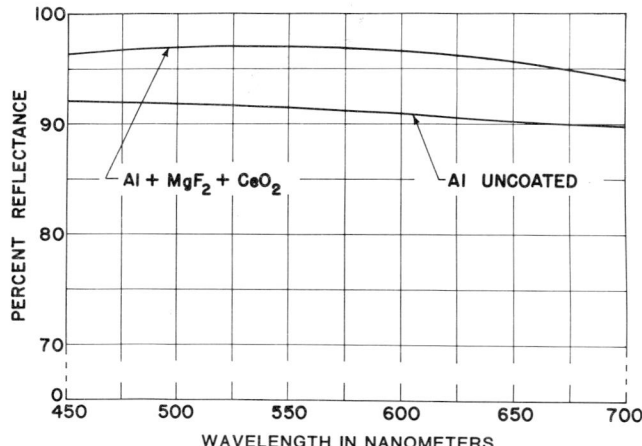

FIG. 28. Visible reflectance of Al with and without a reflectance-enhancing film pair of MgF$_2$ and CeO$_2$.

large difference in their refractive indices to obtain high reflectance with the least number of layers. Because of the demanding requirements for low absorption, low scattering, and high stability, not many film materials have been found practical for preparing laser mirror coatings. Furthermore, these film materials must be selected and tested anew for different lasers. A complete discussion of substrates, film materials, and combinations for laser mirrors as well as their stability and susceptibility to damage is outside the scope of this article. A good survey of this field has been published by Ritter (70). His article cites numerous publications dealing with the above topics.

## V. Metal–Dielectric Mirrors for Use as Reflection-Type Filters

The spectrally selective reduction of unwanted radiant energy in optical instruments by means of reflection-type filters is an important application of modern evaporated front surface mirror technology. Evaporated metal–dielectric films, composed of absorbing and nonabsorbing film materials, can be deposited onto opaque metal mirror surfaces to produce very low reflectance over an extended wavelength region while leaving the high-reflectance region virtually unchanged. Mirrors such as these with low visible and high infrared reflectance, often referred to as "dark mirrors," are especially useful for reducing stray light in IR instrumentation (15, 47). They have also been employed in various types of devices for solar energy conversion (71, 72). The selective reflectance is achieved by using evaporated films whose optical thicknesses are great enough to produce interference effects at visible wavelengths but sufficiently thin to leave the IR reflectances unaffected. Hass *et al.* (15) have described two film combinations that clearly demonstrate this principle and that have served as the prototypes for a number of different dark mirrors with components modified for specific applications (71, 72).

The first design utilizes opaque Al coated with evaporated Ge and SiO, each film approximately one-quarter wavelength thick at visible wavelengths. Germanium was chosen because it absorbs strongly in the visible, but becomes nonabsorbing in the near IR. Figure 29 exhibits the construction and reflectance characteristics of such a filter mirror. Germanium is first evaporated onto opaque Al until the reflectance, controlled at a selected wavelength in the visible region, decreases from 91% to a minimum of about 30–40%. SiO is then added to the Al–Ge combination until the reflectance

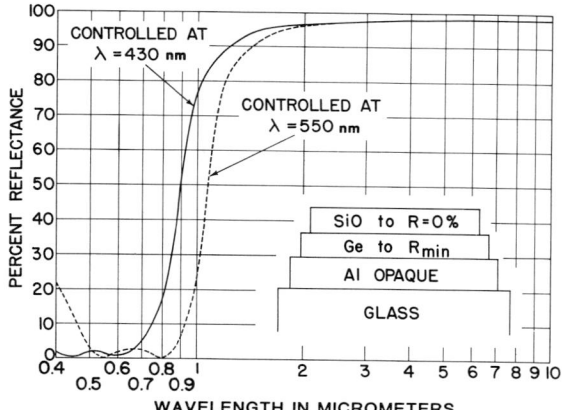

FIG. 29. Construction and reflectance of two Al–Ge–SiO filter mirrors as a function of wavelength in the region 0.4–10 μm. Layer depositions controlled at $\lambda = 430$ nm and $\lambda = 550$ nm.

becomes almost zero. If the coating deposition is controlled at $\lambda = 430$ nm, the filter mirror has a visible reflectance of about 2%. Its reflectance starts to rise rapidly at 0.8 μm, reaches 80% at 1.0 μm, 90% at 1.2 μm and is as high as that of uncoated Al for all wavelengths longer than 2 μm. When the coating deposition is controlled at longer wavelengths, the region of low reflectance shifts to longer wavelengths, as shown in Fig. 29. By controlling the deposition at $\lambda = 550$ nm, the steep rise in reflectance can be made to start at about 1.0 μm and to reach 90% at about 1.5 μm.

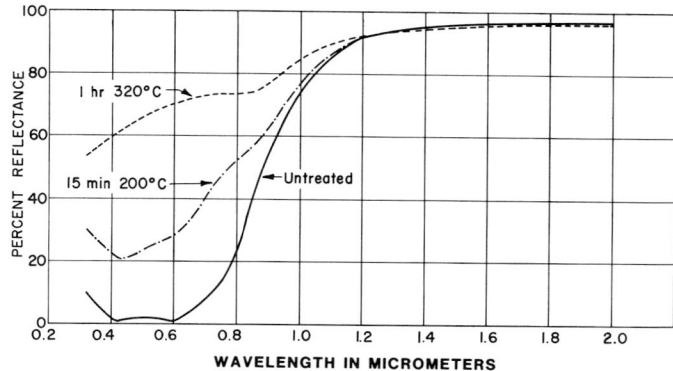

FIG. 30. Reflectance of an Al–Ge–SiO coating before and after heat treatments at 200°C and 320°C as a function of wavelength in the region 0.3–2.0 μm.

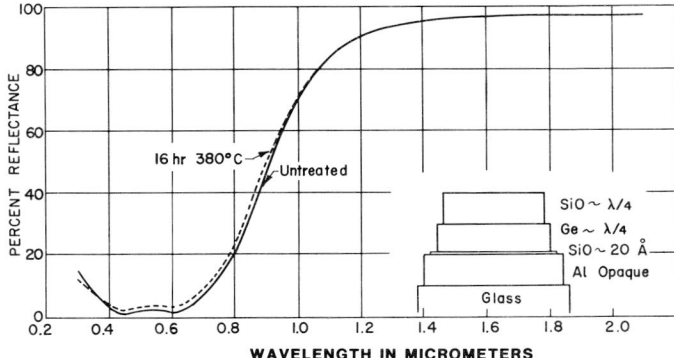

FIG. 31. Construction and reflectance of an Al(SiO)–Ge–SiO coating before and after 16 hr heat treatment at 380°C.

Dark mirrors constructed with the components shown in Fig. 29 are not very temperature resistant. This is important, because mirrors of this type are often used in situations where they receive prolonged exposure to intense light sources. If the temperature of such a mirror reaches 200°C, Al will diffuse into the Ge film and the "dark mirror" will lose its dark appearance. Figure 30 shows the visible and near-IR reflectance of an Al–Ge–SiO coating before and after heat treatments at 200 and 320°C (71). A short heat treatment at 200°C increases the mirror's visible reflectance to more than 20%, and a 1-hr exposure at 320°C results in a visible reflectance exceeding 65%. A minor change in the layer arrangement can greatly improve the temperature resistance of such coatings. This is demonstrated by the results presented in Fig. 31 (71). A thin film of true SiO, about 20 Å thick, is placed between the Al and Ge layers. This extremely thin layer of SiO does not change the reflectance characteristics of the thin-film combination, but forms a diffusion barrier between the Al and Ge films up to temperatures higher than 400°C. Dark mirrors constructed in this way are unaffected by long heat treatments at rather high temperatures, such as might be experienced by prolonged exposure to an intense light source.

The construction of another, more versatile, dark mirror type is presented in Fig. 32 (47). This film design uses two SiO coatings, separated by a semitransparent Al film, on top of opaque Al. The first SiO film is evaporated onto Al until the reflectance reaches a minimum of about 60–65%. Then Al is deposited until the reflectance, after passing through zero, reaches a value of 15–20%. Another film of SiO on top reduces the reflectance again almost to zero. When the deposition is controlled at $\lambda = 550$ nm, the reflectance is low in the visible and near IR, starts to rise rapidly at 1.6 $\mu$m, and reaches

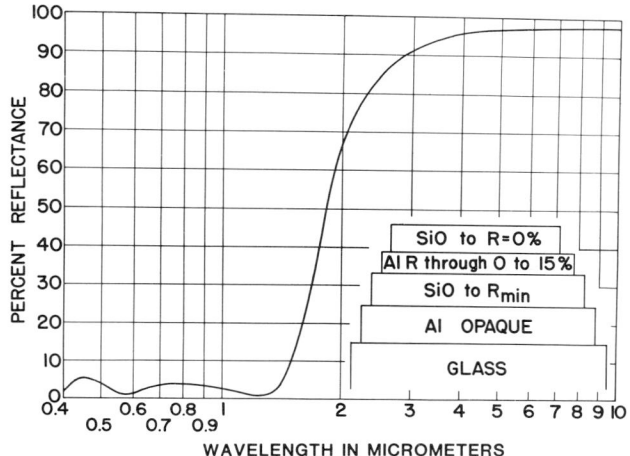

FIG. 32. Construction and reflectance of an Al–SiO–Al–SiO filter mirror as a function of wavelength in the region 0.4–10 μm. Layer depositions controlled at $\lambda$–550 nm.

90% at 2.8 μm, as shown in Fig. 32. The reflectance at normal incidence in the far IR is as high as that of uncoated Al. As in the previous design, the region of low reflectance of this type of dark mirror coating can easily be shifted to shorter or longer wavelengths.

The maximum temperature to which this coating, with opaque Al as the base layer, can be heated before it is destroyed due to crystallization effects is less than 450°C. Thermal stability at higher temperatures can be achieved by substituting higher-melting-point metal–dielectric combinations into the basic design shown in Fig. 32. The most popular usage of dark mirror coatings at the present time, one that also requires stability at high temperatures, is in solar thermal energy conversion systems (72). The low visible reflectance and high IR reflectance of these coatings translates directly into high solar absorptance and low thermal emittance, respectively. Schmidt and Janssen (73) reported that an interference coating in which Mo and $Al_2O_3$, in substitution for the Al–SiO of Fig. 32, are deposited onto a Mo substrate yields a mirror with a solar absorptance of 83% and thermal emittance of 11% at 530°C. This coating was subsequently shown to be unaffected by a 500-hr exposure to a temperature of 930°C and by a simulated exposure to space environment (74).

The low visible reflectance of dark mirror coatings of the metal–dielectric–metal–dielectric type was found to change very little with angles of incidence up to 50°, to increase slightly at 60°, and to become considerably higher at larger incidence angles.

## VI. Water Absorption in Evaporated Dielectric Films

It is well known that many evaporated dielectric film materials, such as $SiO_2$, $SiO_x$, and $MgF_2$, absorb water when exposed to air. This is especially true for coatings deposited at low rates and on unheated substrates. The absorbed water decreases the reflectance of a protected metal mirror significantly in the region close to 3 μm, where water has the highest extinction coefficient. An excellent way to study whether a material absorbs water and whether the water is absorbed at the surface or penetrates through the surface film is to deposit films $\lambda/4$ and $\lambda/2$ thick at $\lambda = 3$ μm and measure the reflectance of each coating at $\lambda = 3$ μm. If there is no water present in the film, only a shallow interference minimum will be observed in the Al + $\lambda/4$ $SiO_x$ spectrum. Figure 33 shows the calculated reflectance of Al coated with $\lambda/4$- and $\lambda/2$-thick films of $SiO_x$ with water of different thicknesses absorbed on the surface only. The optical constants of water published by Hale and Querry, $n \simeq 1.3$ and $k \simeq 0.3$ at $\lambda = 2.95$ μm (75), were used for the calculations. The curves show that the reflectance at $\lambda = 3$ μm of Al coated with $\lambda/4$-thick $SiO_x$ is greatly decreased by water absorbed at the surface, while Al + $\lambda/2$-thick $SiO_x$ shows no change in reflectance even for very thick absorbed water layers. The situation changes completely if the absorbed water penetrates into or completely through the $SiO_x$ protective coatings. This condition is shown in Fig. 34, which exhibits the measured reflectance of Al coated with $\lambda/4$- and $\lambda/2$- thick coatings of $SiO_x$ in the wavelength region 2–4 μm. The measured curves show clearly that the reflectance of Al protected with $\lambda/4$ and $\lambda/2$ $SiO_x$ films is greatly decreased at about 3 μm. This means that $SiO_x$ films absorb water and that the water penetrates into

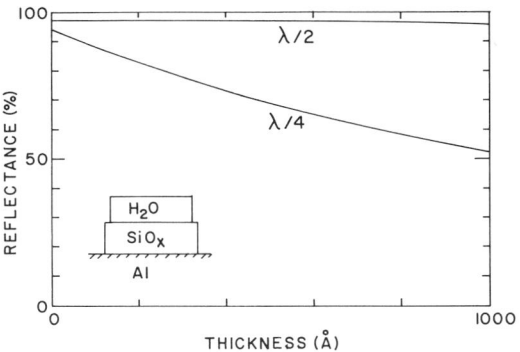

FIG. 33. Calculated reflectance at 3 μm of Al coated with $SiO_x$ films $\lambda/4$ and $\lambda/2$ thick at 3 μm, having water absorbed on the $SiO_x$ surface only.

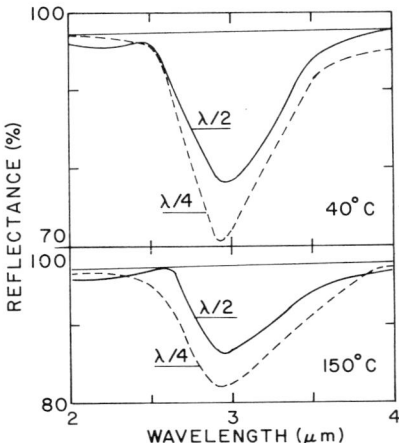

FIG. 34. Measured reflectance of Al coated with $SiO_x$ films $\lambda/4$ and $\lambda/2$ thick at 3 μm in the wavelength region 2–4 μm.

the $SiO_x$ coating. Increasing the substrate temperature during the deposition decreases the amount of absorbed water, which results in a smaller reflectance decrease. In order to determine whether the absorbed water penetrates uniformly all the way through the $SiO_x$ films to the underlying Al layer, calculations of reflectance as a function of the extinction coefficient $k$ for $\lambda/4$- and $\lambda/2$-thick films were made. The curves are shown in Fig. 35. The measured minimum reflectance of Al coated with $\lambda/4$-thick $SiO_x$ was found to be 71%, while that of Al + $\lambda/2$ $SiO_x$ was 77.5% (Fig. 34). Both reflectance

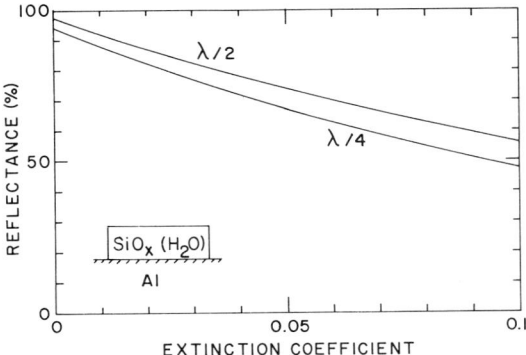

FIG. 35. Calculated reflectance at 3 μm as a function of the extinction coefficient $k$, with $n \simeq 1.50$ for Al coated with surface films $\lambda/4$ and $\lambda/2$ thick at 3 μm.

values agree with those calculated for uniform coatings of $k = 0.04$ and $n = 1.50$. It can, therefore, be concluded that water penetrates through the entire, rather thick $SiO_x$ coatings and forms homogeneous $SiO_x–H_2O$ films.

$MgF_2$ is probably the most common film material for use as a single layer and as a component in multilayer antireflection coatings. It is also employed as a protective coating for Al in the VUV (37). Its water absorption as a function of substrate temperature has been studied by various methods (76–78). Measurements of the refractive index of $MgF_2$ performed in vacuum directly after its deposition and again after exposure to air allow the determination of its packing density and water absorption. It is found that $MgF_2$ films deposited at room temperature have a packing density of 0.82 and a refractive index of only 1.32. Water absorption in air raises $n$ to 1.38, since holes of $n = 1$ are filled with water of $n = 1.33$. Higher substrate temperatures increase the refractive index and bulk values of $n = 1.39–1.40$ are obtained at substrate temperatures of 270–340°C. High-temperature films of this type are hard and do not change when exposed to air. For this reason durable antireflection coatings of $MgF_2$ can be obtained only by evaporation onto heated substrates.

ZnS and true SiO deposited at high rates and low pressures are two of the few film materials that exhibit bulk values of the refractive index and density when evaporated onto substrates at room temperature. Their optical properties measured in vacuum and after exposure to air are practically identical. Films of these materials one-quarter wavelength thick at $\lambda = 3$ μm evaporated onto Al exhibit only the expected shallow interference minimum at $\lambda = 3$ μm both before and after exposure to air. Consequently there is no indication of water absorption. This is not true for most fluoride and oxide films. For many of these coatings, water absorption can be decreased by the use of high deposition rates at very low pressures ($p < 10^{-6}$ Torr). However, this procedure is limited by the fact that at the elevated temperatures required for high deposition rates, many dielectrics, especially oxides, decompose, resulting in films that may be strongly absorbing.

A much better way to decrease or even eliminate the porosity and water absorption of evaporated dielectric films is by deposition onto heated substrates, as has been shown above for $MgF_2$ coatings. For example, $SiO_2$, $Al_2O_3$, and MgO films deposited by electron beam evaporation and $CeO_2$ coatings properly deposited from tungsten boats (49) show very pronounced water absorption after exposure to air when they have been evaporated onto room-temperature substrates. This water absorption disappears if the coatings are deposited onto substrates at 300°C. In addition, the high-temperature depositions produce films with a higher density and, consequently, higher $n$. For films deposited at 50 and 300°C, the $n$ value at $\lambda = 550$ nm of $CeO_2$ increases from 2.18 to 2.40, that of $Al_2O_3$ from 1.60 to 1.63, and that of MgO from 1.70 to 1.74. The high-temperature refractive index of MgO films is identical to that of single-crystal MgO (79). In addition, the high-temperature films are chemically and mechanically more durable than those produced at low temperatures.

Investigations have shown that the high water content of $SiO_x$ and $MgF_2$ films deposited at room temperature after exposure to air consists of two parts. One part, representing about 70%, is reversible, while the remaining part of about 30% is irreversible and chemically bound in the dielectric films (70, 80). This means that about 70% of the absorbed water in the films is released when the coated samples are placed into a high-vacuum system. However, the released 70% is immediately reabsorbed when the films are again exposed to air. These conclusions are supported by measurements of the optical thickness, variations in weight loss determined with a coated quartz crystal oscillator, and observations of the 3 $\mu$m water absorption band of $SiO_x$- and $MgF_2$-coated Al performed in air and after reinsertion into a high-vacuum system. The 70% water loss experienced by $SiO_x$ films in vacuum greatly decreases the 3 $\mu$m absorption of $SiO_x$-protected Al mirrors. This is important when mirrors of this type are used aboard spacecraft, especially since the $SiO_x$ films normally used as protective coatings are much thinner ($t \simeq 1500$ Å) than those used for studying the water absorption of evaporated $SiO_x$ layers ($\lambda/4$ thick at 3 $\mu$m, $t \simeq 5000$ Å). The use of rather thin protective layers, combined with the fact that 70% of the absorbed water departs in the vacuum conditions found in space, should practically eliminate the 3 $\mu$m absorption loss of Al + $SiO_x$ front surface mirrors used in space optical systems.

A more complete discussion of the packing density and refractive indices of many fluoride, sulfide, and oxide films has been published by Ritter (13, 81).

## References

1. G. Hass and R. Tousey, *J. Opt. Soc. Am.* **49**, 593 (1959).
2. R. P. Madden, *in* "Physics of Thin Films" (G. Hass, ed.), Vol. 1, pp. 123–186. Academic Press, New York, 1963.
3. G. Hass and W. R. Hunter, *in* "Physics of Thin Films" (G. Hass and M. H. Francombe, eds.), Vol. 10, pp. 71–166. Academic Press, New York, 1978.
4. G. Hass and N. W. Scott, *J. Opt. Soc. Am.* **39**, 1979 (1949).
5. A. P. Bradford and G. Hass, *J. Opt. Soc. Am.* **53**, 1096 (1963).
6. E. Ritter, Dissertation, University of Innsbruck, Austria, 1958.
7. A. P. Bradford, G. Hass, M. McFarland, and E. Ritter, *Appl. Opt.* **4**, 971 (1965).
8. J. T. Cox, G. Hass, and J. B. Ramsey, *J. Phys.* (*Paris*) **25**, 250 (1964).
9. J. T. Cox, G. Hass, and W. R. Hunter, *Appl. Opt.* **14**, 1247 (1975).

10. J. T. Cox and G. Hass, *Appl. Opt.* **17**, 333 (1978).
11. J. T. Cox and G. Hass, *Appl. Opt.* **17**, 1657 (1978).
12. J. T. Cox and G. Hass, *Appl. Opt.* **17**, 2125 (1978).
13. E. Ritter, *in* "Physics of Thin Films" (G. Hass and M. H. Francombe, eds.), Vol. 8, pp. 1–49. Academic Press, New York, 1975.
14. A. P. Bradford, G. Hass, and M. McFarland, *Appl. Opt.* **11**, 2242 (1972).
15. G. Hass, H. H. Schroeder, and A. F. Turner, *J. Opt. Soc. Am.* **46**, 31 (1956).
16. G. Hass, W. R. Hunter, and R. Tousey, *J. Opt. Soc. Am.* **46**, 1009 (1956).
17. W. R. Hunter, *Appl. Opt.* **6**, 2140 (1967).
18. R. Tousey, *J. Opt. Soc. Am.* **29**, 235 (1939).
19. I. Simon, *J. Opt. Soc. Am.* **41**, 336 (1951).
20. W. R. Hunter, *J. Opt. Soc. Am.* **55**, 1197 (1965).
21. R. P. Madden and L. R. Canfield, *J. Opt. Soc. Am.* **51**, 838 (1961).
22. J. Strong, "Procedures in Experimental Physics," p. 376. Prentice-Hall, New York, 1938.
23. H. E. Bennett and W. E. Koehler, *J. Opt. Soc. Am.* **50**, 1 (1960).
24. H. E. Bennett and J. M. Bennett, *in* "Physics of Thin Films" (G. Hass and R. E. Thun, eds.), Vol. 4, pp. 1–96. Academic Press, New York, 1967.
25. H. E. Bennett, Naval Weapons Center Report TP-6015, p. 87. Naval Weapons Center, China Lake, California, 1978.
26. D. M. Gates, C. C. Shaw, and D. Beaumont, *J. Opt. Soc. Am.* **48**, 88 (1958).
27. L. Harris and P. Fowler, *J. Opt. Soc. Am.* **51**, 164 (1961).
28. D. R. Herriott and H. J. Schulte, *Appl. Opt.* **4**, 883 (1965).
29. D. L. Perry, *Appl. Opt.* **4**, 987 (1965).
30. O. Arnon and P. Baumeister, *Appl. Opt.* **17**, 2913 (1978).
31. G. Hass, *J. Opt. Soc. Am.* **45**, 945 (1955).
32. G. Hass and J. Waylonis, *J. Opt. Soc. Am.* **51**, 719 (1961).
33. H. E. Bennett, M. Silver, and E. J. Ashley, *J. Opt. Soc. Am.* **53**, 1089 (1963).
34. B. Feuerbacher and W. Steinmann, *Opt. Commun.* **1**, 81 (1969).
35. E. T. Hutcheson, G. Hass, and J. K. Coulter, *Opt. Commun.* **3**, 213 (1971).
36. W. Walkenhorst, *Z. Tech. Phys.* **22**, 14 (1941).
37. P. H. Berning, G. Hass, and R. P. Madden, *J. Opt. Soc. Am.* **50**, 586 (1960).
38. R. P. Madden, L. R. Canfield, and G. Hass, *J. Opt. Soc. Am.* **53**, 620 (1963).
39. A. P. Bradford, G. Hass, J. F. Osantowski, and A. R. Toft, *Appl. Opt.* **8**, 1183 (1969).
40. K. H. Behrndt, *in* "Physics of Thin Films" (G. Hass and R. E. Thun, eds.), Vol. 3, pp. 1–60. Academic Press, New York, 1966.
41. H. Herzig, NASA Tech. Note D-3357 (March 1966).
42. H. L. Rook and R. C. Plumb, *App. Phys.* **1**, 11 (1962).
43. R. B. Love and W. K. Bower, *J. Vac. Sci. Technol.* **11**, 1124 (1974).
44. J. K. Coulter, G. Hass, and J. B. Ramsey, *J. Opt. Soc. Am.* **63**, 1149 (1973).
45. G. Hass and E. Ritter, *J. Vac. Sci. Technol.* **4**, 71 (1967).
46. L. Holland, *J. Opt. Soc. Am.* **43**, 376 (1953).
47. G. Hass, *in* "Applied Optics and Optical Engineering" (R. Kingslake, ed.), Vol. 3, pp. 309–330. Academic Press, New York, 1965.
48. G. Hass, J. B. Heaney, H. Herzig, J. F. Osantowski, and J. J. Triolo, *Appl. Opt.* **14**, 2639 (1975).
49. G. Hass, J. B. Ramsey, and R. E. Thun, *J. Opt. Soc. Am.* **48**, 324 (1958).
50. G. Hass, *J. Opt. Soc. Am.* **39**, 632 (1949).
51. G. Hass and C. D. Salzberg, *J. Opt. Soc. Am.* **44**, 181 (1954).
52. J. T. Cox and G. Hass, *J. Opt. Soc. Am.* **48**, 677 (1958).
53. W. Heitmann, *Appl. Opt.* **10**, 2414 (1971).

54. G. Hass, J. B. Ramsey, J. B. Heaney, and J. J. Triolo, *Appl. Opt.* **10**, 1296 (1971).
55. H. E. Bennett, J. M. Bennett, and E. J. Ashley, *Appl. Opt.* **2**, 156 (1963).
56. G. Hass, L. R., Drummeter, and M. Schach, *J. Opt. Soc. Am.* **49**, 918 (1959).
57. A. P. Bradford, G. Hass, J. B. Heaney, and J. J. Triolo, *Appl. Opt.* **8**, 275 (1969).
58. A. P. Bradford, G. Hass, J. B. Heaney, and J. J. Triolo, *Appl. Opt.* **9**, 339 (1970).
59. C. Boeckner, *J. Opt. Soc. Am.* **19**, 7 (1929).
59a. A. H. Lettington and G. J. Ball, Royal Signals and Radar Establishment Memorandum, No. 3295, Ministry of Defense, Worcs., England, Jan. (1981).
60. G. Hass, J. B. Heaney, and J. J. Triolo, *Opt. Commun.* **8**, 183 (1973).
61. L. Harris, *J. Opt. Soc. Am.* **45**, 27 (1955).
62. W. Heitmann, *Appl. Opt.* **12**, 394 (1973).
63. P. Baumeister and O. Arnon, *Appl. Opt.* **16**, 439 (1977).
64. H. E. Bennett, R. L. Peck, D. K. Burge, and J. M. Bennett, *J. Appl. Phys.* **40**, 3351 (1969).
65. D. K. Burge, H. E. Bennett, and E. J. Ashley, *Appl. Opt.* **12**, 42 (1973).
66. S. F. Pellicori, *Appl. Opt.* **19**, 3096 (1980).
67. J. M. Bennett and E. J. Ashley, *Appl. Opt.* **4**, 221 (1965).
68. I. Lubezky, E. Ceren, and Z. Klein, *Appl. Opt.* **19**, 1895 (1980).
69. G. Hass, *Vacuum* **2**, 331 (1952).
70. E. Ritter, in "Laser Handbook" (F. T. Arecchi and E. O. Schulz-DuBois, eds.), pp. 899–921. North-Holland Publ., Amsterdam, 1972.
71. L. F. Drummeter and G. Hass, in "Physics of Thin Films" (G. Hass and R. E. Thun, eds.), Vol. 2, pp. 353–357. Academic Press, New York, 1964.
72. R. E. Hahn and B. O. Seraphin, in "Physics of Thin Films" (G. Hass and M. H. Francombe, eds.), Vol. 10, pp. 1–65. Academic Press, New York, 1978.
73. R. N. Schmidt and J. E. Janssen, in "Thermal Radiation of Solids" (S. Katzoff, ed.), NASA SP-55, pp. 509–524. Natl. Aeronautics Space Admin., Washington, D.C., 1965.
74. P. E. Peterson and J. W. Ramsey, *J. Vac. Sci. Technol.* **12**, 174 (1975).
75. G. M. Hale and M. R. Querry, *Appl. Opt.* **12**, 555 (1973).
76. E. Ritter and R. Hoffmann, *J. Vac. Sci. Technol.* **6**, 733 (1969).
77. D. Hacman, *Opt. Acta* **17**, 659 (1970).
78. W. Heitmann and G. Koppelmann, *Z. Angew. Phys.* **23**, 221 (1967).
79. R. E. Stephen and I. H. Malitson, *J. Res. Natl. Bur. Stand.* **49**, 249 (1952).
80. H. Koch, *Phys. State Solidi* **12**, 533 (1969).
81. E. Ritter, *Appl. Opt.* **15**, 2318 (1976).

# Photoemissive Materials

## C. Ghosh*

*Bhabha Atomic Research Center*
*Optoelectronics Section*
*Bombay, India*

| | | |
|---|---|---|
| I. | Introduction | 54 |
| II. | The Mechanism of Photoemission | 54 |
| | 1. Optical Absorption | 55 |
| | 2. Transport of the Electrons | 59 |
| | 3. Escape from the Surface | 66 |
| III. | The Ag–O–Cs (S-1) Photocathode | 70 |
| | 1. Method of Preparation | 71 |
| | 2. Properties | 76 |
| | 3. Bi–Ag–O–Cs Photocathode | 84 |
| IV. | The Alkali Antimonides | 84 |
| | 1. Preparation of the Cathodes | 85 |
| | 2. Chemical Composition and Crystal Structure | 87 |
| | 3. Photoemissive Properties | 90 |
| | 4. Electrical Properties | 94 |
| | 5. Optical Properties | 101 |
| V. | Negative Electron Affinity Materials | 112 |
| | 1. The Lowering of Electron Affinity by Cs/O Activation | 114 |
| | 2. Carrier Transport | 123 |
| | 3. Diffusion Length and Doping | 125 |
| | 4. Surface Escape Probability | 126 |
| | 5. Activation Techniques | 128 |
| | 6. Fabrication of the Material for NEA Activation | 129 |
| | 7. Secondary Emission | 133 |
| | 8. Cold Cathodes | 134 |
| | 9. Dark Current | 135 |
| | 10. Stability | 135 |
| | 11. Photoemitters beyond 1.1 $\mu$m | 136 |
| VI. | Applications of Photoemissive Materials | 140 |
| | 1. Photomultipliers | 140 |
| | 2. Image Tubes | 145 |
| VII. | Conclusions | 157 |
| | References | 158 |

---

* Present address: ITT Electro-Optical Products Division, Roanoke, Virginia 24019.

## I. Introduction

Photoemission, or the external photoelectric effect as it was called, was discovered in 1887 (*1*) by Hertz, who observed that the length of the spark produced between two electrodes could be increased by exposing the negative electrode to ultraviolet (UV) radiation. After the discovery of the electron, it was realized that incident light of the proper frequency causes emission of electrons from the surfaces of materials. Because the classical electromagnetic theory failed to offer an explanation for some aspects of the phenomenon, Einstein developed an elegantly simple explanation by assuming the quantum nature of electromagnetic radiation. However, the application potential of this phenomenon could not be realized until the discovery of the Ag–O–CS (*2, 3*) photocathode. The discovery of Ag–O–Cs with a maximum quantum efficiency of about $10^{-2}$ electrons/photon opened up various fields of application, such as photometry, television, and night vision. The development of Ag–O–Cs was followed by that of other materials such as $Cs_3Sb$ (*4*), Bi–Ag–O–Cs (*5*), $Na_2KSb(Cs)$ (*6*), and finally the negative electron affinity (NEA) materials (*7, 8*) and the field-assisted photocathodes (*9*). The NEA cathodes increased the maximum quantum efficiency to 0.4 electrons/photon, and field-assisted cathodes extended the spectral response to 2.1 $\mu$m.

This review presents an overview of the physics and materials technology of the important photoemissive materials, including NEA materials and their application in various devices. The understanding of the physics of these materials has increased during the last 50 years and has generated a vast amount of information. A good deal of this information is presented here, to provide a better perspective on the state of the art.

## II. The Mechanism of Photoemission

For a long time photoemission was considered to be a surface phenomenon, with the term "surface" meaning the top few atomic layers in the solid (*10–13*). Subsequently, it was realized that most of the electrons are generated at depths greater than that which is considered to be surface in the solid. The absorption of photons by the surface states is quite small (*14*), and the maximum quantum efficiency of photoemission due to the surface cannot exceed $10^{-3}$ electrons/photon (*15*). Hence, photoemission from the surface is of little significance in the case of high-sensitivity photoemitters, in which the maximum quantum efficiency is usually greater than $10^{-1}$ electrons/photon.

Considering the bulk generation of the photoelectrons, the phenomenon of photoemission can be divided into three steps (16): (1) absorption of photons and consequent excitation of the electrons from the filled to the empty states; (2) the transport of the electrons from the point of generation to the surface; and (3) escape from the surface. To obtain a better insight into the phenomenon, these steps should be examined more closely. It should be remembered, however, that this three-step process is rather arbitrary and classical in origin. From a purely quantum mechanical standpoint, the event as a whole should be considered, rather than its somewhat arbitrarily defined parts.

## 1. Optical Absorption

When a beam of light is incident on a solid, a portion of it is reflected and the remainder is absorbed in or transmitted through it. The intensity of light $I$ at a depth $x$ from the surface is given by

$$I = I_0(1 - R)e^{-\alpha x} \tag{1}$$

where $I_0$ is the incident intensity, $R$ is the reflectivity, and $\alpha$ is the coefficient of absorpion, which is a constant for the medium for a particular frequency. The optical absorption coefficient, along with other constants, determines the optical behavior of the material. The other important parameters are the complex refractive index $N$ and the complex dielectric constant $\epsilon$ defined as

$$N = n + ik, \qquad \epsilon = \epsilon_1 + i\epsilon_2 = \epsilon_1 + (4\pi\sigma/\omega) \tag{2}$$

where $n$ is the real part of the refractive index, $k$ is called the attenuation index, $\epsilon_1$ and $\epsilon_2$ are the real and imaginary parts of the dielectric constant, $\sigma$ is the ac conductivity, and $\omega$ is the angular frequency. The complex refractive index is related to the complex dielectric coefficient by $N = \epsilon^{1/2}$, which gives

$$\epsilon_1 = n^2 - k^2, \qquad \epsilon_2 = 2nk \tag{3}$$

and the absorption coefficient is related to the attenuation index by

$$\alpha = 4\pi k/\lambda \tag{4}$$

where $\lambda$ is the wavelength of light. The reflectivity from a material at normal incidence is determined from the real and imaginary parts of the refractive index:

$$R = [(n - 1)^2 + k^2]/[(n + 1)^2 + k^2] \tag{5}$$

For high photoemission, a material should not have very high reflectivity. The metals, which generally have reflectivities between 80 and 100% in the visible and near-UV regions, are thus unlikely to be good photoemitters. The semiconductors and insulators, on the other hand, generally have reflectivities in the range of 0–30% and thus are better suited to become good photoemitters.

The absorption that gives rise to photoemission usually results from interband transitions from occupied to empty states at higher energies. The transition rate per unit volume for this kind of transition, as obtained from time-dependent perturbation theory, is given by (17)

$$P_{ij} = 4\pi^2 \hbar n^{-2}(e/m\omega)^2 |\langle j|\exp(i\mathbf{q}\cdot\mathbf{r})E_q \cdot \mathbf{V}|i\rangle|^2 \rho(\omega) \delta(E_j - E_i - \hbar\omega) \quad (6)$$

where $n$ is the real part of the refractive index, $m$ and $e$ are the mass and charge of the electron, $\rho(\omega)$ is the density of the electromagnetic energy in the medium, and $E_i$ and $E_j$ are the energies of the states $|i\rangle$ and $|j\rangle$. The rate of removal of energy from a beam of intensity $I$ can be expressed as

$$-dI = \hbar\omega \sum P_{ij} \, dx \quad (7)$$

Recalling that the intensity of the light beam is simply $I = \rho(\omega)c/n$ and that absorption at a particular depth of the material is proportional to the intensity at that depth, we obtain for the absorption coefficient

$$\alpha(\omega) = \frac{\hbar^2}{ne\omega}\left(\frac{e}{m}\right)^2 \sum_{ij} \langle j|e^{i\mathbf{q}\cdot\mathbf{r}} E_q \cdot \mathbf{V}|i\rangle \, \delta(E_j - E_i - \hbar\omega) \quad (8)$$

In the case of interband transitions, if it is assumed that the wave functions of the initial and final state are Bloch type with wave vectors $\mathbf{k}_i$ and $\mathbf{k}_j$, then the matrix element

$$\langle \mathbf{k}_j | e^{i\mathbf{q}\cdot\mathbf{r}} \nabla | \mathbf{k}_i \rangle \quad (9)$$

is zero unless

$$\mathbf{k}_i - \mathbf{k}_j + \mathbf{q} = \mathbf{g} \quad (10)$$

where $\mathbf{g}$ is the principal vector of the reciprocal lattice and can be taken to be zero if a reduced zone scheme is adapted for the energy bands. However, since $\lambda \gg a$, the lattice parameter, the photon wave vector $\mathbf{q}$ would be negligible compared to $\mathbf{k}_i$ or $\mathbf{k}_j$. Thus if the optical matrix element is not to vanish, the initial- and final-state wave vectors must be equal: $\mathbf{k}_i = \mathbf{k}_j$. Because of this condition the summation in Eq. (8) is reduced to an integration over the first Brillouin zone, and thus

$$\alpha(\omega) = \frac{\hbar^2}{nc\omega}\left(\frac{e}{m}\right)^2 \int \frac{d\mathbf{k}}{4\pi^3} \langle j|e^{i\mathbf{q}\cdot\mathbf{r}} E_q \cdot \mathbf{V}|i\rangle \, \delta(E_j - E_i - \hbar\omega) \quad (11)$$

The integral in Eq. (11) can be easily calculated if one assumes that the matrix element varies slowly with $\mathbf{k}$. Recalling the properties of the $\delta$ function we note that the integral

$$J(\omega) = \int \frac{d\mathbf{k}}{4\pi^3} \delta(E_j - E_i - \hbar\omega) \quad (12)$$

called the joint density of states, represents the number of states per unit energy whose energy difference $E_j - E_i$ equals $\hbar\omega$. In three dimensions the joint density of states is

$$J(\omega) = \frac{1}{4\pi^3} \int \frac{ds}{|\nabla_k(E_j - E_i)|_{E_j - E_i = \hbar\omega}} \quad (13)$$

where $ds$ is the element of surface of constant energy defined by $E_j - E_i = \hbar\omega$. Points in $k$ space for which $|\nabla_k(E_j - E_i)| \to 0$ are called critical points or Van Hove singularities (18). They were first discussed by Van Hove in connection with neutron scattering and later applied by Philips (19) to a discussion of electronic energy bands and optical properties. For the joint density of states to be large or for the existence of critical points, the two bands should be almost parallel at that point. There are four different types of critical points, each having a different behavior of the joint density of states in the surrounding region. A single critical point in the joint density of states at a particular energy would only produce an edge in the absorption spectrum. For the production of a peak in the optical spectrum, there should be superposition of two saddle-point-type critical points back to back. Figure 1 shows a certain portion of the band structure of Ge (20) and the observable critical points in it. Figure 2 shows the absorption spectrum of Ge (20) and compares it with the joint density of states spectrum. This very

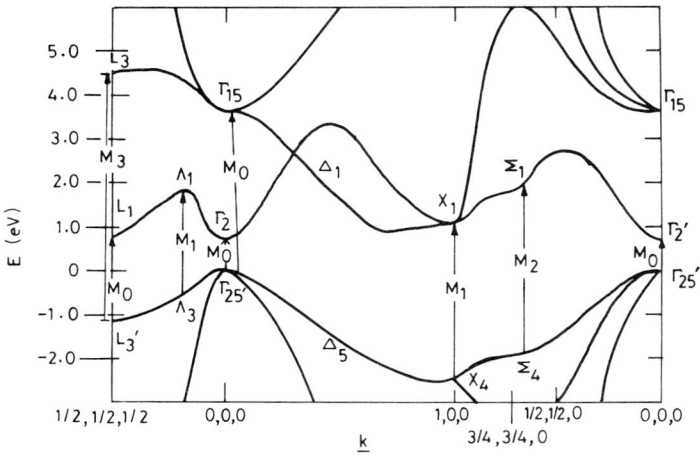

FIG. 1. The band structure of Ge. From Brust et al. (20).

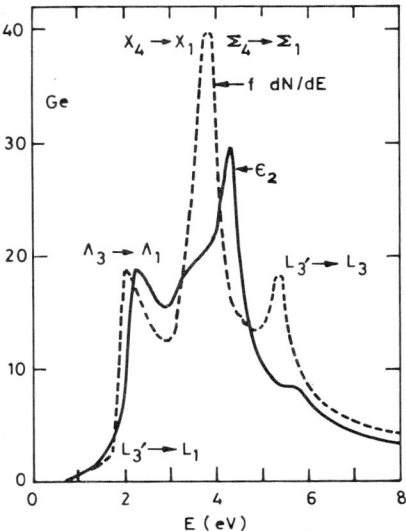

FIG. 2. Observed dielectric constant ($\epsilon_2$) and calculated joint density of state spectra for Ge. From Brust et al. (20).

clearly illustrates how the joint density of states is responsible for the optical absorption in a solid. The strong absorption bands observed at certain energies in some materials are due to the existence of a number of critical points in the same energy region. A very high absorption coefficient due to a few critical points at a particular energy is observable only when a good portion of the filled and empty bands is exposed to the incident photon energy. This is not possible in semiconductors near the band-edge region because only a small depth of energy in both bands is available for the excitation to take place. In semiconductors the absorption near the band edge is low and increases as one proceeds toward higher energy, because of higher density of states. Consequently, in the case of photoemission from conventional photoemitters, the quantum efficiency follows the same trend.

In many solids (e.g., Si, Ge, AgBr) the maximum of the valence band and the minimum of the conduction bands occur at different points in the Brillouin zone. The minimum energy of excitation across the gap then corresponds to an indirect transition with $\mathbf{k}_i \neq \mathbf{k}_j$. Such transitions are observed, and the conservation of the wave vector of the electron can occur through the emission or absorption of phonons.

The transition probability for indirect transitions is different from zero only at the second order, and is then much smaller than for direct transitions. This effect is partly compensated for by the relaxation of the selection rule for direct transitions in the evaluation of the joint density of states. A direct transition can take place from a given $\mathbf{k}_i$ to the states vertically above, while for indirect transitions many $\mathbf{k}_j$ are allowed, provided that energy and momentum are conserved. However, the absorption coefficient corresponding to indirect transitions is usually within the range $0–10^2$ cm$^{-1}$, while for direct transitions it is usually about $10^4$ cm$^{-1}$ and above. For this reason the optical attenuation length in the case of indirect transitions is often on the order of tens of microns; hence for conventional photoemitters, in which the photoemission is due to hot electrons with small escape depth of only a few hundred angstroms, indirect transitions make little contribution to photoemission. However, for negative electron affinity (NEA) materials, in which the photoelectrons are mostly thermalized and the escape depth is consequently very large, indirect transition can and do contribute significantly photoemission.

In the case of interband transitions, $\mathbf{k}_i = \mathbf{k}_j$ would be a selection rule if the initial- and final-state wave functions can be represented as Bloch wave functions. But if these states are relatively localized, rather than Bloch-like, because of, for example, electron–electron scattering, then $\mathbf{k}$ conservation ceases to be an important selection rule. Transitions that do not follow $\mathbf{k}$ conservation had been first identified by Spicer (21). He called these nondirect transitions, to distinguish them from indirect transitions. These transitions do not involve phonons, because optical absorption was found to be large ($>10^5$ cm$^{-1}$) and independent of temperature. As in indirect transitions, for a given $\mathbf{k}_i$, many $\mathbf{k}_j$ are allowed, provided that energy is conserved. However, the optical matrix elements between different coupling states would not be constant; they would reach a maximum in the region where the wave vectors of the two states were almost equal and would decrease as the difference between the initial and final wave vectors increased (22). Such absorption was found in most of the conventional alkali antimonide type of photoemitters, such as $Cs_3Sb$, $K_3Sb$, $Na_3Sb$, $K_2CsSb$, and $Cs_3Bi$ (23–25). Similar types of transitions are observed in amorphous materials (26).

## 2. Transport of the Electrons

The electrons that are excited to higher energy by absorption of photons in the material should move to the surface. There are various mechanisms by which an electron can lose energy in the solid. The most important of these loss processes is electron–electron scattering, for which the mean free path is low and energy loss is high. Electrons also lose energy by lattice scattering, a process in which the energy loss is smaller. In polycrystalline materials, such as conventional photoemitters, there is also energy loss by scattering at grain boundaries and other structural imperfections. If an electron's energy is reduced to below the vacuum level, there is very little probability of its being emitted from the surface. Therefore, for good photoemission it is necessary that the loss of energy suffered by the electron in the solid should be small. In electron–electron scattering a high-energy electron excites an electron from the occupied states. In semiconductors there is a threshold energy for this process, because to excite a valence band electron an amount of energy at least equal to the band gap must be given to the electron. This process has been investigated in semiconductors both experimentally (27–35) and theoretically (36) by a number of workers. The generation of secondary carriers in high fields is due to electron–electron scattering, and has been studied in many semiconductors. Wolff (36) calculated the rate of generation of electron–hole pairs by this process. His method of calculation consists essentially of solving Boltzmann's equation for the motion of electrons in a high field, taking into account the effect on the distribution function of electron–phonon and pair-producing collisions.

Because of the requirement of energy and crystal momentum conservation, the threshold of pair production is usually higher than the band-gap energy. If one assumes spherical nondegenerate energy bands with equal mass for electrons and holes, then this threshold could be estimated at about 1.5 times the band gap, taking into account the conservation laws. The band structures of such semiconductors as Si and Ge are, however, more complicated than this; in addition, such an estimate takes no account of the Umklapp process. The latter would certainly permit some pair production at somewhat lower energies, but this effect would probably be of little importance in determining the energy at which the mean free path for pair production becomes smaller than that for electron–phonon interaction. The band structure has an important role in determining this threshold, as different band shapes have different effective mass ratios. Considering the details of band shape, etc., this threshold in silicon was calculated to be 2.0 eV. An experiment (27) with bombardment of $\alpha$ particles gave a threshold of 3.6 eV, whereas measurement of the quantum yield of the photovoltaic effect (37) gave a threshold of less than 2.15 eV. Measurement of avalanche breakdown (15) gave a value of 1.8 eV for electrons and 2.4 eV for holes. It appears that the experimental results are close to theoretical predictions. The mean free path calculated by taking a value of 2 eV for this threshold would be about 200 Å (36). The calculated rate of pair production rises sharply from a small value near threshold and tends to saturate at higher energies. Correspondingly, the mean free path should also initially decrease rapidly, but should saturate at higher energies. This type of behavior of mean free path was observed from photoemission data for alkali antimonide photocathodes (38, 39).

Since the minimum energy loss in a pair-production collision in a semiconductor is equal to the band gap, the position of the threshold energy should determine the photoemission from a material to a great extent. Sommer and Spicer (15) classified materials on the basis of the relative position of the pair-production threshold and the position of the vacuum level or the magnitude of the electron affinity. They divided the semiconductors into three general classes (Fig. 3). In the first group are the materials in which the threshold for pair production is so large compared with electron affinity (the difference in energy between the bottom of the conduction band and the vacuum level) that even after a pair-production collision the energy of the electron would be above the vacuum level. Therefore, the electron would a good probability of escape even after creation of one electron–hole pair. If the photoemission threshold is small, then an electron from the valence band may be excited to an energy above the vacuum level by scattering. An electron generated this way could contribute to photoemission. The escape depth from these materials would be very large and they would have the highest photoelectric yield.

In the second group of materials, the threshold for pair production has a value lower than the electron affinity. Therefore, the electrons that could

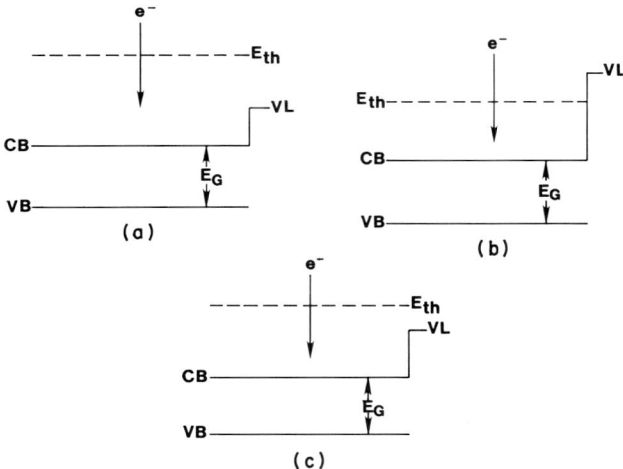

FIG. 3. Scheme of classification of various photoemitters. (a) Type 1: The threshold of pair production $E_{th}$ is very large, so that the electron can escape even after creating an electron–hole pair. This type of material would make good photoemitters. (b) Type 2: The electron affinity $E_A$ is greater than $E_{th}$, so that any electron that can escape would have high probability of creating an electron–hole pair. This type of material would make poor photoemitters. (c) Type 3: $E_{th} > E_A$, but the difference is small, so that only in a narrow energy range would the electrons that might escape not lose their energy by forming pairs. This type of material is also not likely to make good photoemitters.

escape would also have a good probability of taking part in pair-production collisions. If such a collision occurred, the electron would have very little probability of escape unless its energy were very large. Such materials would have very poor escape probability for the electrons. In a third and intermediate group of materials, the threshold of pair production has a value relatively close to, but larger than, the electron affinity. For these materials, the electrons with energy greater than the electron affinity but smaller than the threshold for pair production would have large escape depths, whereas those electrons with energy greater than the pair-production threshold would have a small escape depth. In effect, these types of materials would have intermediate photoemissive quantum efficiency.

It should be remembered that the undesirable effects of electron–electron scattering on photoemission are significant only for conventional photoemitters. In the case of NEA emitters, since the electrons thermalized at the bottom of the conduction band can also escape, pair production should actually increase the quantum efficiency. Although this aspect has not been investigated so far for photoemission from NEA materials, it seems quite reasonable to assume that a low threshold energy and a small mean free path for electron–electron collision would be quite helpful for photoemission.

In the case of metals, there would be no threshold energy for electron–electron scattering, and this scattering would therefore be a dominant mode of energy loss. There is another mode of energy loss for metals that results from the interaction of the excited electron with a set of conduction electrons. This type of excitation becomes important if a plasma resonance can be excited and the loss of energy is quantized into surface or volume plasmons. Quinn (40) found that for plasmon scattering in a free-electron metal the mean free path is

$$l_{pe} = \frac{A_0 p^2}{m\hbar\omega_p} \left[ \ln \frac{(P_F^2 + 2m\hbar\omega_p)^{1/2} - P_F}{P - (p^2 - 2m\hbar\omega_p)^{1/2}} \right]^{-1} \quad (14)$$

where $\hbar w_p$ is the plasmon energy, $P_F$ is the momentum of an electron at the Fermi energy, and $A_0$ is the Bohr radius. Figure 4 represents the variations of mean free path $l_{pe}$ calculated from this equation and with the electron energy above Fermi energy, as calculated by Smith and Fisher (41) for cesium. Figure 4 also shows the variations of mean free path $l = l_{pe}l_{ee}/(l_{pe} + l_{ee})$ that result from the combination of plasmon and electron–electron scattering. In this case the mean free path for electron–electron scattering plays the dominant role, but the situation can be different in other metals. For these strong energy loss processes in the metals, in addition to their high reflectivity, they are not likely to be good photoemitters. Figure 5 shows the quantum efficiency of cesium films (42) and Fig. 6 shows the quantum efficiency of a clean copper sample (43). It can be observed that the maximum quantum efficiency is less than $10^{-2}$ electrons/photon.

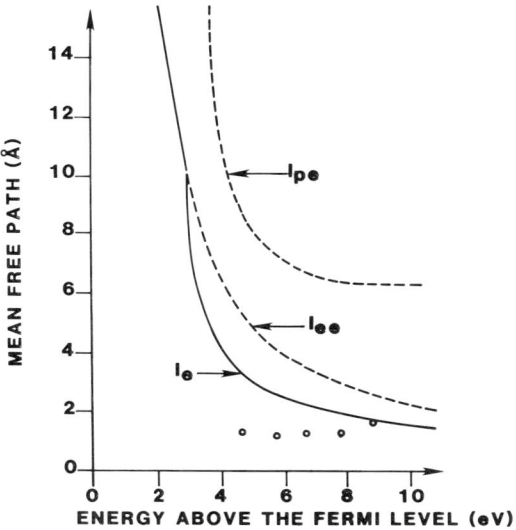

FIG. 4. Theoretical estimate of mean free path of electrons in Cs for plasmon scattering ($l_{pe}$), for pair creation ($l_{ee}$), and for the combined effect of the two ($l_e$). The circles represent the experimental estimate of elastic escape depth. From Quinn (40).

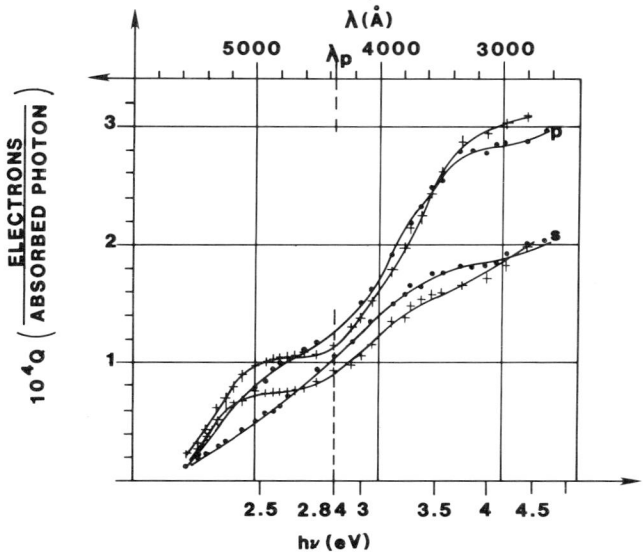

FIG. 5. Spectral distribution of the "true quantum yield" $Q$ of an opaque film of Cs after deposition at 77 K (+) and after reheating at 195 K (●) for s and p polarizations, respectively. From Monin (42).

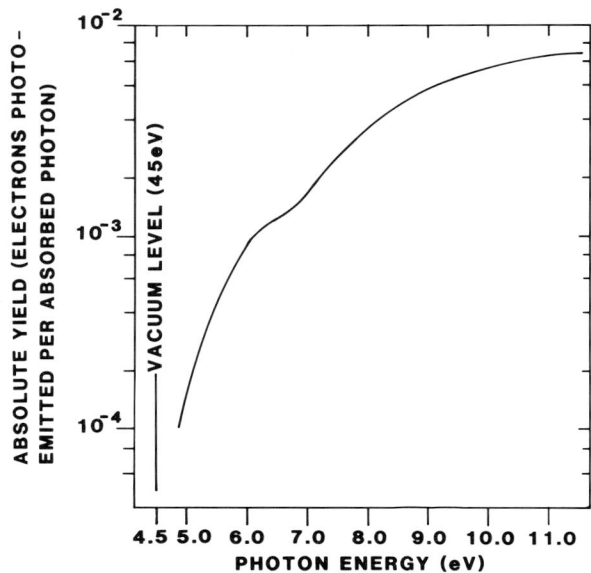

FIG. 6. Quantum yield from a clean Cu sample. From Krolikowski and Spicer (43).

The other mechanism of energy loss is by means of lattice scattering. This would be the dominant loss mechanism for electrons with energy below the pair-production threshold. Electrons with energies approximately 1 eV above the thermal electron would lose their energies primarily to optical phonons (44). The energy loss per collision with optical phonons would be about 0.1 eV, and the mean free path would be about 100 Å, a value that does not vary appreciably with energy (45, 46).

Scattering at grain boundaries and other structural imperfections would be of importance in polycrystalline materials, but it would not be of much importance in single-crystal photoemitters. Almost all the conventional photoemitters are polycrystalline materials. An increase in the grain size appears to increase the photoemission from the alkali antimonide photoemitter $K_3Sb$ (47). If polycrystalline GaAs is treated with Cs–O, the resulting photoemitter does not have good quantum efficiency. This is evidently due to scattering and recombination at the defects and grain boundaries. Similarly, amorphous materials are also not expected to be good photoemitters, because of the poor mobility of the electrons in the materials.

The probability of electrons reaching the surface after encountering energy-loss processes has been derived by a number of workers. The Fermi age theory was used by Hebb (48) and Lye and Dekker (49) to calculate this probability for secondary electrons. They assumed phonon loss as the only loss mechanism and that an electron of energy $E$ above the bottom of the conduction band $E_c$ loses a fraction $\xi$ of $E - E_c$ in each phonon creation. They found the probability of reaching the surface to be

$$F(x) = \text{erfc}(x/2) \tag{15}$$

where

$$\text{erfc}(x) = (2/\pi)^{1/2} \int_x^\infty \exp(-x^2)\, dx$$

and $\tau_F$ is the age of the electron after excitation:

$$\tau_F = \int_{E_v}^{E} Dl_p\, dE/\xi E \tag{16}$$

where $D$ is the diffusion coefficient due to phonon scattering, and $l_p$ is the mean free path due to this scattering. However, it was found that the photoelectric data could be fitted reasonably well by a function

$$F(x) = F(0)e^{-x/L} \tag{17}$$

where $F(0)$ is the probability of escape of an electron generated right at the surface and $L = \tau_F^{1/2}$.

A similar result (50) was arrived at by using Fermi age equations and assuming loss of energy by electron–phonon and electron–electron interactions (23). In this case $L$, which is called the escape depth, should be represented by

$$L = [l_e^2 l_p/3(l_e + l_p)]^{1/2} \tag{18}$$

where $l_e$ is the electron–electron and $l_p$ the electron–phonon mean free path. Spicer (16) used a similar function for deriving the theoretical expressions of quantum efficiency for alkali antimonide photocathodes and found agreement with experimental results.

The transport of electrons from the point of generation to the surface is esssentially hot electron transport, and all of the above discussion pertains to that particular problem. In the case of NEA emitters however, the whole character of this electron transport changes. Since the escape of thermalized electrons is possible, the transport mechanism changes from hot electron transport to minority carrier diffusion, and very large escape depths are possible. The escape depth for a positive electron affinity photoemitter is typically a few hundred angstroms. The equivalent lifetime during which a hot electron can escape is on the order of $10^{-13}$ sec. In a NEA material the electrons can come to equilibrium at the conduction band minima, where the lifetime is limited only by trapping and recombination with holes. These lifetimes are much longer, typically $10^{-9}$ sec in the $\Gamma$ minimum for III–V compounds and $10^{-7}$ sec in the X minimum for silicon. The electron velocities are still large, making much greater escape depths possible. The escape depth can be identified with the minority carrier diffusion length $L = (D\tau)^{1/2}$, where the diffusion constant $D = \mu kT/q$, $\mu$ being the electron mobility. Taking $D = 100 \text{ cm}^2 \text{ sec}^{-1}$ and $\tau = 10^{-9}$ sec, we obtain $L = 3 \mu\text{m}$. Usually the escape depth varies between 0.1 and 10 $\mu$m for NEA materials. We examine the transport equations for NEA material in Section V.

### 3. Escape from the Surface

In the case of positive electron affinity photoemitters, the photoexcited electrons encounter a barrier at the surface, which has to be overcome for emission into vacuum. In the case of a semiconductor, the surface barrier is expressed in terms of electron affinity, which is the difference in energy between the vacuum level and the bottom of the conduction band. Its magnitude is usually calculated by subtracting the band gap from the threshold energy for photoemission. However, in the case of direct transitions the apparent electron affinity, which would be obtained by subtracting the band gap from the threshold energy, could be larger than the actual electron affinity. As can be seen in Fig. 7, for the electrons to be excited to vacuum level and above, the initial energy need not be at the top of the valence band. Therefore, the true electron affinity would actually be less than the apparent

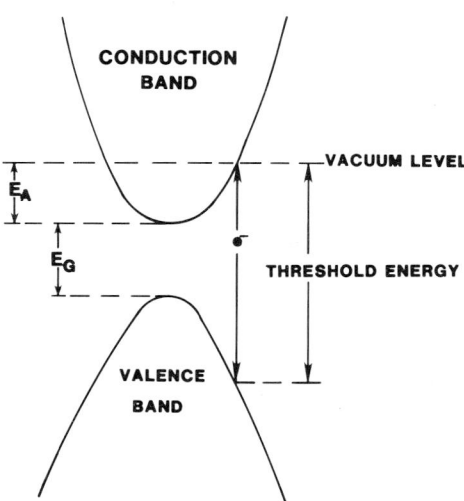

FIG. 7. In the case of direct transitions and positive electron affinity, the threshhold energy for photoemission is often higher than the sum of the band gap ($E_G$) and real electron affinity ($E_A$).

electron affinity by an amount equal to the difference in energy between the top of the valence band and the initial energy of the electrons excited to the vacuum level.

The probability of escape of the electrons would increase if the potential barrier at the surface or the electron affinity were reduced. The electron affinity of a material is strongly dependent on the conditions of the surface and the doping in the material. If the electron affinity can be reduced, the threshold energy for photoemission will also decrease. A low threshold energy can be obtained by choosing a material with smaller band gap, but small band gaps give rise to the higher associated noise or dark emission that results from thermionic emission. Reduction of the surface barrier by some surface treatment has the advantage of maintaining the position of the bands at the same energy inside the volume but still lowering the surface barrier by bending the bands at the surface.

In a crystalline solid the surface can be considered to be a discontinuity in the periodic potential of the crystal. The atoms at the surface give rise to bound states that may lie in the same regions of energy as the volume states or in the forbidden energy range for the volume state. The surface states that are in the same energy range as the bulk states are not easily distinguishable from the volume states, but the surface states in the forbidden gap dominate the behavior of the semiconductor surface. The density of surface states in the forbidden gap may be comparable to much less than the "one per surface atom" sometimes postulated.

If the density of surface states is large, these states pin the Fermi level at the surface. Changing of the type of conductivity or the position of the Fermi level inside the bulk results in filling or emptying of the surface states, a change in the net charge residing on the surface, and formation of a depletion or accumulation layer near the surface. In the case of p-type material with n-type surface states, the bands bend downward at the surface, causing a decrease in apparent electron affinity, while p-type states on n-type material bend the bands upward, causing an increase of apparent electron affinity (15).

It is possible to determine the amount of band-bending at the surface by solving the Poisson equations with appropriate boundary conditions (51). Assuming that the space charge is equal to $eN_a$ from $x = 0$ to a depth $d$ in the depletion region, at the surface the boundary conditions would be $E = 0$ at $x = d$ and $E = -eN_s/\epsilon$ at $x = 0$, where $\epsilon$ is the dielectric constant and $N_s$ is the number of surface states above the Fermi level. The solution is

$$E = -(eN_a/\epsilon)(d - x) = -(eN_s/d\epsilon)(d - x) \tag{18}$$

For the electrostatic potential $V$ we have, taking $V = 0$ at $x = d = N_s/N_a$,

$$V = -(eN_a/2\epsilon)(d - x)^2 \tag{19}$$

The surface potential is then (at $x = 0$)

$$V_s = eN_s^2/2\epsilon N_a \tag{20}$$

This is the depth of band-bending at the surface. If we take a typical case of $\epsilon = 10\epsilon_0 \simeq 10^{-10}$ F m$^{-1}$, $N = 2 \times 10^{17}$ cm$^{-3}$, $N_s = 10^{11}$ cm$^{-2}$, we would have $V_s = 0.5$ eV. In the case of a semiconductor in which the surface state density is not sufficient to produce the desired band bending, a low work-function material like Rb or Cs can be added which would produce n-type surface states, increase the density of surface states, and cause more band-bending in accordance with Eq. (20).

There is another way of reducing the effective electron affinity at the surface, namely by adding a thin layer of n-type material on p-type bulk. The bands bend at the surface, thus lowering the electron affinity. The surface layer should be thin enough for the photoemitted electrons to pass through without much energy loss. An example of this type of band-bending is the generation of negative electron affinity by using a thin layer of n-type $Cs_2O$ on the surface of p-type GaAs. In the case of a multialkali photocathode, it is believed that a thin layer of intrinsic $K_2CsSb$ on p-type material produces band-bending at the surface (52).

If the electron arrives at the surface with energy higher than the vacuum level, it may have a reasonable probability of escape, however, there is another factor, which is the escape cone consideration (53).

If we consider an electron with momentum $\hbar k$ traveling toward the vacuum surface at an angle $\theta$ with respect to the surface normal, then the component of energy corresponding to the perpendicular component of momentum should be greater than the surface barrier. That is,

$$\hbar^2 k_x^2/2m \geq E_A \tag{21}$$

where $k_x$ is the normal component of momentum and $\cos\theta = k_x/k$. The escape cone for electrons with energy $E = \hbar^2 k^2/2m$ above the bottom of the conduction band is defined by the critical angle $\theta_c$, which is the maximum value of $\theta$ that will permit an electron to escape. The critical angle is found with the help of the two conditions given above to be

$$\theta_c = \cos^{-1}(E_A/E)^{1/2} \tag{22}$$

If the electrons are isotropically distributed after generation, then the probability that an electron is within the escape cone is

$$T_0(E) = \tfrac{1}{2}[1 - (E_A/E)^{1/2}] \tag{23}$$

This result is found simply from the ratio of the solid angle included by the escape cone to the total angle $4\pi$. A look at the escape function reveals that the probability of escape is very small when $E \simeq E_A$, but it approaches a value of 0.5 at electron energies much higher than photoemission threshold.

We can derive an expression for photoemission quantrum efficiency by taking into consideration the factors discussed in the foregoing analysis. If we consider a thin layer of a photoemitter of thickness $d$, then the intensity of light at a depth $x$ from the surface when the light is incident from vacuum side would be

$$I = I_0(1 - R)\exp(-K_T x)$$

where $K_T$ is the total absorption coefficient, $I_0$ is the incident intensity, and $R$ is the reflectivity. The absorption of light that can give rise to photoemission in an infinitesimal thickness $dx$ at depth $x$ would be (*16*)

$$dI(hv) = \frac{K_p}{K_T}\frac{dI}{dx}dx$$

where $K_p$ is the fraction of the absorption coefficient due to the transitions for which the final states are above the vacuum level. Now considering the transport and escape functions as described earlier, we have for the photoemissive current for monochromatic light of photon energy $hv$

$$i(hv) = \int_0^d dI(hv)F(hv)T_0(E)\,dx$$

The quantum efficiency in terms of electrons per photon would be

$$Y(hv) = \frac{i}{I_0} = \frac{1}{2}(1-R)\left[1 - \left(\frac{E_A}{E}\right)^{1/2}\right]$$

$$\times \frac{K_p(hv)F_0(hv)}{K_T(hv) + g_0(hv)}[1 - \exp\{-[K_T(hv) + g(hv)]d\}] \quad (24)$$

where $g(hv) = [L(hv)]^{-1}$. This equation highlights the importance of various parameters that are related to photoemissive efficiency of the material, particularly the parameter $K_p/K_T$. Apparently, if $K_p/K_T$ is very small, which happens near the threshold region if the electron affinity is large, the photoemission quantum efficiency will be poor. The ratio of $K_p/K_T$ is generally dependent on the ratio $(hv - E_g - E_A)/E_A$, and $E_A$ should be as small as possible for this factor to be high, particularly in the low-energy region. It is found that materials with low $E_A$ have high quantum yield compared to materials with high $E_A$. This is true not only of alkali antimonides with low $E_A$ but also of materials such as cesium and rubidium tellurides, cesium iodide ($E_A < 0.5$ eV for all), and barium oxide ($E_A = 1$ eV). All have very high quantum efficiency in the UV region.

The tremendous increase in quantum efficiency in the threshold region resulting from the application of a high field and the consequent lowering of electron affinity have been observed for multialkali and other photoemissive materials (54–56). It is observed that the increase in photoemission is principally in the threshold region, in which the lowering of electron affinity can affect the photoemission most. An increase of photoemission by a factor of more than 2 with a field of $2 \times 10^4$ V cm$^{-1}$ has been observed, which shows the effect on photoemission of lowering of electron affinity

The important general conclusions that can be arrived at from this discussion on the phenomenon of photoemission are the following:

(1) The metals are not likely to be good photoemitters because of their high reflection and large electron–electron and electron–plasmon interaction cross sections.

(2) The material (semiconductor or insulator) should have a high absorption coefficient due to direct or nondirect transitions for positive electron affinity photoemitters. However, for NEA materials the absorption coefficient need not be as high.

(3) The electron–electron scattering threshold energy should be well above the vacuum level for good photoemission and the electron affinity should be as small as possible. However, for NEA materials this threshold is immaterial.

(4) The reduction of electron affinity can be achieved by p-type doping of the material and having n-type surface states or by having an n-type surface layer on p-type material. In either case, a p-type material would generate fewer of the thermal electrons that cause dark emission.

## III. The Ag–O–Cs (S-1) Photocathode

As mentioned earlier, the Ag–O–Cs or S-1 photocathode was the first material to be discovered that had high sensitivity and could be used for device applications. The maximum quantum efficiency was about $10^{-2}$ electrons/photon, and even now this is the most commonly used commercial photoemitter for wavelengths up to 1.2 μm. The present-day use of this photocathode is primarily because of its longer threshold wavelength compared to other photoemitters.

The discovery of the Ag–O–Cs photocathode followed the experimental observation that the work function of metals such as tungsten can be reduced drastically if they are superficially oxidized and, subsequently, a thin layer of cesium is deposited on them (57). Working on such combinations with silver, Koller (2) and Campbell (3) observed that Ag–O–Cs had much higher sensitivity than any previously known material.

1. METHOD OF PREPARATION

After the discovery by Koller and Campbell, an important improvement in the technique of preparation was made by Asao and Suzuki (58). Other workers (59–64) have made a number of small alterations to the processing method, but it is basically a process in which a silver substrate is oxidized and then activated by cesium vapor, often followed by the evaporation of a thin film of silver and superficial oxidation.

The silver substrate used to process the photocathode can be chemically deposited by reduction of silver nitrate by an organic reducing agent (65). The film produced by this method, however, is thick, and though photocathodes of good sensitivity can be prepared in this way, they can be used only in the reflection mode. Even a plate or a thin sheet of silver can be used for preparing Ag–O–Cs cathode (66). For processing semitransparent cathodes, which are used in most of the devices, the initial layer should be very thin and can be formed only by evaporation. The thickness of the silver layer should correspond to about 60% loss of transmission for white light. The condition of the substrate, the rate of evaporation, angle of incidence, etc. influence the growth of silver films and the light transmission through them. The growth of thin films of silver or any other material starts after nucleation on the substrate. Initially the film is of island type; the islands grow larger as their thickness increases, and after a certain stage they coalesce to form a continuous film (67).

The growth of silver films and their optical properties have been studied by a number of workers (68–71). Sennett and Scott studied extensively the growth of silver films at different rates of evaporation, using electron microscopy, and measured their optical transmission and reflection. They observed that for a high rate of evaporation (175 Å sec$^{-1}$), the film coalesces at a thickness of 120 Å, while for a low rate of evaporation (0.08 Å sec$^{-1}$), the film does not coalesce even at a thickness of 300 Å. The thin film shows a type of absorption behavior that probably could be explained by Maxwell Garnett's theory of absorption by an aggregate system of metallic particles (72). The thin films show an absorption band in the visible region, which can not be accounted for by interband transitions. The peaks in the absorption for thicknesses of 30, 70, and 110 Å are at 4500, 5000, and 5500 Å, respectively. This is consistent with Maxwell Garnett's theory that the peak in the absorption band due to the system of aggregate metallic islands should shift toward longer wavelengths with increase of island size. For silver film this type of absorption has been identified by others (71, 73, 74). Since different rates of evaporation produce different sizes of islands, at a particular thickness the absorption due to islands is very different. Therefore, in the determination of thickness by measurement of light transmission, one would arrive at different thicknesses of film depending on the rate of evaporation and the nature of the substrate. This has in fact been observed. Figure 8 shows the variation of transmission of silver films with thickness for two rates of evaporation, as observed by Philip (70). The increase in transmission with increase in thickness, which appears somewhat paradoxical, is probably caused by higher absorption at smaller thickness due to Maxwell Garnett's type of absorption. The curves of Fig. 8 give an indication of how much difference there can be in thickness between films with the same transmission but different rates of evaporation. For 50% loss of transmission, the film thickness can be between 100 and 500 Å. Even with this variation, it appears that various workers, starting with the same light transmission for the initial film, arrive at ultimate sensitivities that are not too different (66). This shows that the thickness of the initial silver layer is not so critical for the processing of high-sensitivity Ag–O–Cs cathodes. Of course, some change in spectral response has been reported with different rates of evaporation of the initial film (64). It has also been observed by Hoene and Saggan (75), who

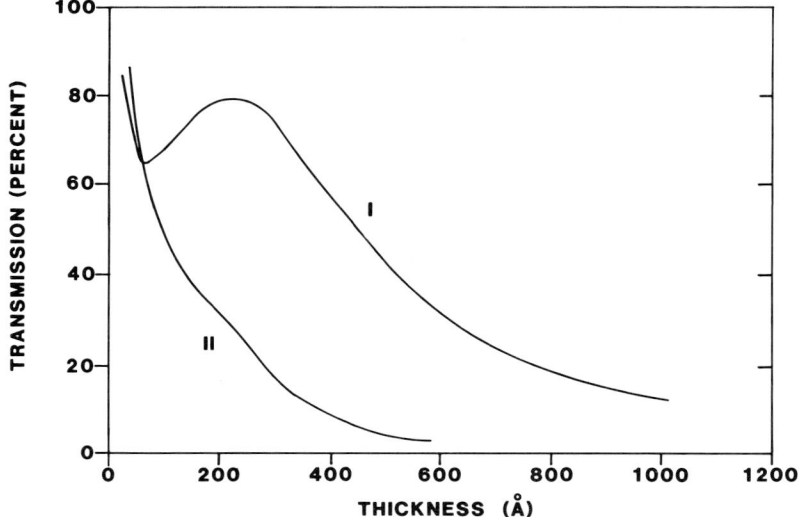

FIG. 8. Light transmission ($\lambda = 4800$ Å) of thin silver films as a function of thickness for two different rates of evaporation. Curve I, 500 Å min$^{-1}$; curve II, 5 Å min$^{-1}$. From Philip (70).

used single crystals of silver for processing of this cathode, that there is no observable effect of the crystal face on the emission property of the cathode. It has generally been observed in our laboratory that a loss of transmission of 60% achieved in about 10 sec gives the best sensitivity for the ultimate cathodes. According to Sommer (66), a film deposited in about 30–60 sec with a transmission of 50% of white light would give good sensitivity.

The silver layer thus evaporated is usually oxidized by means of a glow discharge. The silver layer is evaporated in a vacuum of $10^{-7}$–$10^{-8}$ Torr, and then pure oxygen is admitted through a leak valve at a pressure in the range 1–0.1 Torr. The oxygen could also be generated by heating potassium chlorate or other compounds, such as potassium permanganate or mercury oxide, in a side arm by means of a flame. A rf generator of about 1 kW power is used for oxidation of the silver film by glow discharge. The uniformity of oxidation is quite important for a uniform final sensitivity of the cathode. The oxidation is continued until the silver layer becomes completely transparent. It is observed (75) that if the oxidation is continued, the oxide layer becomes translucent beyond a certain maximum transmission. The oxidation can also be done with a dc discharge by applying a voltage of about 500 V in the case of chemically deposited silver layers and silver films or plates. In the case of semitransparent cathodes, only a rf discharge is used because the silver oxide is an insulator, and its resistance would therefore increase too much with oxidation to sustain the discharge, resulting in nonuniform oxidation.

The formation of the silver oxide layer was investigated by Prescott and Kelly (59), who determined volumetrically the quantity of oxygen absorbed in a gas discharge by silver foil. Tjapkina and Dankov (76) oxidized silver as an anode in a glow discharge, and they were able to detect the compound $Ag_2O$ by electron diffraction. According to Wilman (77), the oxidation occurring in a glow discharge of a silver film grown epitaxially on mica is an epitaxial film of $Ag_2O$. Suzuki (78) also observed the presence of $Ag_2O$ on oxidized film. It was generally believed that in a dc glow discharge, the compound $Ag_2O$ is formed (79, 80) regardless of whether silver is exposed to positive or negative discharge. However, Heimann et al. (81) have made a precise determination of the amount of silver and oxygen during oxidation by quartz crystal oscillator, and their result is different from others. For example, they found that in most cases the compound $Ag_2O_2$ was formed, with an oxygen excess of 25%. In such films the existence of $Ag_2O_2$ was confirmed with the help of electron- and X-ray diffraction measurements. However, they found the compound $Ag_2O$ at certain conditions in the negative glow discharge, which they ascribed to the fact that the higher oxides could not be formed due to intense ion bombardment. In the high-frequency inductive discharge they found a ratio of oxygen and silver that may correspond to the compounds $Ag_2O$, $Ag_2O_2$, and $Ag_2O_2$ with oxygen excess. $Ag_2O_3$ could also be produced during the discharge, depending on the energy and the type of the reactive oxygen components in the plasma (O, $O_2^+$, $O^-$).

After the oxidation of silver, another layer of silver is often deposited on the silver oxide (62, 81). This silver layer is thought to have a beneficial effect on the final cathode sensitivity. For semitransparent cathodes, the silver evaporation is continued until the loss of transmission is about 50% (66, 81). In the case of opaque cathodes, the silver deposition is continued until the oxide color changes to dark purple. The evaporation of metallic silver on the oxide layer reduces the $Ag_2O_2$ to $Ag_2O$ (81), and the excess silver remains in the metallic state. The electron diffraction pattern of silver shows diffused rings (81). This probably indicates that the remaining metallic silver is in the form of very fine droplets, which for many metals give diffuse rings because of the smallness of the microcrystals (82).

After the second silver evaporation, the silver oxide film with silver is treated with cesium vapor. The temperature of activation is usually maintained at about 160°C. The temperature can be in the range of 150–200°C. At higher temperatures the reaction is faster, but because of the higher supersaturation vapor pressure the rate of cesium generation has to be higher. At the higher reaction rate it becomes difficult to control and stop the reaction at the optimum level of cesiation. Thus, it is advisable to activate at lower temperatures (150–160°C). With the introduction of cesium the photosensitivity begins to develop. It increases to a maximum and then falls from the peak value by about 20%. If the photosensitivity is monitored by a dc method, then the thermionic emission can also be monitored; the activation with cesium should be discontinued when the thermionic emission has fallen considerably from its peak value. Sayama (83) reported that the peaks in thermionic emission and photoemission do not coincide and the peak in thermionic emission occurs after the peak in photoemission. However, it was found in our laboratory that the peaks in thermionic emission and photoemission occur at the same time. After the cesium is stopped, the sensitivity and thermionic emission again increase, and often both exceed the peak value obtained earlier. Addition of too large an amount of cesium reduces the sensitivity permanently. The reaction of silver oxide with cesium can be represented by (81)

$$Ag_xO + yCs \rightarrow xAg + Cs_yO$$

As is well known, a number of oxides of cesium can exist (84, 85), such as $Cs_2O$, $Cs_2O_2$, $Cs_2O_3$, $CsO_2$, $Cs_3O$, and $Cs_7O_2$. Borzyak et al. (60) observed the presence of $Cs_2O$ in the Ag–O–C photocathode. Heiman et al. (81)

found out through detailed and careful experiments that the compound $Cs_2O$ is ultimately formed, but that at initial stages, peroxides such as $Cs_2O_2$ and $Cs_2O_4$ are formed. The ratio of cesium to oxygen at the peak of photosensitivity was about 2.03, and when the sensitivity began to decrease this ratio reached a value of 2.12. Whether at this stage (when the cesium generation is stopped) there is free cesium or the excess cesium is in the form of lower oxides was not established by these experiments. The existence of lower or higher oxides can reduce the work function of the cathode and decrease the photoemission, as it has been found that for a particular composition the cesium oxide has a minimum work function of 0.6, which is the lowest of the oxides of cesium (84). Sommer suggests that a mixture of $Cs_2O_2$ in $Cs_2O$ might be responsible for lowering the work function (86). Another study, by X-ray photoelectron spectroscopy (87), showed that in S-1 photocathode, $Cs_2O$ and another suboxide of cesium, $Cs_{11}O_3$, are present, and that the presence of $Cs_{11}O_3$ in $Cs_2O$ is responsible for the high quantum efficiency and low work function of this material.

It was found by Asao (88) and Asao and Suzuki (58) that an additional silvering process at the end of cesium activation improves the sensitivity of the photocathodes. The silver is evaporated slowly at room temperature, during which process the sensitivity falls after an initial small rise. The sensitivity, which falls from one-half to one-third of its initial value after cesium activation, is improved by a subsequent bake at 120°C. It was found at our laboratory (89) that good enhancement is obtained if the sensitivity is monitored with a visible light cutoff filter and IR sensitivity is brought down to zero by silver addition. Visible light sensitivity meanwhile first rises and then falls to the initial value on baking. This is done in cathodes in which no silver is added after oxidation of the silver film. In cathodes in which good sensitivity is already obtained after cesium activation, addition of silver does not increase the sensitivity very much; but in the case of cathodes that do not give high sensitivity after cesium activation, the silver may enhance integral sensitivity by up to 3 times its value before silver addition. The amount of silver added at this step is much less than the initial layer of silver, which implies that the silver is essentially granular at this stage of addition.

In cathodes that do not give good sensitivity up to this stage of processing, a final, carefully controlled superficial oxidation (32) often leads to higher sensitivity and longer threshold. In cathodes that have already attained high sensitivity, the oxidation is not much help. Oxygen is added carefully by means of a leak valve or by resistive heating of metal channels containing pure barium peroxide to about 800°C. However, if the cathode sensitivity decreases due to overexposure to oxygen, it can be recovered by a bake at 150°C; and very often the sensitivity goes to a higher value after this bake.

## 2. Properties

*a. Electrical.* Ag–O–Cs photocathode is found to be composed of $Cs_2O$ and particles of metallic silver (*66, 81*). Asao (*88*) observed that thin films of Ag–O–Cs prepared from $Ag_2O$ of a thickness of about 70 molecules had a high conductivity and a positive temperature coefficient of resistivity. This was attributed to the overlapping of the silver particles in the film. However, later experiments by Harper and Choyke (*90*) showed that the electrical conductivity of thin layers of Ag–O–Cs showed semiconducting behavior, and the activation energy for the temperature dependence of conductivity was found to be 0.2 eV. $Cs_2O$ films formed by direct oxidation of cesium also had similar conductivity and activation energy. Subsequently, Davey (*91*) observed that electrical conductivity has an activation energy of 0.1 eV for some films, which was the same value he had calculated from Harper and Choyke's data. In some films with a higher cesium content, he observed a break at lower temperatures and noted that the activation energy dropped and the work function increased. With a very large quantity of cesium—to the extent of almost shorting between anode and cathode—the behavior became metallic.

The electrical conductivity data obtained by different authors is perhaps not so confusing as it was previously considered to be. Since the cathode contains both metallic silver particles and $Cs_2O$, and the ratio of these components may vary depending on the conditions of processing, the electrical conductivity data are quite reasonable. If the fraction of metallic silver is higher, then metallic behavior would result due to overlapping of the islands. Similarly, a large amount of cesium would produce a layer of metallic cesium on the surface and give rise to metallic behavior. If the amount of silver is less, so that the silver islands are separated and electron tunneling cannot take place between them, the electrical conductivity would be entirely dominated by $Cs_2O$, an effect that has been observed by Harper and Choyke. Cesium forms donor levels, which would have characteristic activation energies, and the changes in activation energy and work function are probably due to the filling of the surface states, as has been suggested by Davey.

*b. Thermionic Emission.* Among the commercial photoemitters, Ag–O–Cs photocathodes have the highest thermionic emission. Typical values of thermionic emission lie in the range $10^{-11}$–$10^{-14}$ A cm$^{-2}$. The thermionic works functions measured from Richardson plots are in the range 0.7–1.0 eV (*81, 88, 91, 92*). The work function of the cathode depends in general on the method of processing and the surface treatment. For example, the final silvering process that reduces the threshold wavelength of the photocathode also reduces the thermionic emission, and superficial oxidation increases the threshold wavelength, with a consequent increase of thermionic emission.

The dependence of the thermionic emission on the threshold wavelength is found to very linearly, as can be seen in Fig. 9. However, measurement of the Kelvin work function of $Cs_2O$ layers on Ag by Uebbing and James (92) shows that the photoelectric threshold varies almost linearly with the Kelvin work function above about 1 eV (they are almost the same), but that the Kelvin work function decreases to a value as low as 0.6 eV, while the optical threshold does not decrease below 1 eV. Uebbing and James observed that the work function determined from Richardson's equation and the Kelvin work function are not the same for $Ag-Cs_2O$ layers, as a layer with a Kelvin work function of 0.6 eV had a thermionic work function of 0.95 eV. In addition, the layers with a low Kelvin work function, between 0.6 and 0.7 eV, did not have high thermionic emission as expected. Moreover, thick $Cs_2O$ layers gave dark currents of less than $2 \times 10^{-14}$ A cm$^{-2}$, which is much less than that of most Ag–O–Cs photocathodes. All these results suggest that the thermionic emission in Ag–O–Cs photocathodes is due to thermionic emission of the electrons from the silver across the Schottky barrier to the $Cs_2O$. Thus the work function obtained from thermionic emission data gives this barrier height, which is about 1 eV, according to Uebbing and James. However, this height depends on the concentration of Cs in $Cs_2O$ and on

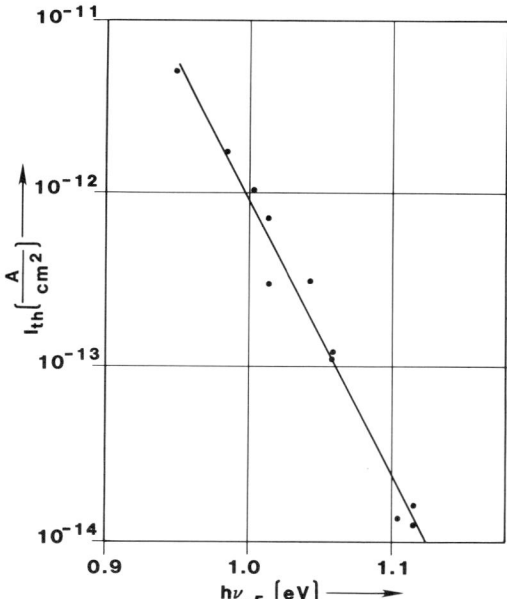

FIG. 9. Relationship between thermionic emission $I_{th}$ and long-wavelength threshold limit $hv_{-5}$ for semitransparent Ag–O–Cs at $T = 293$ K. From Heimann et al. (81).

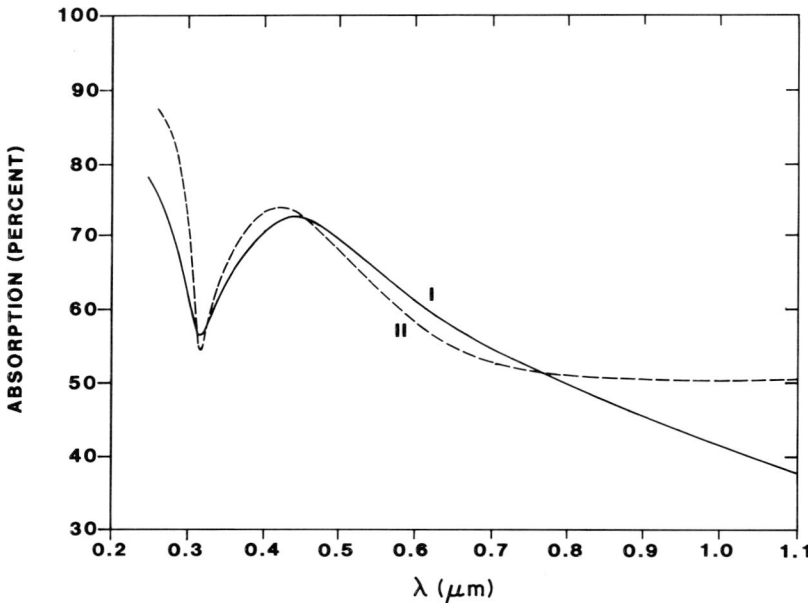

FIG. 10. Light absorption of Ag–O–Cs cathode before (curve I) and after (curve II) additional silver deposition, as derived from data of Asao (88).

the interface of $Cs_2O$ and Ag. Gregory et al. (85) observed that the work function of Cs during oxidation passes through a sharp minimum at 0.7 eV. The minimum is believed to be a point at which there is plenty of unreacted cesium in $Cs_2O$.

c. *Optical Properties.* The optical absorption of the Ag–O–Cs photocathode was first measured by Asao (88); the absorption curves are shown in Fig. 10. The absorption shows a minimum at 3200 Å and a broad maximum at about 4500 Å. The position of this maximum changes with the treatment of the photocathode with silver. It is generally believed that the Ag–O–Cs photocathode is composed of particles of elementary silver embedded in the $Cs_2O$ matrix. Thus this material should have the optical properties of either $Cs_2O$ or silver or both. The light absorption of $Cs_2O$, as measured by Borzyak et al. (60), showed little variation in between 1 and 2 eV, and rose quite sharply at higher energies. The absorption of bulk silver was measured by Ehrenreich and Phillipp (93). There is a sharp minimum in the absorption coefficient curve at 3.85 eV (Fig. 11). On the higher-energy side of this minimum the absorption is due to interband transitions and at lower energy the absorption rises monotonically due to free-electron

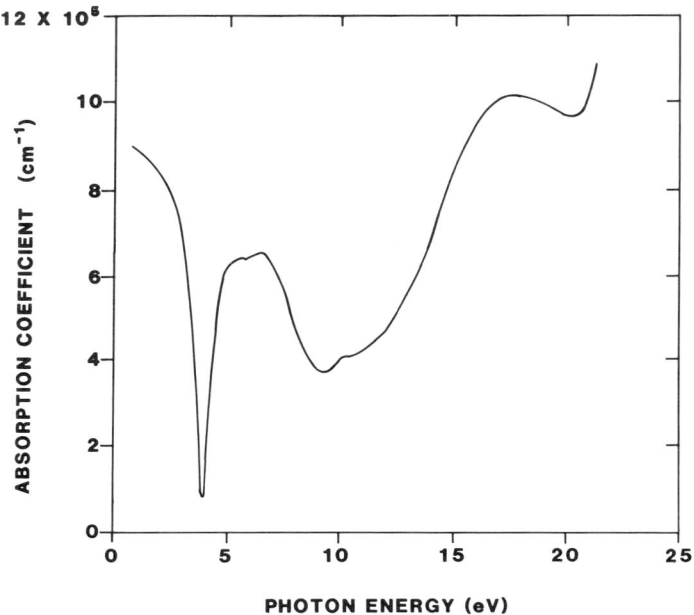

FIG. 11. Absorption coefficient of silver. From Ehrenreich and Philipp (93).

absorption. The minimum in absorption coefficient in silver coincides with the minimum in the absorption of Ag–O–Cs, and it is mainly on this basis that Sommer (94) suggested that absorption in this material is predominantly due to silver. However, neigher $Cs_2O$ nor Ag has the broad maximum in absorption observed with Ag–O–Cs at longer wavelengths. This absorption may be due to the silver defect levels in cesium oxide. The absorption coefficient corresponding to the peak position would be about $3 \times 10^5$ cm$^{-1}$, assuming a thickness of 300 Å for the photocathode. This is much higher than the absorption coefficient that would normally be due to the defect levels. Moreover, the absorption due to this defect level should give rise to photoemission, but the quantum efficiency plot (see Fig. 12) does not show a peak corresponding to this absorption peak. The absorption of the silver film prior to oxidation was measured by Asao (88), who obtained an absorption behavior similar to that of Ag–O–Cs, although the absorption peak was at a longer wavelength of about 0.6 $\mu$m. This absorption is probably the result of anomalous or Maxwell Garnett absorption due to the aggregate of metallic silver, islands, as observed by many experimentalists (71–74). In Ag–O–Cs film also the absorption peak at longer wavelengths could be due to the aggregate of silver granules embedded in the $Cs_2O$ matrix. This absorption would not contribute to photoemission.

*d. Photoemission from Ag–O–Cs Photocathodes.* Figure 12 shows the spectral response of Ag–O–Cs photocathode from the near UV to the near IR. The cutoff wavelength is near 1.2 μm (1 eV), and a sharp minimum at 0.325 μm and a peak at 0.350 μm were observed. This led Borzyak *et al.* (*60*) and Sommer (*94*) to conclude that the photoemission in Ag–O–Cs in the region between 0.300 and 0.600 μm is due to elementary silver. The minimum coincides with a minimum in the absorption of silver (Fig. 11) at 3.85 eV. However, the minimum in quantum yield is not due entirely to a decrease in absorption, as the plot of quantum efficiency in terms of electrons per absorbed photon also shows a minimum at the same energy. According to Berglund and Spicer, the yield minimum is due to two factors. One is that due to the small absorption coefficient of electrons, more electrons would be excited in silver at depths greater than escape depth. This happens if the absorption depth $1/\alpha$ is much larger than the escape depth. Moreover, since the yield minimum is at the energy in which strong plasma resonance is expected (*93*), a large fraction of absorbed photons would contribute to plasma oscillations rather than to photoemission. The peak at 3500 Å is, however, due to high optical absorption, because in the yield plot of electrons per absorbed photons this peak does not appear. It was pointed out by Ehrenreich and Phillipp (*93*) that optical absorption in silver is dominated

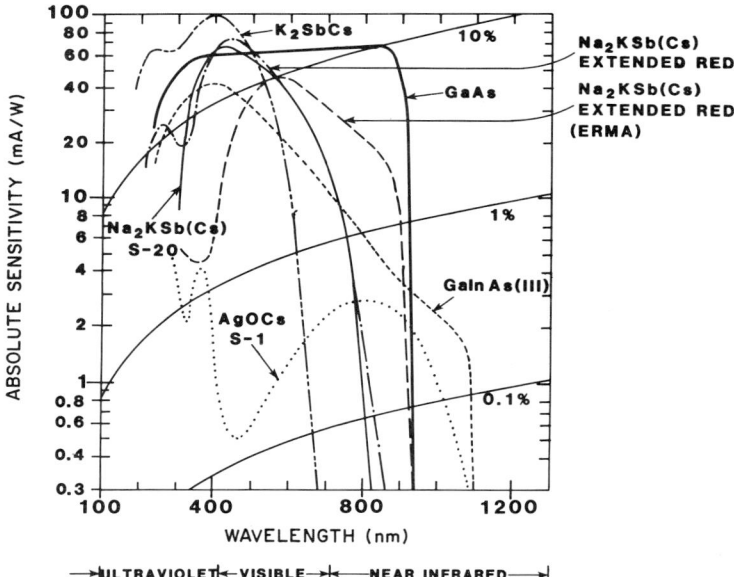

FIG. 12. Spectral response of a number of photoemissive materials.

by interband absorption above 3.5 eV, and below this energy it is due to free-electron absorption. This absorption rises quite sharply at lower energies.

It was proposed by Borzyak et al. (60) and later supported by Sommer (66, 94) that photoemission in Ag–O–Cs photocathodes at wavelengths greater than 0.300 μm is due to emission of electrons from metallic silver into $Cs_2O$. For wavelengths less than 0.300 μm, $Cs_2O$ photoemission is considered to predominate. $Cs_2O$ is an n-type material with a band gap of 2 eV (92) and electron affinity of 0.6 eV (94). The work function is also about 0.6 eV (84, 92). The potential barrier between a metal and a semiconductor would be the difference of the work function between the two. The work function for cesiated silver is 1.65 eV (95). Therefore, if there is free cesium at the interface, the interfacial barrier could be 1.0 eV. This barrier would determine the photoelectric threshold and thermionic work function, and thus the observed anomaly between thermionic and Kelvin work functions could be explained (92).

Strong evidence of photoemission from silver has been obtained from photoelectron spectroscopy of cesiated silver films by Berglund and Spicer (95) and Neil and Mee (96) with Ag–O–Cs cathodes. The remarkable similarity between the two sets of spectra up to 5 eV photon energy is probably due to the fact that elementary silver is responsible for photoemission from both. The photoemission at higher photon energies could not be compared because data for Ag–O–Cs were not available. In the low excitation energy range in silver, the energy distribution shows a peak at about 0.3 eV below the maximum electron energy (Fig. 13). This peak corresponds to a peak in the density of states in the valence band of silver at 0.3 eV below the Fermi level. There is a group of slow electrons which increase in number as the photon energy is increased. This group appears as a large peak in kinetic energy near 0.5 eV. This peak is due to the electrons scattered by electron–electron scattering and the Auger electrons produced due to the hole created in the d band of silver due to optical excitation (95). However, this Auger process would be insignificant below photon energies of 3.8 eV because the d band of silver states is at 3.8 eV below the Fermi level. Thus the slow electrons produced below this photon energy would be the electrons excited to lower-lying states and due to scattered electrons. There is no threshold for electron–electron scattering and the fraction of slow electrons at lower energies is considerable; for example, at 3.5 eV it is about 40%.

A similar situation is observed in the case of Ag–O–Cs photocathodes (96). A fast-electron peak 0.3 eV below the maximum electron energy is observed, the intensity of which increases with evaporation of additional silver onto the photocathode. Therefore, this peak is quite likely to be due to silver, generated by the peak in the density of states at an energy 0.3 eV below the Fermi energy. On the lower-energy side also a peak appears at

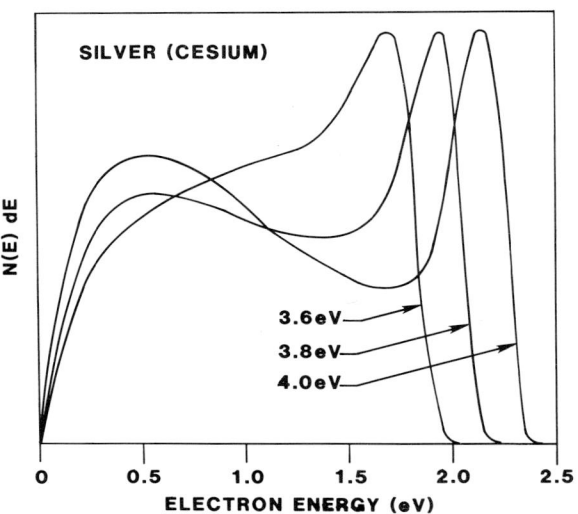

FIG. 13. Energy distribution of photoemitted electrons from silver at low excitation energies. From Berglund and Spicer (95).

a kinetic energy of ~0.5 eV. However, Neil and Mee consider that the slow-electron energy peak arises because of pair production by fast electrons in $Cs_2O$, a process having a threshold of 3.4 eV. But in their experimental curves the existence of a good proportion of slow electrons is evident even at 2.85 eV (Fig. 14). Of course, pair production in $Cs_2O$ is quite probable at higher energies, but at lower energies the behavior is similar to that of silver, suggesting that the photoemission at lower energies is mainly from silver.

There is another model of photoemission from Ag–O–Cs, originally proposed by Eckart (97), according to which silver forms donor levels in $Cs_2O$, and the photoemission in the IR region is from these donor levels, which would be about 1.1 eV below the vacuum level. The photoemission in $Cs_2O$ could be considered to be due to the valence band of $Cs_2O$. However, only about 1% of the silver atoms would form an impurity band in $Cs_2O$, and it is not understood why 50% of Ag–O–Cs cathodes should be free silver (66). Even at the stage of final silver addition it is seen that for good increase in sensitivity a substantial amount of silver should be added and not the minute quantities that would have been necessary if silver were acting as a dopant. Moreover, photoemission from impurity levels is unlikely to produce a quantum yield near $10^{-2}$ electrons/photon. In fact, cesium forms donors in $Cs_2O$, and $Cs_2O$ doped highly with cesium has a quantum efficiency in the range of $10^{-4}$ electrons/photon (81, 92), about two orders of magnitude lower than that of Ag–O–Cs. Some photoemission in the

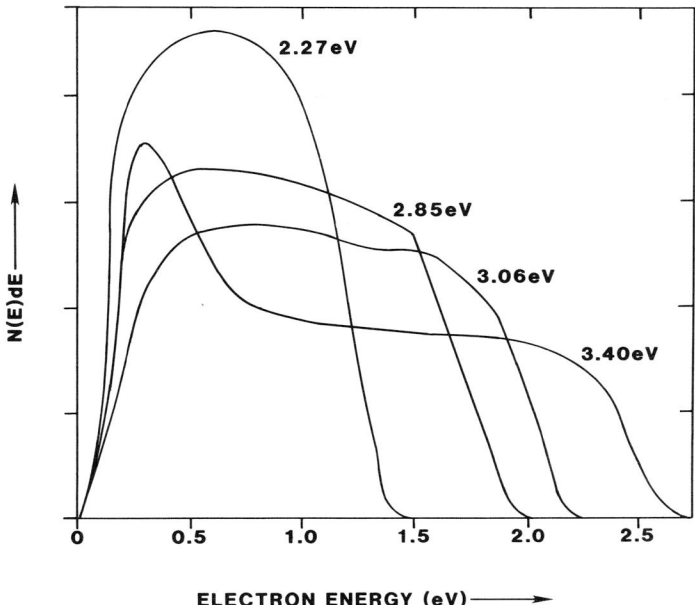

FIG. 14. Energy distribution of photoelectrons from Ag–O–Cs. From Neil and Mee (96).

threshold region could be due to the defect levels, but in general this source would contribute little to the higher quantum efficiency away from threshold. Neil and Mee suggested that the fast-electron peak in the energy distribution 0.3 eV below the maximum energy is due to the defect level. However, the width of this peak is more than 1 eV, which should correspond to the width of the defect band. From most considerations it appears that the photoemission at longer wavelengths may not be due to the defect level or band in $Cs_2O$ due to silver atoms.

The question may arise, however, of how photoemission from metallic silver accounts for a quantum efficiency nearing $10^{-2}$ electrons/photon, while most metals have a lower quantum efficiency near the threshold region. The answer to this question is not straightforward. Two factors contribute to the low quantum efficiency of metals. One is their large reflectivity, and the other is their small escape depth because of electron–electron collision. In the case of Ag–O–Cs, when silver particles are embedded in $Cs_2O$, the reflection at silver would be much smaller than the reflection at the silver–vacuum interface because the refractive index of $Cs_2O$ is much higher than that of air; in fact it is greater than 2 (81). The small escape depth in metals is mainly due to electron–electron scattering. If electron affinity is high, the escaping electrons will have high energy. But for electrons with low energy, the escape depth may be large. For example, for gold, Sze et al. (98) found an escape depth less than 70 Å for ~5 eV electrons, whereas the escape depth was more than 1000 Å for electron energies of less than 1 eV. Corresponding high-energy values for silver are not known, but Crowell et al. (99) reported an escape depth of 440 Å for electrons in the 1 eV range. Thus the electrons that would have sufficient energy to cross the Schottky barrier and escape into $Cs_2O$ may have a large escape depth in silver, and in $Cs_2O$ their escape depth would also be very large, enabling them to escape. This may be the reason for the higher quantum efficiency of Ag–O–Cs compared to metals near threshold. Another factor that may contribute to their high sensitivity is the increase in absorption due to multiple reflection by the colloidal silver particles (36).

The high quantum efficiency of Ag–O–Cs at wavelengths below 3000 Å has been ascribed to photoemission from $Cs_2O$ (60, 66). Borzyak et al. (60) and subsequently Heimann et al. (81) measured quantum efficiency of $Cs_2O$, and they observed that the quantum efficiency is quite large at shorter wavelengths. Silver also has high quantum efficiency at shorter wavelengths in the UV (95), comparable to that of $Cs_2O$. Thus the photoemission at wavelengths below 3000 Å could be due to both silver and $Cs_2O$ instead of only $Cs_2O$, as postulated earlier.

### 3. Bi–Ag–O–Cs Photocathode

Sommer and Spicer (100) developed Bi–Ag–O–Cs cathode by first depositing a layer of bismuth, then depositing silver, and subsequently oxidizing and cesiating the layer in a manner similar to the technique used for the Ag–O–Cs cathode. This cathode has panchromatic response together with high quantum efficiency, and an application was found in image orthicon tubes. It also, however, has a high dark current, similar to that of Ag–O–Cs cathodes. After the discovery of high-sensitivity multialkali cathodes, the use of this cathode was largely discontinued.

## IV. The Alkali Antimonides

The alkali metals, on reaction with antimony, produce semiconducting materials that have good photoemissive properties. The mono-, bi-, and trialkali antimonides have photoemission thresholds ranging from the UV to the near IR, and some of them are used for a variety of applications. They are, in fact, the most widely used photocathodes at present.

The chemical composition of an alkali antimonide can be generally represented by $M_3Sb$, M representing one or a combination of alkali elements. All the alkalis from lithium to cesium react with antimony to form photoemissive compounds with properties some of which follow the order in the periodic table. For example, $Cs_3Sb$ has the longest threshold wavelength, followed by $Rb_3Sb$, and so on. The bi- and trialkali antimonides also have well defined chemical compositions.

## 1. Preparation of the Cathodes

*a. The Monoalkali Antimonides.* The cesium antimonide $Cs_3Sb$ was the first alkali antimonide to be discovered (*4*). It has a maximum quantum efficiency of about 0.15 electrons/photon, with a threshold at about 5800 Å. This cathode is prepared by exposing a thin film of antimony to cesium vapor at an elevated temperature. The thickness of the film is usually monitored by measuring the amount of loss in a light beam transmitted through the substrate. A loss of transmission of 15–30% gives good sensitivity (*66, 101*). The thickness that gives highest sensitivity is believed to be quite critical; however, the different values of loss of transmission reported by different authors may be due to the fact that the same thickness of antimony film can give different loss of transmission, depending on the angle of incidence, substrate preparation, source of evaporation, etc. (*66, 102–105*). The thickness of antimony film that gives good sensitivity corresponds to a mean thickness of about 2–3 $\mu g\ cm^{-2}$, corresponding to an average thickness of about 45–60 Å. The term "average thickness" is used because at this thickness the film will be of island type.

The antimony film, on reaction with cesium, forms the semiconducting compound $Cs_3Sb$, which is photoemissive. The photosensitivity is usually monitored during cesium activation. Photosensitivity to visible light starts sometime after the beginning of activation with cesium. The sensitivity increases to a maximum and then starts falling when the activation is stopped. At this point the cathode has more cesium that is needed for maximum sensitivity; the excess, however, is distilled out later and a higher sensitivity is obtained on cooling. Superficial oxidation, produced by admitting very small amounts of oxygen into the vacuum system, increases the sensitivity and extends the cutoff to longer wavelengths (*4, 106, 107*).

A substantially enhanced sensitivity and extended cutoff wavelength are obtained by depositing the $Cs_3Sb$ cathode on a manganese oxide substrate (*108*). The substrate is prepared by depositing a thin layer of manganese on the glass substrate and then oxidizing it in a glow discharge. The reason for this increase in sensitivity is not known (*109*). It has been suggested (*110*) that band-bending at the $MnO_2$–$Cs_3Sb$ interface may be responsible for it.

The other antimonides, such as $Rb_3Sb$, $K_3Sb$, $Na_3Sb$, and $Li_3Sb$, can be prepared by similar methods. The temperature of activation is different in each case because of different vapor pressure of the alkali elements. These compounds have less photosensitivity and shorter cutoff wavelengths than $Cs_3Sb$ and hence are not used for photocathode applications. Consequently, they have received much less attention from investigators. However, $K_3Sb$, being the base layer for multialkali photocathodes has been investigated in greater detail. $K_3Sb$ is found in general in hexagonal and cubic crystalline phases. The cubic variation has higher photosensitivity. Sommer and Mc-Carroll (*111*) observed that if the thin film of Sb is slightly oxidized in a glow discharge of oxygen before the exposure to K vapor, the resulting $K_3Sb$ layer is of cubic type.

A more recent method for processing these monoalkali antimonides (*52, 112*) is to deposit the alkali metal and antimony simultaneously in controlled amounts on the substrate. This method usually gives higher sensitivity than does activating the antimony thin film with alkali vapors.

*b. Bi- and Trialkali Antimonides.* Sommer observed that a combination of some alkali elements activating antimony produces higher photosensitivity than a single alkali. He first developed the multialkali photocathode (*6*) consisting of potassium, sodium, cesium, and antimony. It has been established that the chemical composition is $(Na_2KSb)Cs$. The bulk of the compound has the chemical composition of $Na_2KSb$ and cesium is believed to play some role at the surface. The amount of cesium is very small compared to the other two alkali elements.

The original method (*6*) for processing multialkali photocathode developed by Sommer is as follows: antimony is evaporated to a light transmission loss of 30% and the film is exposed to potassium vapor at 160°C. After the sensitivity reaches a maximum the film is activated with sodium at 220°C. The next step is to add K and Sb alternately at 160°C in small steps until the increase of sensitivity is stopped, after which the film is exposed to Cs and Sb alternately. The sensitivity reaches a maximum and is allowed to fall to some extent with addition of excess of Cs, after which the cathode is allowed to cool to room temperature. There have been numerous modifications of this processing technique (*113–116*), mainly involving temperature and sequence of activation. A more drastic modification, which does not require the deposition of the initial antimony film, has also been reported (*52, 112*). Both potassium and antimony are released simultaneously in controlled quantities, so that they deposit on the substrate as $K_3Sb$. At other stages also the antimony is evaporated along with the alkali metals whenever the deposition of both is necessary. All these variations of processing technique result in essentially the same composition of the cathode, but the sensitivity and the threshold wavelengths differ depending on the processing technique and other experimental conditions.

A variant of this cathode with much higher red sensitivity and extended response is sometimes called extended red multialkali (ERMA or S·25) photocathode. This is prepared by making a layer of $Na_2KSb$ with thickness in the range 1000–1300 Å. In this cathode a higher red sensitivity is achieved at the expense of blue sensitivity in the transmission mode (Fig. 12). A higher red or IR sensitivity is a much desired characteristic for certain applications, such as in night vision devices.

The bialkali photocathodes exist in such compositions as $K_2CsSb$ ([117]), $Na_2KSb$, and $Rb_2CsSb$ ([118], [119]). The $K_2CsSb$ cathode is prepared by first forming a $K_3Sb$ layer and subsequently activating it with Cs. $Na_2KSb$ is the material which on activating with Cs and Sb gives multialkali photocathode. Rb–Cs–Sb cathode is processed in a similar way, and the composition corresponding to maximum sensitivity is $Rb_2CsSb$.

## 2. Chemical Composition and Crystal Structure

Though the alkali antimonides with highest photosensitivity have the general composition $M_3Sb$, where M is an alkali metal, a number of intermediate compounds are known to exist ([120]). Sommer first attempted to determine the composition of cesium antimonide by determining the weight of reacting elements. He found the composition to be $Cs_3Sb$, though the accuracy of his measurements was not better than $\pm 10\%$. Garfield and Thumwood ([121]) determined the composition by the vacuum microbalance method, and Hagino and Takahashi ([122]) determined the composition by flame photometry. Both observed the compound to be $Cs_3Sb$, but since the quantities involved were very small these methods could give only a rough estimate of the chemical composition.

More accurate determination of the chemical composition was made by crystal structure determination of these compounds. The crystal structure of cesium antimonide was determined by Jack and Wachtel ([123]), Gnutzman ([124]), and Scheer and Zalm ([125]). They prepared the sample by reacting the individual elements in bulk quantities. The chemical composition of the resulting compound was found to be $Cs_3Sb$ by means of X-ray diffraction. However, the preparation technique differed radically from that used to prepare the photocathodes, and there was no way of knowing whether the compound that was formed was photoemissive. Subsequently, McCarroll ([126]) investigated the crystal structure of the actual photoemissive films by scraping the film off the substrate and putting it into a capillary to study the X-ray diffraction. Robbie and Beck (127) and Dowman et al. (128) studied the alkali antimonides by means of scanning electron diffraction. All of them found the composition of the compound of good photosensitivity to be $Cs_3Sb$.

$Cs_3Sb$ has been found to exist in bcc structure with a lattice constant of 4.56 Å (126). McCarroll (126) observed other lines in addition to those corresponding to $Cs_3Sb$, which he ascribed to lower antimonides. The higher the sensitivity of the cathode, the lower was the intensity of the lines due to lower antimonides; in the case of the most sensitive cathodes the lines attributable to other phases were almost absent.

There is some controversy about the degree of ordering in the $Cs_3Sb$ cathodes. While McCarroll (126) did not find any evidence of ordering, Jack and Wachtel (123) observed superlattice lines that suggested a partially ordered sodium thallide type of structure. Gnutzmann (129) and subsequently Robbie and Beck (127) found evidence for a completely ordered $Cu_3Al$ type of structure with a lattice constant of 9.12 Å.

$Rb_3Sb$ was found to exist both in hexagonal and cubic structures (130). Chikawa et al. (131) found that when antimony film is activated with rubidium vapor, a cubic material is formed, which on subsequent exposure to Rb changes partially to hexagonal form. In the final cathode, the cubic and hexagonal phases exist together. Dowman et al. (128) did not find any hexagonal phase. They observed a fully ordered $DO_3$ structure with lattice constant of $8.80 \pm 0.01$ Å. The films they prepared were much thinner than those prepared by others.

Early experiments (132) on the crystal structure of $K_3Sb$ found it to be hexagonal. Sommer and McCarroll (111) first found a cubic modification, which was normally stable but transformed into the hexagonal variety when heated to 180°C. The cubic compound has a $DO_3$ structure similar to those of $Rb_3Sb$ and $Cs_3Sb$ with a lattice constant of 8.50 Å. The hexagonal modification has a $Na_3As$ structure with $a = 6.03$ Å and $c = 10.69$ Å. The cubic modification has much higher photosensitivity than the hexagonal modification.

$Na_3Sb$ has been observed only in the hexagonal modification both in the case of films (131) and in bulk (132). There is a relationship between the crystal structure of the monoalkali antimonides and the position of the alkali elements in the periodic table (133). For example, $Cs_3Sb$ exists only in cubic form and $Na_3Sb$ exists only in hexagonal form. $K_3Sb$ and $Rb_3Sb$ exist in both cubic and hexagonal form. The sensitivity of the photocathodes is also apparently linked to the crystal structure. It is observed in general that the cubic modifications of compounds have higher photosensitivity than the hexagonal ones even in the case of materials that exist in both variations. The reason for this is not well understood.

Chemical analysis and crystal structure investigation of the bialkali antimonides has determined the chemical composition of the bialkalis as $K_2CsSb$ (134), $Rb_2CsSb$ (128), and $Na_2KSb$ (135). McCarroll (134) determined the crystal structure of $K_2CsSb$ to be cubic $DO_3$ type with a lattice constant of 8.61 Å. Dowman et al. (128) found that during processing of the cathode, which they did by activating a $Cs_3Sb$ cathode with potassium vapor, potassium goes on replacing cesium atoms in the lattice and forming the solid solution $K_xCs_{3-x}Sb$. The structure changes continuously until $x$ reaches a value of 2. They obtained a crystal structure and lattice constant similar to those obtained by McCarroll. The crystal structure of $Rb_2CsSb$ was found to be similar (128). The compound has its highest photoemission when the chemical composition is $Rb_2CsSb$, but the $Rb_xCs_{3-x}Sb$ system gave a solid solution over the complete range $x = 0$–3.

Scheer and Zalm (125) and McCarroll (136) investigated the crystal structure of bulk-synthesized NaKSb system. It was found that $Na_2KSb$ has a cubic $DO_3$ structure with a lattice constant of $7.7235 \pm 0.0005$ Å. McCarroll observed that the cubic phase is obtained only when the Na:K ratio is close to 2:1. At other compositions, such as $Na_{0.99}K_{2.01}Sb$, the material had a hexagonal structure similar to that of $Na_3Sb$ with $a = 5.61$ Å and $c = 10.932$ Å (136). Dowman et al. (128) also observed a cubic structure for $Na_2KSb$ with $a = 7.74$ Å. They observed that during activation of the $K_3Sb$ with Na vapor the hexagonal $NaK_2Sb$ phase is formed first, which with further addition of Na changes to cubic $Na_2KSb$. However, this conversion is not usually complete and in the final cathode $Na_2KSb$ and $NaK_2Sb$ phases may coexist. Investigations on multialkali photocathodes by Garfield (116) by means of the microbalance technique show that the amount of Na is less than required to form $Na_2KSb$, and the available amount of Na and K can form a compound of composition between $Na_2KSb$ and $NaK_2Sb$. In electron diffraction studies, Garfield observed the rings corresponding to cubic $Na_2KSb$ with lattice constant of 7.73 Å. However, the samples on which microbalance and electron diffraction studies were made were different. Hoene (137) measured the chemical composition of multialkali cathode by chemical analysis and found it be be $Na_xK_yCs_zSb$, where $x + y + z = 3$ with $1.2 < x < 1.7$ and $z < 0.21$. All these results support the coexistence of $Na_2KSb$ with other phases such as $NaK_2Sb$. It was observed (128) that the higher-sensitivity samples contain mostly $Na_2KSb$. In multialkali cathodes, the existence of a $K_2CsSb$ phase at the surface has been observed, which could explain the band-bending at the surface that results in a lower threshold energy (52).

## 3. Photoemissive Properties

For monoalkali antimonides the threshold energy is a minimum for $Cs_3Sb$ and increases in the sequence of $Rb_3Sb$, $K_3Sb$, $Na_3Sb$ and $Li_3Sb$. The band gaps of these compounds are not very different (16, 52), but the electron affinities also increase in that order. An increase of electron affinity brings about a decrease in the quantum efficiency in the energy ranges near threshold, but at considerably higher energy the difference in electron affinity does not affect the photoemission from these materials. Instead, the photoemission is determined primarily by the absorption and loss mechanism encountered by the photogenerated electrons.

In the case of bi- and trialkalis, such as $K_2CsSb$, $Na_2KSb$, $Na_2KSb(Cs)$, and $Na_2KSb(Rb)$, the threshold energy is a minimum for $Na_2KSb(Cs)$ or S·20 (Fig. 15). At the shorter wavelengths, the bialkali $K_2CsSb$ shows higher photosensitivity. $Na_2KSb$ and $Na_2KSb(Rb)$ have intermediate sensitivities.

Photoemission from these materials is a strong function of the processing conditions. For example, almost identical processing conditions can produce S·20 cathodes with cutoff wavelengths between 8000 and 9000 Å, with integrated sensitivities ranging from 150 to 250 $\mu A\, lm^{-1}$. Similar variation occurs for other alkali antimonides. The reason for this is not likely to be

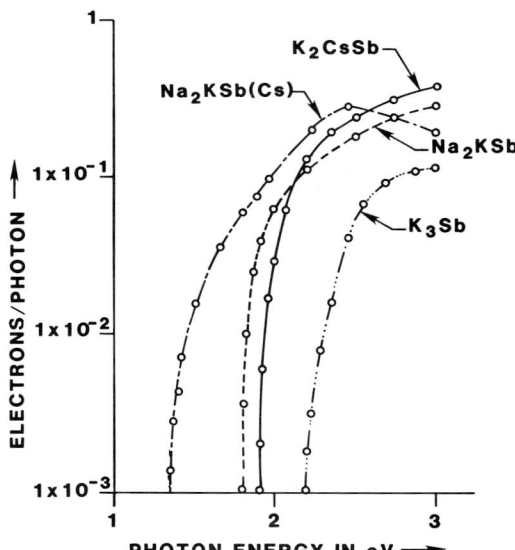

FIG. 15. Spectral response of some alkali antimonide photocathodes. From Ghosh and Varma (52).

a drastic change in the composition of the material, which would give rise to a change in such bulk properties as optical absorption, but a small variation in the processing condition could give rise to change in doping and surface composition, thereby affecting the electron affinity and emission from the defect levels.

The photoemission very near to the threshold region in p-type materials is considered to be due to the defect levels in the material (*16*). In p-type materials the shallow acceptor states would be occupied at room temperature and can contribute to photoemission. This contribution would decrease when the material is cooled because of the reduced occupancy of the levels. Figure 16 shows the variation of photoemission with cooling in case of a $Na_2KSb(Cs)$ photocathode, which is a p-type material (*47*). The reduction in photoemission in the threshold region may also take place due to an increase of band gap on cooling and an unfavorable change of band-bending on cooling (*138*). The increase that has been observed at shorter wavelengths is probably the result of increased escape probability due to reduction in lattice scattering (*138–140*).

Most of the alkali antimonides are unstable at high temperatures. The photocathodes containing cesium, such as $Cs_3Sb$, $K_2CsSb$, and $Na_2KSb(Cs)$, start losing their sensitivity at temperatures above 100°C. The sensitivity can often be restored by addition of further cesium. The loss of sensitivity is apparently due to loss of cesium atoms from the cathode, as cesium has the highest vapor pressure of the alkali metals. Cathodes without cesium have

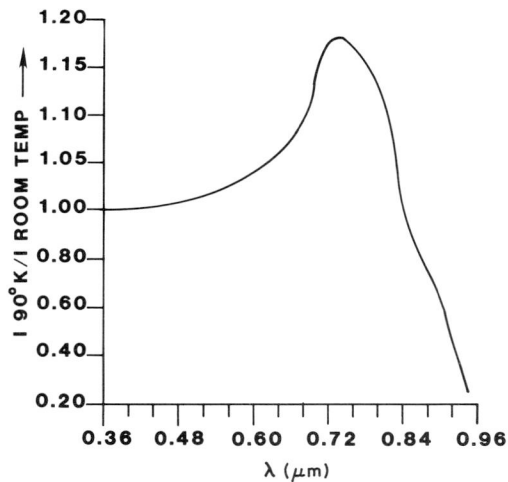

FIG. 16. Effect of cooling on photoemissive quantum efficiency in the case of multialkali photocathode. From Ghosh (*47*).

a better stability at higher temperature—bialkali $Na_2KSb$, for example, is stable up to about 160°C—and they are therefore used in high-temperature applications.

There are other factors that reduce the sensitivity of the photocathodes during its shelf life or during its operation. If during operation a high current is drawn from a high-resistivity photocathode, a potential gradient is created across the photocathode film, which causes electrolytic decomposition of the material (141). This effect is more prominent in higher-resistivity cathodes such as $K_2CsSb$ and $Cs_3Sb$. For most applications, however, the current drawn from the cathodes is normally below $10^{-8}$ A, and in this range the effect is not very serious. The sensitivity can decrease during operation due to positive ion bombardment resulting from ionization of the residual gases. This effect also would be minimized for smaller operating currents and if the metal parts are thoroughly degassed and good gettering is used, so that the residual gas pressure is small. The decrease in sensitivity during storage (shelf-life stability) could be due to a number of factors (142): a fine leak in the tube; an imperfect reaction of the alkalis with the antimony; nonequilibrium of the alkali vapors with certain parts of the tube, causing loss of alkali vapors from the cathode; sudden exposure to strong light, possibly causing heating of the cathode.

Superficial oxidation of some photocathodes appears to enhance their sensitivity and decrease the threshold energy for photoemission (4). In the case of $Cs_3Sb$ the superficial oxidation may increase the sensitivity to almost double its value before oxidation (142). In the case of bialkali $K_2CsSb$, a similar effect is observed (66). The effect of oxidation is apparently a surface effect, because with a larger exposure to oxygen the cathode is permanently damaged. The reduction of surface barrier with oxygen at the surface is probably due to formation of cesium oxide, which has been found to lower the surface barrier in a large number of materials by formation of dipoles or heterojunctions at the surface. However, the increase of sensitivity of the cathodes on oxidation is not universal for alkali antimonides. In the case of trialkali $Na_2KSb(Cs)$, superficial oxidation does not increase the sensitivity (66) except for cathodes that have poor sensitivity to begin with. This may be because in the case of $Na_2KSb(Cs)$ there is probably no free cesium at the surface.

The spectral response of a photocathode depends much upon the mode of light incidence on it. It has generally been observed that the difference in quantum efficiency between semitransparent and reflection modes of operation increases at longer wavelengths (143). The spectral response may also differ greatly if the photocathode is deposited on a reflecting substrate (144). It can be seen (Fig. 17) that the spectral response of thin $Na_2KSb(Cs)$ film on a reflecting substrate is very much higher than that of a semitransparent

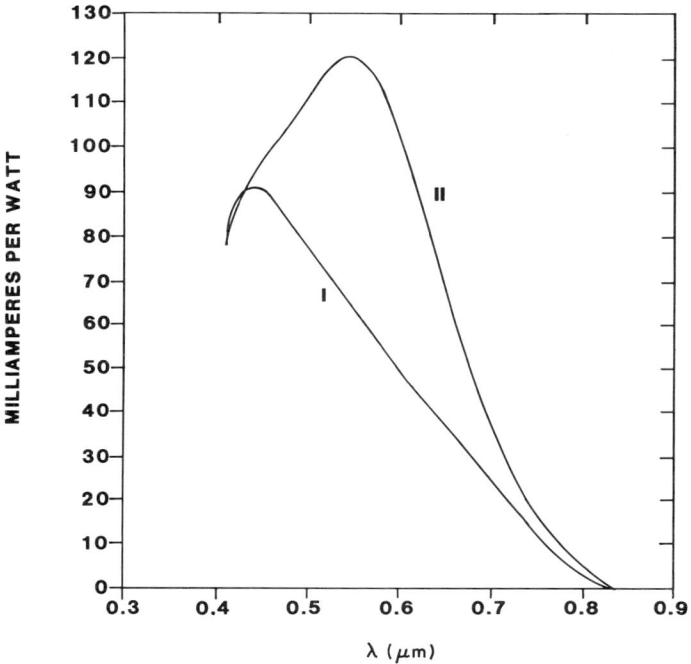

FIG. 17. Spectral response of very thin Na$_2$KSb(Cs) cathode deposited on reflecting substrate (II) compared with standard semitransparent cathode (I). From Sommer (66).

cathode. Deposition on a reflecting substrate effectively doubles the thickness of the cathode. It can be seen that the increase is more on the long-wavelength side, because at the short wavelength the absorption coefficient is high, and most of the light is absorbed before it reaches the reflecting substrate, while at the long wavelength, light that has been transmitted once traverses the cathode again and is absorbed, causing an enhancement in photoemission. The extension of red response has been employed to obtain ERMA or S·25 photocathodes, which have very high sensitivity in the red region (Fig. 12). The blue sensitivity of these cathodes is less in the semitransparent mode because in a thick cathode the blue light would be absorbed more in the layers close to the substrate and the photogenerated electrons would consequently have reduced escape probability.

For a particular composition the photoemission has been found to vary with the crystal structure. Figure 18 shows the spectral response of the hexagonal and cubic modifications of K$_3$Sb (*111*). The hexagonal variation has higher threshold energy and lower quantum efficiency. The band gaps for these two phases have not been measured, but the difference in quantum

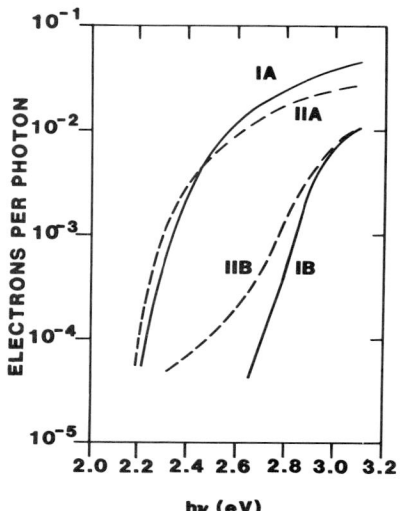

FIG. 18. Spectral response of two $K_3Sb$ cathodes, I and II, of cubic (A) and hexagonal (B) crystal structure. A and B of each cathode are formed from Sb films of identical thickness. From Sommer and McCarroll (*111*).

efficiency could be due to the difference in the type of conductivity in these two materials and the consequent difference in band-bending at the surface.

## 4. Electrical Properties

It was first pointed out by Sommer (*106, 145*) and Borzyak (*146*) that the cesium antimonide is not an alloy but is a semiconducting compound of composition $Cs_3Sb$. Subsequently all other alkali antimonide materials that were developed were found to have similar semiconducting behavior. It was observed that when cesium reacts with the antimony film to form the photocathode film the resistivity of the film increases by more than a factor of $10^6$ (*147*). When cesium reacts with antimony the most drastic change of resistance occurs when the Cs:Sb ratio in the material changes from 0.91 to 1.02 (*147*). The resistivity of the film with a Cs:Sb ratio of 0.91 was $10^{-3}\,\Omega\,cm$ and it was metallic, where as with a Cs:Sb ratio of 1.02 the resistivity was $10^3\,\Omega\,cm$ and the film was semiconducting. The resistance of the film during activation has been reported by a number of authors (*106, 148, 149*) and it was found that the resistance of the materials goes through several peaks corresponding to stoichiometric compounds such as $CsSb$, $Cs_3Sb_2$, and $Cs_2Sb$, finally reaching the peak corresponding to $Cs_3Sb$. Not all authors reported all of the peaks, because the observation of the resistance peaks would be dependent on a number of factors, such as speed of reaction and layer thickness. Apparently the variation of resistivity rather low among the antimonides of cesium (*147*), and measurement of activation energy shows that these compounds also do not have very different activation energies. However, photoemission increases steadily as one goes from lower-cesium to higher-cesium compounds, and reaches a maximum just before stoichiometric $Cs_3Sb$ is formed.

A well defined value for the band gap of $Cs_3Sb$ was not available from the photoconductivity measurements by which the band gaps of similar materials are obtained, but it could be estimated at 1.6 eV (*16, 150*) from optical absorption data. This establishes the effective electron affinity of the material as 0.45 eV. The large band gap implies that the intrinsic material should be a poor conductor at room temperature. The conductivity was found to be quite high ($10^{-2}$–$10^{-3}$ $\Omega^{-1}$ cm$^{-1}$), which indicates extrinsic conduction. It was found by means of thermopower and Hall measurements (*135, 151–156*) that the material had p-type conduction. Sakata's (*151*) Hall data indicate a mobility of about 10 cm$^2$ V$^{-1}$ sec$^{-1}$ and thus the free-carrier density would be about $5 \times 10^{16}$ cm$^{-3}$. By assuming this carrier density to be due to a single carrier without compensation, Sakata obtained a value of $10^{20}$–$10^{21}$ cm$^{-3}$ for defect concentration. However, this estimate could be in considerable error if the scheme of the defect level is different. Nevertheless, it is beyond doubt that these materials contain a high density of defects. These defects are not the result of the impurities, but are due to deviation from stoichiometry. Borzyak (*157*) found that the normal p-type films could be changed to n type by adding excess cesium. Sommer (*133*) observed that the conductivity of $Cs_3Sb$ could be increased by antimony addition. He pointed out that in $Cs_3Sb$ the photoemission reaches a peak during cesium addition before the resistance peak due to formation of stoichiometric $Cs_3Sb$ is reached. Thus the peak sensitivity is obtained in films that are slightly deficient in cesium. In fact Sommer found a relationship between the peak photoemission and the type of conductivity for the alkali antimonides. If the photoemission reaches its peak after the resistance peak is reached, then the material contains an excess of alkali atoms over the stoichiometric composition and is n type. Conversely, if the photoemission attains a peak before the resistance peak is reached, then the material is slightly deficient in the alkali and is p type. Materials such as $Cs_3Sb$ and cubic $K_3Sb$ were p type, whereas hexagonal $K_3Sb$ and $Na_3Sb$ were n type. In general, the materials with cubic lattice were found to be p type and materials of hexagonal lattice were found to be n type. Hagino *et al* (*156*) observed that by application of a higher field across the film the transport of cesium is observed in $Cs_3Sb$. Due to the transport of cesium ions towards the negative electrode the material nearer the negative electrode becomes richer in cesium and becomes n type, whereas the other side becomes deficient in cesium and becomes p type. They also observed a p–n junction-like behavior of the $I-V$ characteristics. Thus it appears that a deficiency of alkali atoms from the stoichiometric ratio in $Cs_3Sb$ produces acceptor-type defects. Spicer (69) suggests from the number of defects present in $Cs_3Sb$ that it is highly unlikely for a close-packed cubic lattice to incorporate such a high density of interstitials, and since no impurities are likely to act as substitutional defects, the p-type conductivity in $Cs_3Sb$ should arise from the alkali atom vacancies. In hexagonal lattices, however, due to more openness, the interstitials may be accommodated, and the excess alkalis may give rise to n type defects by acting as interstitials. This is probably how cubic materials act as p type and hexagonal as n type.

The electrical behavior of other alkali antimonides is similar to that of $Cs_3Sb$. For example, monoalkali antimonides such as hexagonal $K_3Sb$, $Na_3Sb$, and $Li_3Sb$ also exhibit similar peaks in resistivity during activation of the Sb film with the alkalis (133, 149, 159, 160). However, the final compound giving maximum photoemission is $M_3Sb$, where M represents the alkali material. In the case of $K_3Sb$ both p- and n-type conductivity were found by Sommer and McCarroll (111) for cubic and hexagonal materials. The band gaps of some of these compounds have been measured by photoconductivity. Spicer (16) reported a band gap of 1.1 eV for n-type material and Ghosh and Varma (52) obtained a band gap of 1.8 eV for a material whose conductivity type was not determined. Spicer suggests that the electron affinity is between 1.1 and 1.8 eV, whereas the other experiments suggest a value of 0.4 eV. Conductivity versus temperature plots give the activation energies of the material as 0.7 and 0.2 eV (52) and 0.79 and 0.23 eV (159). The measured values of the activation energies of these materials have been widely different in the alkali antimonide materials, but assuming these sets of results to be closer to actual values it is possible that 0.2 eV is the position of the n-type defect level below the conduction band and 0.7 eV is the band gap. The difference between the optical and electrical gap probably implies an indirect gap.

For $Na_3Sb$ the band gap determined (16) by photoconductivity is 1.1 eV and the electron affinity is between 2.0 and 2.4 eV. Imamura (154) found the n-type behavior of this cathode from the polarity of thermo-emf. In the case of $Rb_3Sb$, Sommer (133) did not observe a peak in resistivity during activation with rubidium before or after the peak in photoemission, in contrast to observations with other alkali antimonides. He ascribes this absence to the fact that $Rb_3Sb$ has been found (131) to exist in a cubic variety, but that with excessive exposure to rubidium the material changes to the hexagonal form. In a film the two forms can exist together, so that p- and n-type materials may coexist in different regions, making it difficult to obtain the characteristics of a particular material in the resistivity measurement. But this would be the result of nonuniform activation by rubidium on antimony film, because, in the case of uniform activation the whole film under investigation should change from p type to n type. The band gap and electron affinity, as determined by Spicer (16), are 1.0 and 1.2 eV, respectively.

In the case of $Li_3Sb$ the activation energy obtained from an electrical conductivity versus temperature plot was found to be 1.0 eV (160), which was considered to be the band gap. The electron affinity of this material then turns out to be 2.9 eV, highest in the alkali antimonide family.

In the case of bi- and trialkali cathodes the electrical properties are not very different from those of monoalkalis. The band gap is almost the same and the resistivity is also similar, except for $K_2CsSb$, which has higher resistivity. In the case of $Na_2KSb$ and $Na_2KSb(Cs)$, Spicer (16) has found the band gap to be the same: 1.0 eV. A similar value of 1.1 eV for band gap was found in a photoconductivity experiment by Ghosh and Varma (52). Electron affinities of 1.0 eV ($Na_2KSb$) and 0.55 eV ($Na_2KSb$–Cs) were observed by Spicer; the results of other experiments (52) are 0.7 and 0.24 eV, respectively. Both these materials are p type. The type of conductivity changes during processing of the material (116, 161). During processing of $Na_2KSb$, an n-type $K_3Sb$ film is initially formed. When sodium is added to the material the conductivity remains n type, but when potassium and antimony are then added to convert the material to $Na_2KSb$, the conductivity changes from n type to p type. This is illustrated in Fig. 19. During initial stages of potassium–antimony addition the resistance tends to increase with addition of antimony. This is because the antimony reacts with the excess alkalis to bring about a reduction in the number of interstitial defects and thus increases the resistivity. At the later stages the reverse happens; i.e., with increase of antimony the number of vacancies for alkalis increases thereby increasing the acceptor-like defects and increasing the conductivity. When alkalis are added the vacancies are filled and resistivity increases. The activation energies for $Na_2KSb$ and $Na_2KSb(Cs)$ have been measured. They seem to be identical, suggesting that the addition of cesium and antimony to $Na_2KSb$ probably does not change the defect level scheme. The resistance, however, decreases during addition of cesium and antimony by a factor of 5–10, depending on the number of alternations required for maximization of sensitivity. The activation energies of these two cathodes are almost the same: 0.34, 0.21, and 0.1 eV for $Na_2KSb(Cs)$ and 0.36, 0.22, and 0.1 eV for $Na_2KSb$ (Fig. 20).

There has been some controversy about the role of cesium in multialkali $(Na_2KSb)Cs$ photocathodes. Spicer (16) observed that the band gaps of the two materials were the same. Addition of cesium and antimony changed the threshold energy but not the band gap. Also, the optical absorption was

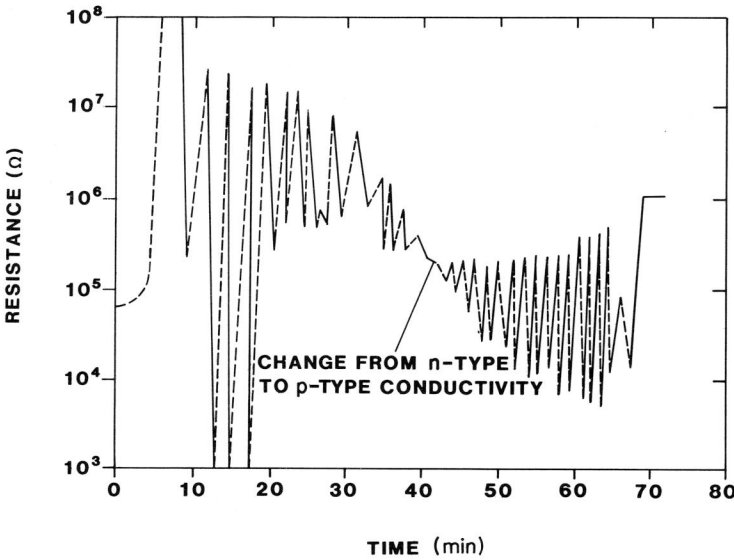

Fig. 19. The change of the type of conductivity from n type to p type during K–Sb addition in the processing of Na$_2$KSb(Cs) photocathode: (---) Sb on; (—) K on. From Garfield (*116*).

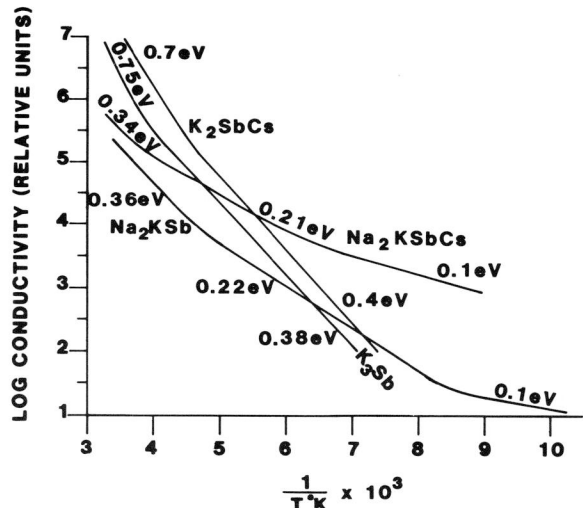

Fig. 20. Log of conductivity versus $1/T$ for a few alkali antimonide photocathodes. (The curves are displaced vertically with respect to one another.) From Ghosh and Varma (*52*).

almost the same for both materials, a point that was later verified (*39*). Spicer concluded from his observation that the cesium deposits as a monolayer on the surface and lowers the electron affinity, and its effect is only on the surface.

McCarroll *et al.* (*162*), using careful X-ray analysis, observed that the lattice constant of $Na_2KSb$ was $7.727 \pm 0.003$ Å, which after cesium addition increased to $7.745 \pm 0.004$ Å. They concluded that some cesium must have been incorporated in the volume of the material, causing an increase in lattice constant. By using Vegard's law the quantity of cesium was estimated to be 1% of the volume.

Ninomiya *et al.* (*113*) made electron diffraction studies in a modified transmission electron microscope and they estimated that $Na_2KSb$ cathode has a lattice similar to that of $Cs_3Sb$, but in which 30% of the cesium atoms are replaced by potassium and sodium atoms.

Hoene (*137*) reported a chemical analysis of high-sensitivity photocathodes. He found that the cesium content was between 3 and 7%. Garfield (*116*) also found, by using a microbalance technique, that about 10% of the weight of the whole cathode was due to the last few additions of cesium and antimony. This means that the cathode contained a much higher quantity of cesium than was necessary to make a monolayer at the surface.

Oliver (*163*) reported that the vapor pressure of cesium over the S·20 photocathode was about four orders of magnitude lower than it should be if one assumed the cesium was at the surface. Again Dowman *et al.* (*128*) observed from their scanning electron diffraction studies that good multialkali photocathodes are composed primarily of $Na_2KSb$ with a layer of $K_2CsSb$ at the surface. They also observed that the surface layer of $K_2CsSb$ was formed during the last stage of photocathode processing when cesium and antimony were added. Since no potassium was added at the last stage, it was inferred (*52*) that the potassium atoms required to form the $K_2CsSb$ layer at the surface must have migrated from the volume of $Na_2KSb$ as a result of substitution of cesium atoms for potassium atoms. Cesium atoms can replace the potassium atoms quite easily, as the reaction constant for this reaction is very high (*164*). The lattice would expand due to this substitution, because the cesium atoms are larger than the potassium atoms. Also, the vapor pressure of cesium and the $K_2CsSb$ surface layer would be much lower than could be expected with free cesium at the surface. A scheme of band-bending on the surface of $Na_2KSb$ has been proposed (*52*) (Fig. 21). $Na_2KSb$ is a highly p-type material and $K_2CsSb$ is nearly intrinsic. Thus band-bending is created at the surface, by which the effective electron affinity is lowered and the theoretical threshold comes down to 1.3 eV, compared to the 1.34 eV that was measured experimentally.

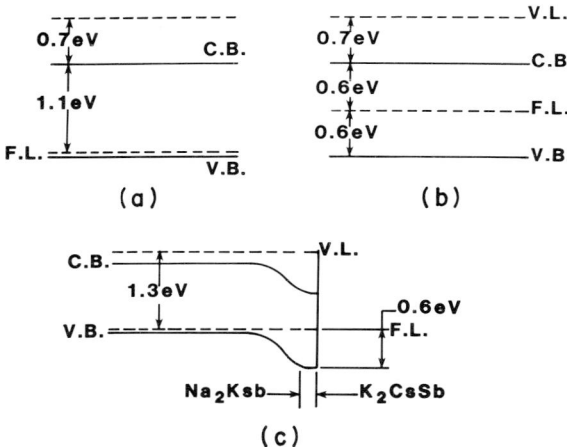

FIG. 21. Band-bending at the surface of multialkali photocathode: (a) $Na_2KSb$; (b) $K_2CsSb$; (c) energy-band diagram of $Na_2KSb$ and $K_2CsSb$. From Ghosh and Varma (52).

$K_2CsSb$ was found by Ghosh and Varma (52) to have a band gap of 1.2 eV and an electron affinity of 0.7 eV. Nathan and Mee (165), from their optical absorption data, determined a band gap of 1.0 eV. An activation energy of 0.7 eV at higher temperature and 0.4 eV at lower temperature has been observed (52). This was the same as in $K_3Sb$ (Fig. 8). Thus substitution of potassium by cesium does not affect the energy level scheme much, as was also observed for $Na_2KSb$ and $Na_2KSb(Cs)$ materials. If 0.7 eV represents the electrical band gap, then $K_2CsSb$ may also have an indirect gap. $K_2CsSb$ has the highest resistivity of all the alkali antimonides. A typical value is $5 \times 10^2$ Ω cm, compared to the value of 0.5 Ω cm for $Na_2KSb$ (166). Compared to other cathodes which are considered to be highly doped, $K_2CsSb$ is considered to be nearly intrinsic, with an acceptor concentration on the order of $5 \times 10^{15}$ cm$^{-3}$. However, the high resistivity makes it incompatible with applications for which currents of more than $10^{-7}$ A are required, unless it is deposited on conducting substrate.

The thermionic emission from different cathodes is one important parameter for practical use of the cathode. Since thermionic emission, which is governed by Richardson's equation, is dependent on the position of the Fermi level, it could be assumed that p-type cathodes in which the Fermi level is close to the valence band would have a lower thermionic emission. The thermionic emissions of most of the cathodes have been measured. The ones like $Cs_3Sb$ have thermionic emissions of less than $10^{-16}$ A cm$^{-2}$ at room temperature. Correspondingly, $Cs_3Sb$ on MnO substrate has a dark current $<10^{-15}$ A cm$^{-2}$, Na$_2$KSb $<10^{-16}$, (Na$_2$KSb)Cs $<10^{-15}$, and K$_2$CsSb $<10^{-17}$, which is the lowest. Oxidized K$_2$CsSb has a dark current $>10^{-16}$ A cm$^{-2}$ (66).

The alkali antimonide photocathodes are known to be good secondary emitters. In fact, all good photoemitters are and should be good secondary emitters. This is because the mechanism of electron emission is the same in both the cases, the only difference being the mode of excitation. Cs$_3$Sb has been found to be of practical importance and is often used in photomultipliers. Sommer (167) studied the secondary emission of some of the alkali antimonides. For Cs$_3$Sb he observed a maximum yield of about 12 at a primary energy of 500 eV. The yield saturated at a va'ue of 12. For K$_2$CsSb he obtained a value of 20 at 600 eV. The yield was measured up to that energy, and saturated at the maximum of primary energy. For K$_2$CsSb(O) yield was about 30 at 800 eV and was still rising. For S·20 the yield was about 32 at 900 eV and was still rising. Mostovskii et al. (168) measured the yield for S·20 and found a $\delta_{max}$ of about 30 at $\sim$1400 eV. Ghosh and Varma (169) obtained a maximum yield of 39 at a primary energy of 1800 eV. Thus K$_2$CsSb and multialkalis could be used as very efficient secondary emitters, but their use in photomultipliers is limited by the difficulty of processing them.

In secondary electron emission the primary electrons excite from the valence band electrons which then escape to the vacuum. The increase of $\delta$ with $E_p$ happens when the range of primary electrons is smaller than the escape depth (170). The maximum in $\delta$ usually corresponds to the primary energy, at which the range of primary electrons is equal to the escape depth of electrons, because beyond this energy the primary beam would generate some secondary electrons, which cannot escape. Using the phenomenological theory (170) based on such assumptions, the escape depth for multialkali antimonide was found to be about 400 Å. Materials with greater escape depths would give higher values of $\delta_{max}$. Since $\delta_{max}$ for Cs$_3$Sb is the lowest of all the photocathodes for which $\delta_{max}$ has been measured (Fig. 10), it must have the smallest escape depth. In fact the escape depth of Cs$_3$Sb was estimated by Burton to be 250 Å (171).

## 5. Optical Properties

The optical properties of photocathodes have received considerable attention because they determine the excitation of the photoelectrons. The determination of optical constants has been followed by the investigation of the optical transitions by photoelectron spectroscopy.

Among alkali antimonides the cesium antimonide was the first material whose optical properties were investigated quite extensively (16, 34, 35, 97,

*107, 150, 171, 172*). Figure 22 shows the value of optical absorption coefficient measured by different workers. Wallis (*150*) and Kunze (*107*) have measured the thickness of the photocathode film to determine the absolute value of the absorption constant, and others have used their value at some point to determine the thickness of their film. The values of absorption coefficient determined by different workers are remarkably consistent. This is because of the simplicity of producing the stoichiometric material and because the optical absorption in the range of measurement is due to the interband transition between the valence and conduction bands and is a bulk property. This is unlikely to be much affected by slight changes of defect concentration and surface states, etc., which may have great influence on photoemission. Wallis (*150*) and Eckart (*97*) fitted an equation of the form $\alpha = c(hv - E_g)^{1/2}$ where $\alpha$ is the absorption coefficient and $E_g$ the band gap. Wallis obtained a value of 1.6 eV for the band gap and Eckart a value of 1.7 eV. These values probably indicate an upper limit of the band gap energy (*16*), which could not be determined through photoconductivity measurements. Taft and Philipp (*35*) observed a number of structures in the absorption coefficient

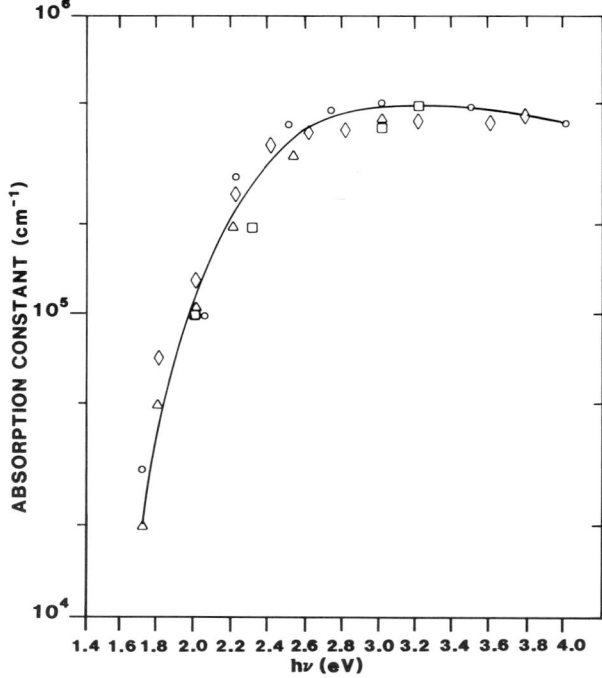

FIG. 22. The absorption constant of $Cs_3Sb$ measured by different workers: (□) Kunze (*107*); (○) Taft and Philipp (*35*); (△) Wallis (*150*); (◇) Spicer (*16*). From Sommer (*66*).

spectrum at 90 K, which were at $hv \simeq$ 1.6, 2.0, 2.25, and 2.45 eV. In the data of Wallis (58) a peak is observed at 1.6 eV. The small peak at 2.25 was observed by Borzyak (173), who attributed it to the formation of excitions in $Cs_3Sb$. This is supported by the energy distribution measurements of Taft and Apker (35), who did not observe any structure corresponding to this excitation energy. Wallis (150) also measured the refractive index of $Cs_3Sb$ between 0.4 and 1.7 μm, in which range the maximum value was 3.3 at 0.6 μm and the minimum was 1.8.

The optical properties of other mono- and bialkali antimonides have also been studied extensively (16, 35, 39, 115, 174, 175). Spicer (16) investigated the optical absorption coefficient of most alkali antimonides in the photon energy range of 0.5–4.0 eV. The $K_3Sb$ absorption coefficient spectrum had two peaks in this region, at about 2.6 and 3.7 eV, and a minimum at 3.0 eV. Taft and Philipp (35) also observed two peaks in the absorption coefficient, near 2.4 and 3.4 eV, and a minimum at 3.0 eV. The first peak at smaller energy has split, the weaker component in it being at 2.2 eV. This weaker component became quite prominent when the material was cooled to 90 K. They concluded from the energy distribution data that these peaks are due to the structures in the valence band density of states.

Sommer and McCarroll (111) measured the optical absorption of hexagonal and cubic variations of $K_3Sb$ and observed that the optical absorption of their hexagonal variation was exactly similar to the spectra obtained by Spicer (16) and Taft and Philipp (35) which were probably also from hexagonal variations. The absorption spectrum of the cubic modification was not very different as the numbers of peaks and shoulders were almost the same, except for one additional shoulder for the cubic phase at 4.5 eV. The positions and relative strengths of the peaks were, however, considerably different. Ebina and Takahashi (174) investigated the optical constants of the purple and brown variations of $K_3Sb$, which were considered to be the hexagonal and cubic modifications although the crystal structures were not directly determined. The purple variation had optical properties almost identical to those of hexagonal $K_3Sb$ (111) and the brown variation was similar to cubic $K_3Sb$. Their measured optical constants were highly reproducible from sample to sample. They calculated $n$ and $k$ from their transmittance data by means of Kramer's Kronig relationship. $n$ was found to vary between 0.5 and 2.2 in the energy range of 0.5–6 eV. The $\epsilon_2$ spectra looked the same for both variations, and both had two strong bands of almost equal magnitude with a sharp minimum in between. The low-energy band comprised two peaks in both cases and was in the range of photon energy of 2–3 eV, while the high-energy one comprised a peak and small shoulder at higher energy and appeared between 3.5 and 4.2 eV. The position of the minimum was at about 3 eV for both types of film.

Spicer measured the absorption coefficient of $Na_3Sb$, and observed that it increased quite sharply from an energy of $\sim 1.2$ eV and started levelling off at $\sim 2.8$ eV. Ebina and Takahashi (174) extended their measurement up to 6.2 eV and observed that the absorption of $Na_3Sb$ consists of two bands, as in $K_3Sb$. The important difference between $Na_3Sb$ and $K_3Sb$ spectra was that in the $\epsilon_2$ spectrum the two levels in $K_3Sb$ were of almost the same magnitude, while in $Na_3Sb$ the lower-energy band was much stronger than the higher-energy one. The $\epsilon_2$ spectrum of $Na_3Sb$ was found to be similar to that of $Na_2KSb$ (Fig. 23). In $Na_2KSb$ samples that had high sensitivity, both the absorption bands had two peaks. The two absorption bands in $Na_3Sb$ had three peaks each. In low-sensitivity $Na_2KSb$ samples that were prepared by activation of $Na_3Sb$ films with potassium vapor, three peaks were observed in both bands. This is thought to be due to the presence of $Na_3Sb$ in the films. In high-sensitivity samples, however, there were only two peaks in both bands. The $n$ and $k$ spectra of the two films were also similar for these two films. The variation of the position of some of the peaks can be traced and appears to be composition dependent (174).

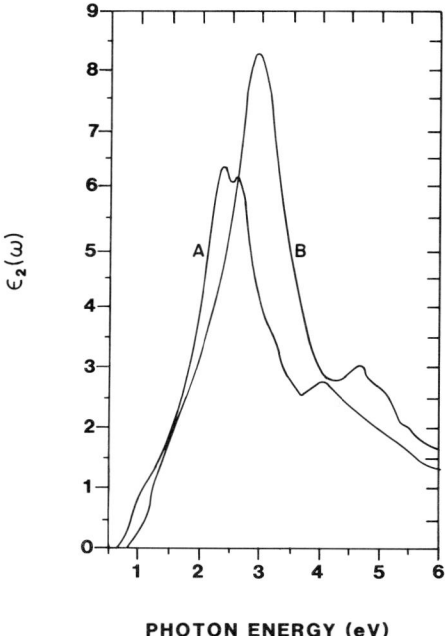

FIG. 23. The $\epsilon_2(\omega)$ spectra of (A) $Na_3Sb$ and (B) $Na_2KSb$. From Ebina and Takahashi (174).

The optical absorption coefficient of $Rb_3Sb$ was measured by Spicer (*16*). As in other alkali antimonides, the optical absorption coefficient rose sharply from about 1 eV to about 3 eV. The peak at 3.4 eV had a low-energy shoulder. The optical absorption of $K_2CsSb$ was measured by Natnan and Mee (*165*) and by Ghosh and Varma (*175*). This material also had two absorption bands, like all other alkali antimonides. The optical constant of multialkali $Na_2KSb(Cs)$ was measured by Spicer (*16*). He found that the absorption coefficient of $Na_2KSb(Cs)$ was identical with that of $Na_2KSb$. Ghosh (*39*) determined optical constants that were almost the same as those of $Na_2KSb$, as determined by Ebina and Takahashi (*174*). The shapes of the $\epsilon_2$ and $n$ and $k$ spectra were identical. The positions of the higher-energy peaks were the same, and in Ebina and Takahashi's experiments that did not vary from sample to sample.

Thus the optical properties of all the alkali antimonides are almost identical in behavior, except for the change in position of the peaks, which is a result of the change in crystal structure and replacement of one alkali atom by another.

The band structures of the alkali antimonide compounds have been calculated by pseudopotential (*176*) and empirical pseudopotential (*177*) methods, making it possible to identify the transitions in the Brillouin zone that are responsible for the observed spectra. In the pseudopotential calculations the Heine–Abarenkov model potential (*178*) was used to establish the qualitative nature of the band structure. In the empirical pseudopotential method the crystal potential is constructed from the ionic potentials, the Fourier components of which are determined after appropriate interpolation to the new lattice period. The interpolation is done in such a way that the width of the forbidden zone of this compound is as close as possible to that observed experimentally. The calculated band structures of these compounds are very similar. Figure 24 shows the band structures of cubic $K_3Sb$, $Na_2KSb$, and $K_2CsSb$ as obtained from pseudopotential calculations (*176*). The band structures of other compounds are similar.

The significant feature of the band structure of these compounds is that the valence band comprises two narrow subbands separated by a large gap of about 6 eV. This implies that these compounds have chemical bonds with highly ionic nature. The forbidden band is determined by the gap at the $\Gamma$ point, for example between $\Gamma_{15}$ and $\Gamma_1$ in cubic $K_3Sb$ and $Na_2KSb$, between $\Gamma_6$ and $\Gamma_{1+}$ in hexagonal $K_3Sb$, and between $\Gamma_{2-}$ and $\Gamma_{1+}$ in hexagonal $Na_3Sb$. In certain compounds, such as cubic $K_3Sb$, there is a forbidden gap in the conduction band between lower- and higher-energy subbands. If the spin–orbit interaction is taken into account, the splitting of the levels of alkali atoms is less by an order of magnitude than those of antimony atoms and hence can be neglected. It is found for example, that for $K_3Sb$ the top

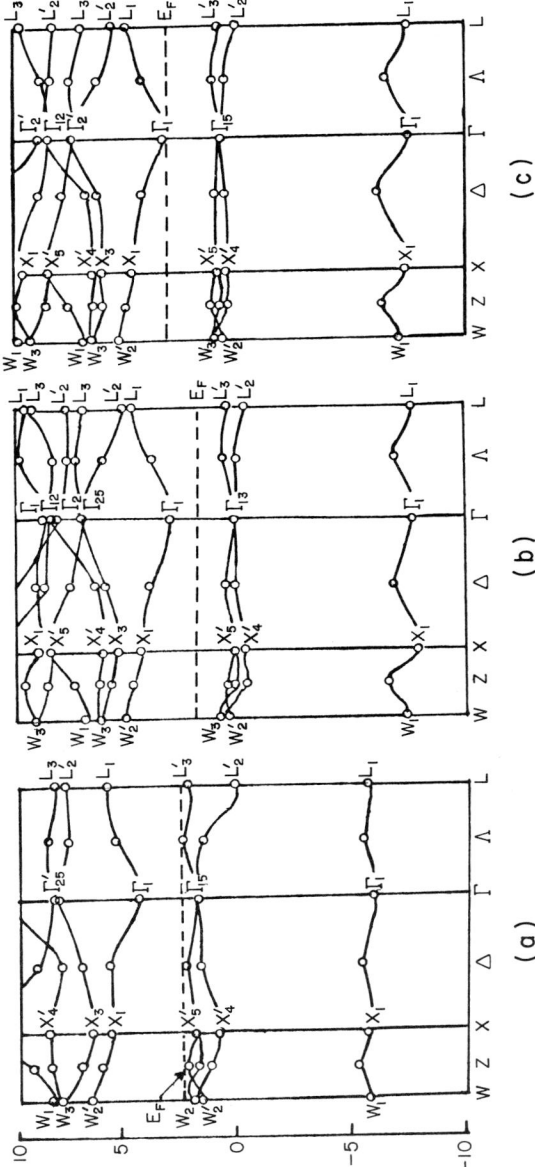

FIG. 24. Band structure of: (a) $Na_2KSb$; (b) $K_2CsSb$; (c) $K_3Sb$. From Mostavskii et al. (176).

of the valence band at Γ which is due to the antimony orbital would split into two bands with an energy separation of 0.97 eV due to spin–orbit splitting. The upper band would move compared to the position of the top of the valence band, without taking into account the spin–orbit splitting, by an amount equal to 1/3 of the spin–orbit splitting, and the lower band would move down by 2/3 of this splitting energy.

The comparison of optical properties of $(Na_2KSb)Cs$ with the band structure of $Na_2KSb$ (39) shows that the energy separation of the subbands in the valence band that are below the energy gap in the valence band are more than 8 eV from the bottom of the conduction band. Therefore they cannot take part in the optical transition below an energy of 8 eV. Thus the optical properties in the visible and UV regions below 8 eV must be due to transitions between the top of valence band and the conduction band. At the lower-energy end of the $\epsilon_2$ and k spectra a small bump at 1.5 eV has been assigned to transitions between the lower level of the spin–orbit split $\Gamma_{15}$ band and the bottom of the conduction band, the calculated energy separation between which appears to be about 1.7 eV. The optical transition strength spectrum rises sharply above an energy of 2 eV. This is apparently due to the transitions in the Δ and Λ directions between the valence and conduction bands. The first shoulder and peak in the optical transition strength spectrum (Fig. 25), at 3.3 and 3.6 eV, are probably due to transitions at both the L and X points of the Brillouin zone. From the band structure calculations, done without taking into account spin–orbit splitting, strong transitions are indicated between $L'_3$ and $L_1$, with an energy gap of 3.21 eV. At the X point the $X'_5$–$X_1$ gap is 3.16 eV, the $X'_4$–$X_1$ gap is 3.82 eV, and the $X'_3$–$X_3$ gap is 3.92 eV. The peak and shoulder in the first band may be the

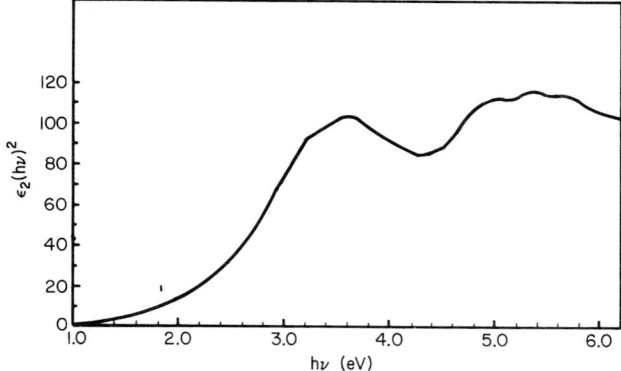

FIG. 25. The optical transition strength $\epsilon_2(h\nu)^2$ spectrum of $Na_2KSb(Cs)$. From Ghosh (39).

result of these transitions. After 3.6 eV the optical transition strength falls, probably because of the gap in the conduction band along the $\Delta$ and $\Lambda$ directions, so that only the Z direction is available for optical transitions. The optical transition strength again starts rising beyond 4.5 eV photon energy and there are three peaks at the energies of 4.85, 5.25, and 5.7 eV. The energy gap between $\Gamma_{15}$ and $\Gamma_{25}$ corresponds to an energy of 4.91 eV. The separation between the first and third peaks in the observed optical transition strength at the higher-energy band is 0.85, while the calculated spin–orbit splitting at $\Gamma_{15}$ is 0.97 eV. The first and third peaks are considered to be due to the transitions at the $\Gamma$ point from the top two valence subbands. The middle peak at 5.25 eV is assigned to the parallel bands in the Z direction, making a strong contribution to the joint density of states. Similar assignments could be made for other cathodes.

Photoelectron spectroscopy has been used by a number of investigators (*21*, *34*, *35*, *39*, *179*) to find out the nature of transitions in the alkali antimonides. In photoelectron spectroscopy the energy distribution of the electrons photoemitted from the solid is measured. In the absence of strong scattering this distribution should be proportional to the number of transitions to the final electron states. Thus while optical properties such as $\epsilon_2$ give information about the total transition probability between different bands at a particular energy, the photoelectron spectrum resolves the transition probabilities to various final states. Taft and Philipp (*35*) measured the energy distribution of $K_3Sb$ and $Cs_3Sb$. They found two peaks separated by 0.4 eV in the energy distribution spectra of $Cs_3Sb$, and concluded that these were due to two peaks in the densities of states in the valence band, because with a change in incident photon energy the peaks moved by an amount equal to the change in photon energy. The two peaks in optical absorption, at 2.0 and 2.4 eV, were also assumed to be due to these peaks in the valence band. A similar observation was made for $K_3Sb$, but three peaks were observed in energy distribution curves (EDCs).

Spicer (*21*) concluded from his measurement of the energy distribution spectra of $Cs_3Bi$ and $Cs_3Sb$ that the optical transitions in these materials do not require the conservation of crystal momentum. He observed two peaks in the energy distribution spectra of $Cs_3Bi$ and $Cs_3Sb$, which moved with change of photon energy by an amount equal to the change of photon energy. He thus concluded that these peaks were due to the density of states in the valence band arising due to the spin–orbit splitting of 5 p and 6p orbitals in antimony and bismuth, respectively. If the two peaks were due to spin–orbit splitting, then in $K_3Sb$ the valence band would be split into three bands because of hexagonal symmetry. In cubic symmetry the spin–orbit splitting should give two peaks, as has been observed in $Cs_3Sb$. Actually three peaks have been observed for $K_3Sb$ (*35*). If the transitions involved emission or absorption of phonons, the absorption coefficient would be affected by a change of temperature, which changes the phonon density, and the absorption coefficient would not be as large as has been observed. Because of the low Debye temperature of these materials the probability for transition involving a phonon should increase by a factor of 3–4 between 77 and 300 K. However, there was no significant change of absorption due to this change of temperature (*180*). Thus, it was concluded that the transitions do not involve phonons and do not have $k$ conservation as a selection rule. This could be a result of low hole mobility and also of strong interaction of the hole with the lattice. Materials with similar hole mobility should behave likewise, and in fact similar behavior has been observed in alkali halides (*104*). Similar behavior was observed for other photocathodes, such as $K_2CsSb$ (*165, 175, 181*) and $Na_3Sb$ (*182*).

In the case of nondirect transitions it is possible to determine the effective density of states in the valence and conduction bands by looking at the energy distributions (*95*). The normalized energy distribution can be represented as

$$N(E)/N_{ph} = (B/\omega^2\epsilon_2)M^2N_c(E)N_v(E - h\nu) \qquad (25)$$

The constant $B$ contains the probability of escape functions. $B$ can be considered to be a step function such that the probability of escape is zero at an energy below the vacuum level and is constant above it. $N(E)$ is the number of electrons at a particular energy, $N_{ph}$ is the total number of photoelectrons emitted, and $M$ is a matrix element combining the initial and final states, which are $N_v(E - h\nu)$ and $N_c(E)$. Now if the matrix element is not a rapidly varying function of energy, then the valence band density of states could be determined by looking at a particular final state energy as

$$M^2N_v(E_c - h\nu) \propto \omega^2\epsilon_2(N_E/N_{ph})|_{E=E_c} \qquad (26)$$

Thus a plot of the magnitude of the normalized energy distribution at a particular final state energy for all photon energies would give a plot of the valence band effective density of states. Similarly, the conduction band density of states can be obtained by holding some point in the valence band as a reference, changing the photon energy, and observing the distribution at the energy $(E_v + h\nu)$ for a range of $h\nu$. This expression and the method of analysis would be valid in the absence of electron–electron and electron–lattice scattering. Wooten et al. (*179*) determined the density of states in $Cs_3Bi$ by computing the contribution due to both electron–phonon and electron–electron scattering by the Monte Carlo method. While in the case of $Cs_3Bi$ they could find the fit with experimental data without assuming any interaction with the defect levels, in the case of $Cs_3Sb$ this was not possible. They had to assume a defect level corresponding to a cesium vacancy concentration of about 1% of $Cs_3Sb$ in order to match the calculated energy distribution curves with experiment. They assumed the level at 0.4 eV above the valence band, but its position was not important in the calculation except for the fact that it had to be in the lower half of the forbidden band. The defects could even be assumed to be spread uniformly in the lower half of the forbidden band. The resulting density of states is shown in Fig. 26. The scattering cross sections at different energies were chosen to fit the experiments.

The nondirect transitions were identified for all other alkali antimonides except multialkali antimonides. In the case of multialkali antimonides, the equal-increment rule was not observed (180). Figure 27 shows the positions of the peak, shoulders, and valleys in these photocathodes. By comparison of the energy distributions at different energies with the band structure (180), it is observed that initially the peak in the energy distribution near threshold closely follows the change of photon energy because the slope of the band at the bottom of the conduction band is high around the $\Gamma$ point. Subsequently, however, the position of the peak changes more slowly with change of photon energy, because as the X and L points are approached, the slope of the band is smaller. Beyond the X point the peak again moves faster as the transitions start taking place in the Z direction. In the Z direction the transitions would give rise to two peaks at about an energy of 5 eV, which has in fact been observed. The energy separation of these two peaks is $\sim 0.5$ eV and is in reasonable agreement with the expected energy separation due to two parallel bands in the conduction band separated by about 0.7 eV. At still higher energies more peaks are observed, because subbands lying higher in the conduction band are exposed and transitions take place to these.

The threshold for electron–electron scattering and the mean free path for pair production were determined by means of photoelectron spectroscopy for most of the alkali antimonides. Due to electron–electron scattering the maximum energy lost by the photoelectron is equal to the band gap energy. However, because of $k$ conservation the threshold energy for pair production is much higher than $2E_g$ above the top of the valence band. If the photoelectrons lost energy by electron–electron scattering they would appear as slow electrons. In the energy distribution curves a group of slow electrons appear at a particular energy, their number increasing with increase of photon energy. These electrons are thought to be generated by electron–electron scattering. The threshold energy for this phenomenon is considered to be the energy at which the slow group of electrons first appears. This energy is about 3.7 eV from the bottom of the conduction band for hexagonal $K_3Sb$, 3.0 eV for $Rb_3Sb$, 2.0 eV for $Cs_3Sb$, $<1.3$ eV for $Cs_3Bi$ (38), 2.3 eV for $(Na_2KSb)Cs$ (39), and 3.3 eV for $K_2CsSb$ (183). The electron energy is

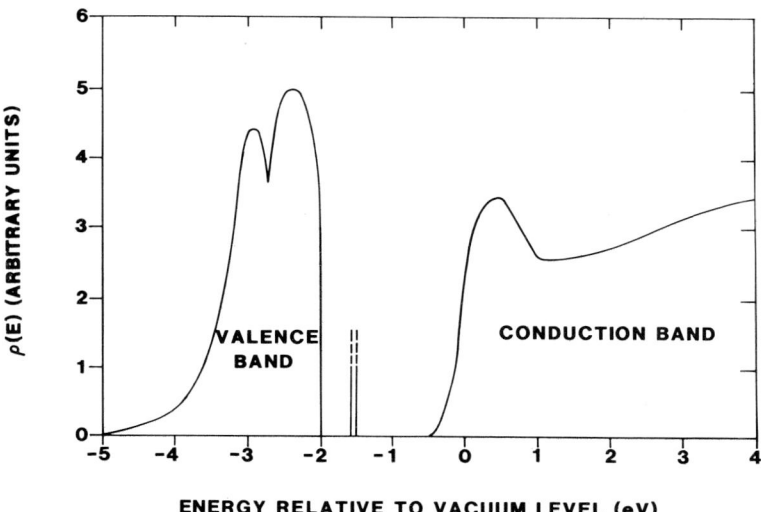

FIG. 26. Density of states for $Cs_3Sb$. The position of the defect levels is 0.4 eV above the valence band maximum. From Wooten et al. (179).

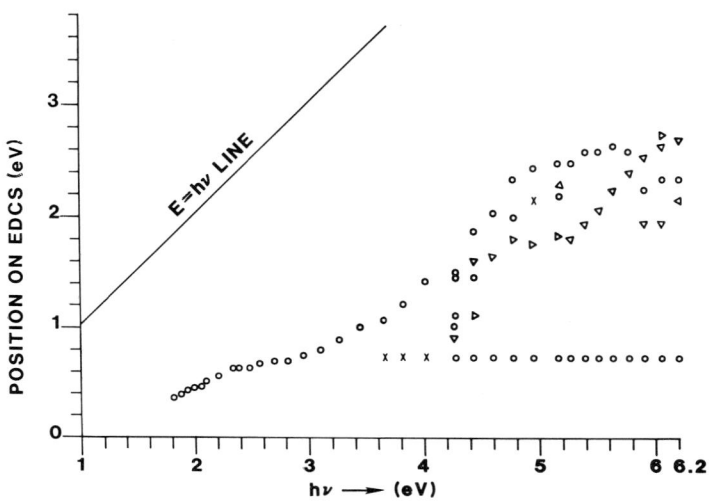

FIG. 27. Plot of the positions of the peak (○), shoulder (×), and valley (△) points in the energy distribution curves (EDCs) versus photon energy in $Na_2KSb(Cs)$ cathode. From Ghosh and Varma (180).

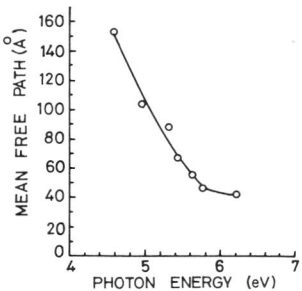

FIG. 28. Mean free path for electron–electron scattering in multialkali photocathodes. From Ghosh (39).

lowered by this scattering and in general it causes a decrease in quantum efficiency. However, in $(Na_2KSb)Cs$ cathodes an increase in quantum efficiency has been observed due to the electrons scattered from the valence band having more energy than the threshold for photoemission (39).

The mean free path for electron–electron scattering has been calculated by measuring the relative number of scattered electrons at each energy. This mean free path is large near the threshold, but soon falls with increase of electron energy. Figure 28 shows the change of mean free path for multialkali photocathodes with change of photon energy (39). The mean free path falls sharply from about 150 Å near 4.5 eV to about 40 Å near 6 eV, i.e., within a span of about 1.5 eV. For $Cs_3Sb$ also the scattering mean free path as obtained by fitting to Monte Carlo calculations to obtain the experimental energy distribution changes from about 250 Å at $\sim 1.0$ eV above the vacuum level to $\sim 15$ Å at a value of 3.5 eV, i.e., in a span of 2.5 eV (179).

## V. Negative Electron Affinity Materials

The quantum efficiency of photoemissive materials has been dramatically improved by the discovery of negative electron affinity photoemitters. NEA was first observed by Scheer and Van Laar (7) with highly doped p-type GaAs, the surface of which was activated by cesium atoms. The effective electron affinity is lowered by band-bending at the surface, as can be seen in Fig. 29. A further enhancement of sensitivity was obtained by oxidizing the cesium layer at the surface (184). In the case of positive electron affinity, the difference in energy between the bottom of the conduction band and the

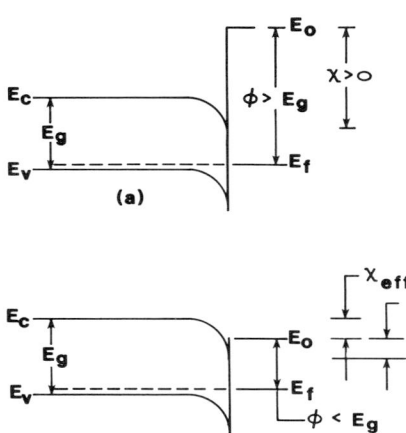

FIG. 29. Band diagrams showing (a) positive electron affinity and (b) negative electron affinity conditions in relation to band gap energy $E_g$ and work function $\phi$. [From Martinelli and Fisher (*185*). Copyright © 1974, IEEE.]

Fermi energy is less than the work function of the material. However, in the case of NEA, this difference

$$E_c - E_f > \phi \qquad (27)$$

where $\phi$ is the work function. Thus, to obtain NEA the quantity $E_c - E_f$ should be maximized (*185*). This could be achieved by doping the material to make it highly *p*-type, so that the Fermi level almost coincides with the valence band maximum. In this case, $E_c - E_f \simeq E_g$, the band gap. So for NEA the condition $E_g > \phi$ should be satisfied. The quantity $E_g - \phi$ is called the degree of NEA, and the photoemission increases with the degree of NEA.

As can be seen, in the case of NEA an electron thermalized at the bottom of the conduction band can escape upon reaching the surface. Thus the electron emission for NEA materials is determined by the lifetime of the minority carriers, and the escape depth of the electrons is equal to their diffusion length. The diffusion length is quite large for good crystalline materials ($\sim 10^{-4}$ cm). Thus, compared to conventional photoemitters, in which the electron emission is due to the hot electrons, the probability of emission is very high in NEA materials.

A number of materials have been found to exhibit NEA. Materials such as GaP (*186*), InP (*187*), Ga(AsP) (*188*), In(AsP) (*189*), (GaAl)As (*190*), Si

(8), (In, Ga, As, P) (191), and Ga(As, Sb) (192) have been found to give rise to NEA when treated with cesium or cesium and oxygen. Surface treatment by Cs and F gives rise to NEA in the case of GaAs (193) and In(AsP) (194). However, it was reported (195) that Si (100) could not be activated to NEA by Cs and F. Si (195) has been found to produce NEA when treated with Rb + O, but while Rb + O on Si (100) gives an activation similar to that of Cs + O, its activation on InAsP (111) B yields photoemission inferior to that of Cs + O activation (194). Cs + O activation gives rise to the best NEA with most of the materials. The clean (111) face of diamond has been found to exhibit NEA without any surface treatment (196).

### 1. The Lowering of Electron Affinity by Cs/O Activation

Lowering of the work function of metals with Cs coverage has been known for a very long time (197, 198). Langmuir (199) first proposed a theoretical explanation for lowering of the work function. According to him, the cesium atoms, being highly electropositive, readily give up their valence electrons to the substrate, which induces an image charge on the substrate, thereby producing a dipole layer and lowering the work function. With the advent of modern surface analysis tools such as AES, LEED, UPS, and ESCA, along with ultrahigh-vacuum (UHV) technology, it has been possible to investigate the surface of many metals and semiconductors. The reduction of surface barrier by surface treatment has also been investigated in great detail, and though a clear picture of the mechanism has yet to emerge, many interesting aspects of the phenomenon have been investigated.

One important requirement for obtaining good NEA on semiconductor surfaces is that the surface be almost atomically clean before activation with Cs or other alkalis (200). The surface cleanliness condition is even more stringent in the case of Si (201, 202). The clean semiconductor surface is necessary so that the extrinsic surface states are not present in large numbers. A surface in a semiconductor, as it terminates the lattice, should give rise to localized states. In the case of GaAs, an ideal surface, in which the surface atoms are in the same positions as in the bulk, should produce a large number of surface states in the energy gap (203). A large number of surface states are likely to pin the Fermi level at a certain energy in the gap and create band-bending at the surface. It has been found that for the vacuum-cleaved p-type GaAs(110) surface the Fermi level is not pinned but is at an energy very close to the bulk position, and there is little or no band-bending (204–206). Similar observations have been made in the case of vacuum-cleaved InP, GaSb, InAs, and p-type GaP (207–210). Only n-type GaP consistently shows Fermi level pinning at an energy 0.55–0.62 eV below the conduction band edge (208, 211). Pinning has often been observed in n-GaAs (212). The

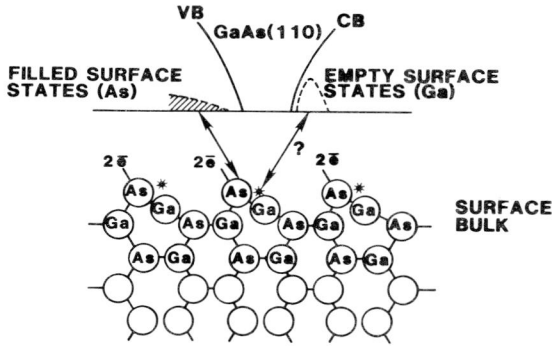

FIG. 30. Schematic diagram of the electronic state distribution near the fundamental band gap of GaAs (110) and the atomic surface reconstruction for this surface. VB and CB refer to the valence and conduction band edges, respectively. Other faces of GaAs may have somewhat different electronic and atomic structures than (110). From Spicer et al. (212).

absence of surface states in the gap has also been observed experimentally (213, 214).

The surface states can be moved away from the band gap to the bands if there is a reconstruction of the surface atoms, which means the surface atoms are displaced with respect to one another (215, 216). A model has been proposed (217) for the (110) face of GaAs, which does not give rise to any intrinsic state in the band gap. According to this model, the surface states have been pulled out of the band gap by electronic and lattice rearrangement at the surface. Such a reconstruction of the surface atoms has been observed directly by LEED (218), which suggests that the surface rearranges, with As atoms moving outward and Ga atoms inward.

The filled surface states in the valence band are associated with the surface As atoms (219). The empty surface states may be due to Ga atoms (Fig. 30). This model for the (110) face may be modified for other faces. In the case of the GaAs(111) As face (B face), the As atoms can contribute the electrons to the filled surface state band, and thus may become more chemically active than the (111) Ga (A face) (220).

The unpinned p-typeGaAs(110) surface is very sensitive to surface treatment by metals and other impurity atoms, such as oxygen. For example, only a tenth of a monolayer of surface coverage by cesium atoms moves the Fermi level to near the middle of the gap (221). Other metals and oxygen also were observed to pin the Fermi level at about the same coverage of about 0.1 monolayer. The pinning position of the Fermi level was almost at the same energy for different metals, which is quite surprising (Fig. 31) if the

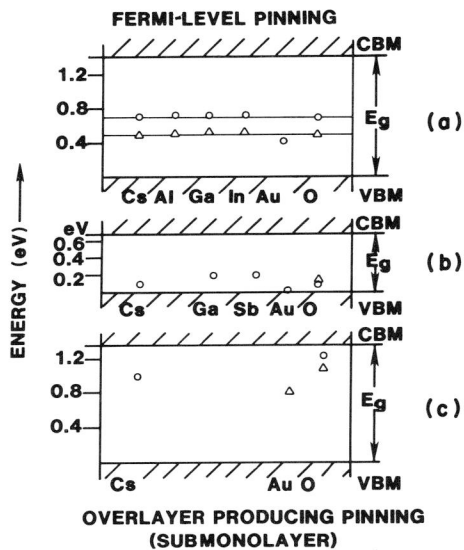

FIG. 31. Approximate Fermi level pinning positions on (○) n- and (△) p-type vacuum cleaved III–V surfaces with submonolayer coverage of adatoms: (a) GaAs (110), (b) GaSb (110), (c) InP (110). From Lindau et al. (222).

extrinsic surface states due to metal atoms themselves are pinning the Fermi level (222). So it is presumed that the pinning states may be produced indirectly by the adatoms (212). The number of extrinsic defect surface states produced by the adatoms corresponding to 0.1 monolayer coverage also may not be sufficient for the pinning of the Fermi level. No pinning states on as-cleaved p-type (110) GaAs material and pinning on the n-type surface at 0.75 eV above the valence band are evidence for the production of a particular type of defect states. A relaxation of the reconstructed surface back toward a bulk-like position could be involved in creating the pinning states. Careful experiments designed to determine whether the intrinsic surface states move back to the band gap and cause the Fermi level pinning with oxidation or metal addition show that the intrinsic surface states do not move, but that new surface states are formed (212). According to a model proposed by Spicer et al., the energy of formation of the defects would be derived from the heat of condensation of the metallic atoms on GaAs. The heat of condensation of Cs on GaAs is about 60 kcal mol$^{-1}$ at lower coverages (223). This energy would produce the defects at the surface by displacing some of the surface atoms into the metal. This process is depicted in Fig. 32. If a few semiconductor atoms are displaced into the metal for each

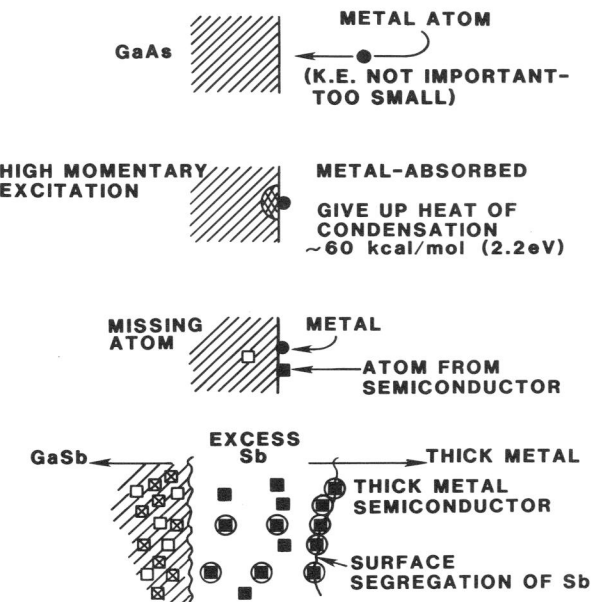

FIG. 32. Schematic of the mechanism of defect formation due to deposition of metal atoms on clean III–V surfaces. (⊠) Sb deficit; (●) Sb in metal; (□) Ga deficit; (■) Ga in metal. This process of defect formation need occur only about once for every hundred metal atoms striking the surface in order to explain the Fermi level pinning. From Spicer *et al.* (*212*).

hundred metal atoms bound to the surface, the resulting departure from stoichiometry and resulting defect states could produce the kind of pinning that has been observed. These defects are antisite defects. There is some support for this model from ESCA and AES experiments and from other empirical observations (*224*).

The pinning of the Fermi level by the surface states that is observed at about 0.1 monolayer coverage does not change much with addition of more cesium or other metal atoms. Figure 33 shows the change of Fermi level position with addition of more of the metal (*221*). Before addition of the metal a clean surface is usually produced by vacuum cleaving, epitaxial growth in ultrahigh vacuum (MBE), ion bombardment annealing (e.g., argon sputter annealing), or simple vacuum heat cleaning. The first two are used more for fundamental studies, whereas ion bombardment and heat cleaning are used for device purposes.

Though addition of metallic cesium on the surface gives rise to NEA, further lowering of electron affinity takes place with oxidation of the cesium

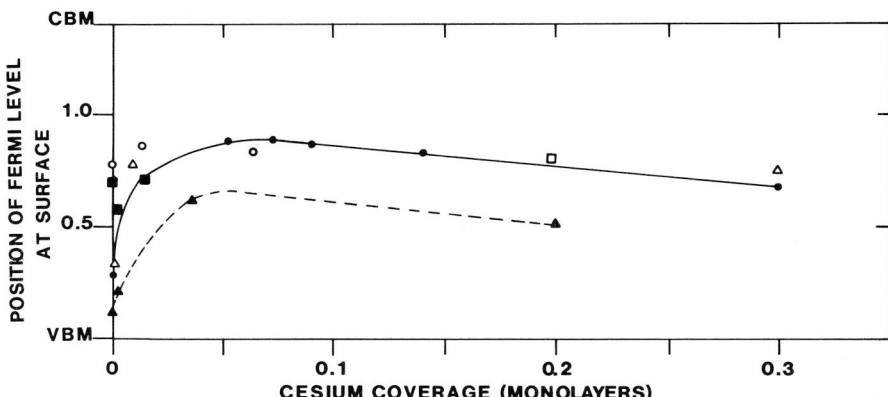

FIG. 33. Experimental measurement of the change of Fermi level position during submonolayer cesium coverage of vacuum-cleaved GaAs (110) surfaces. The Fermi level position does not change significantly beyond 0.1 monolayer of Cs coverage: (●) 17p (first cleave), (△) 17p (second cleave), (▲) 19p, (○) 18n, and (■) 14n refer to $10^{17}$ cm$^{-3}$ p-type, $10^{19}$ cm$^{-3}$ p-type, $10^{18}$ cm$^{-3}$ n-type, and $10^{14}$ cm$^{-3}$ n-type samples, respectively. From Gregory and Spicer (*221*).

layer (*184*). With addition of cesium on the clean surface the work function decreases and the photoemission increases. The maximum in photoemission is reached at about 1/2-monolayer coverage (*225*). Beyond this the work function increases and photoemission starts decreasing. Thermal desorption studies on Cs-covered GaAs surface show (*223, 226*) that the first half-monolayer of Cs is much more tightly bound to the surface than a complete monolayer (Fig. 34). It is observed that the first half-monolayer desorbs at much higher temperature than the second half. This is consistent with the assumption of increasing depolarization with increasing coverage. The first half-layer has greater sticking coefficient and stronger bonding. Further addition of Cs atoms would result in weaker binding because of increasing electrostatic repulsion between the Cs atoms themselves. The less strongly bonded Cs atoms appear even before the formation of the first half-monolayer is complete. Electron-loss spectroscopy (ELS) observation (*227*) suggests that the Ga atoms are the initial binding sites for Cs on the (110) surface.

Oxygen is added to the cesiated surface when maximum photoemission is reached. Usually a few alternations of Cs and $O_2$ give the minimum work function and maximum sensitivity. An oxygen exposure of about 1 langmuir ($1 \times 10^{-6}$ Torr sec) gives the optimum photoresponse, which corresponds to a $Cs_2O$ layer about 10 Å thick (*92, 228–230*). The existence of an interfacial barrier between GaAs and $Cs_2O$ has been observed (*231*). There are two models of surface barrier lowering by means of $Cs_2O$. One is the heterojunction model, (*232, 233*) in which the Cs–O layer has the characteristics

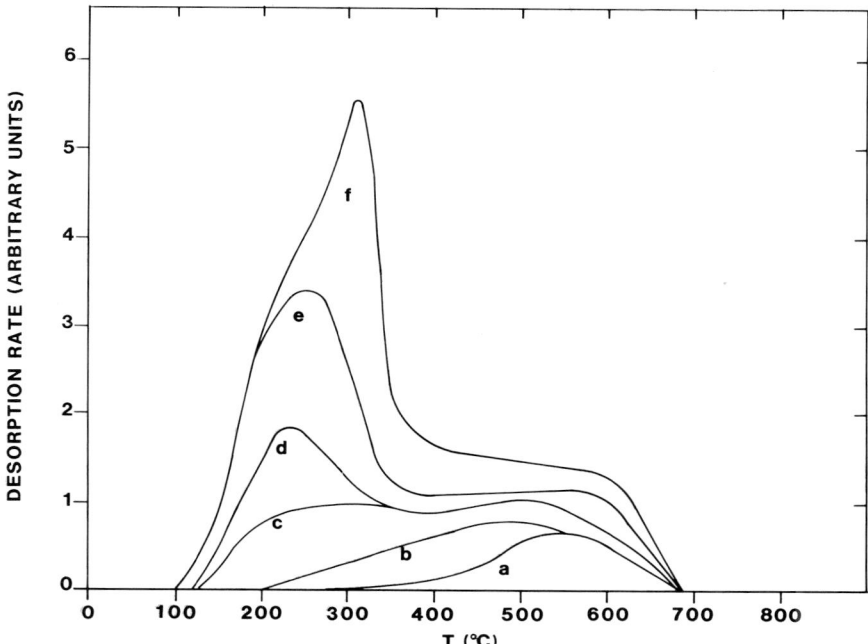

FIG. 34. Flash desorption curves of Cs from S-A GaAs(100) surfaces for different initial Cs coverages, $\theta$; $\theta = 1.0$ is defined as $7.2 \times 10^{14}$ atoms cm$^{-2}$ (from Goldstein and Szostak (226)):

| Curve | Coverage $\theta$ | Curve | Coverage $\theta$ |
|---|---|---|---|
| a | 0.15 | d | 0.58 |
| b | 0.38 | e | 0.77 |
| c | 0.44 | f | 0.88 |

of bulk n-type $Cs_2O$. $Cs_2O$ is an n-type material with band gap of 2 eV and the Fermi level at an energy 0.25 eV below the conduction band minimum (60, 234). If there is a heterojunction formed at the surface, the band profiles adjust so that the bottom of the conduction band in GaAs is above the vacuum level at the surface (Fig. 35). There would be an interfacial barrier at the surface of height of 0.6 eV. The electrons generated in GaAs bulk should either overcome or tunnel through the interfacial barrier. Once in $Cs_2O$, they can be emitted as hot electrons over the small positive electron affinity barrier on $Cs_2O$. The principal objection to this model is that it has

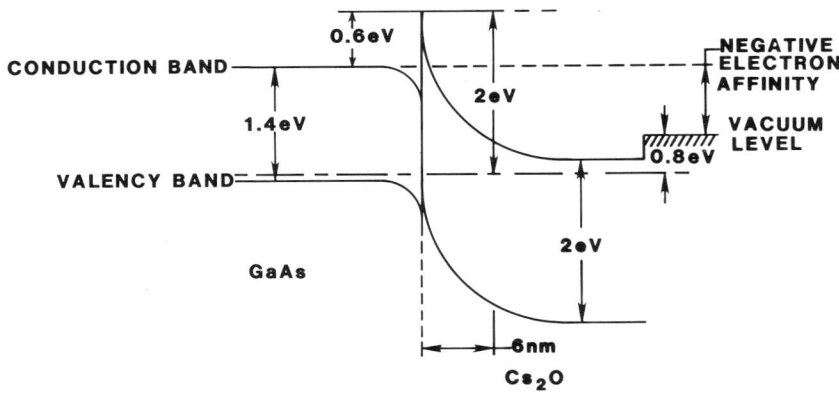

FIG. 35. The heterojunction model of band-bending at the surface of p-type GaAs with n-type $Cs_2O$ layer.

to assume bulk $Cs_2O$ energy bands for almost a monolayer dimension of $\sim 10$ Å.

A second model, shown schematically in Fig. 36, assumes a surface double dipole as the mechanism for activating NEA (229). In this model a dipole thickness of $\sim 8$ Å is consistent with the monolayer dimension of the Cs–O layer. At this thickness, the electron tunneling probability is appreciably less than unity, resulting in an effective barrier above the vacuum level. Thus in terms of experimentally observed barrier characteristics, the dipole model and the heterojunction model give similar predictions.

The composition and atomic arrangement of the Cs–O layer on GaAs are not well established. There are many possible oxides of Cs, ranging from metallic $Cs_7O$ to semiconducting $CsO_3$. UPS experiments with the oxidation of bulk Cs at 100–140 K suggest (235) that the oxygen ions initially dissolve below the Cs surface. Thermal desorption studies on Cs–O-covered GaAs

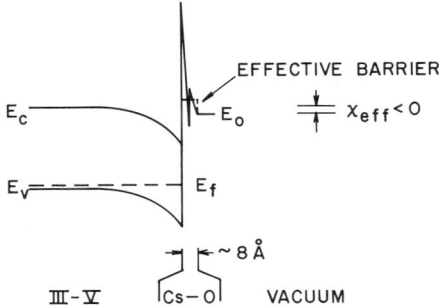

FIG. 36. The surface dipole model of band-bending with Cs–O on GaAs.

(*230*) show that initially the Cs atoms are released from the surface. The fraction of Cs atoms strongly bonded to GaAs increases with addition of oxygen. AES observation (*236*) shows that the oxygen does not start escaping until nearly all the Cs is gone. The desorption of oxygen takes place from the $Ga_2O$ phase, which remains after Cs desorption. Of course, this does not necessarily imply the existence of $Ga_2O$ on GaAs when Cs–O activation is done. It appears that the mechanism of the formation of the interfacial barrier and the arrangement of Cs and O atoms on the surface may be more complex than is assumed in rather simplistic models.

The activation of Si to NEA is similar to that of III–V compounds. Unlike in GaAs, only the Si(100) surface has been successfully activated to NEA. Studies of the vacuum-cleaved Si(111) surface have shown that the as-cleaved surface is pinned at the midgap by the intrinsic surface states. But a slight addition of oxygen and other metals, resulting in submonolayer coverage, removes the intrinsic states from the band gap (*237*). The subsequent pinning of the Fermi level is due to the extrinsic states created by the metals or oxygen (*238*).

The decrease of work function with addition of oxygen after addition of Cs on the surface is observed in the case of activation of the material to NEA. The work function-lowering mechanism appears to be better understood in the case of Si than in that of GaAs. A model has been proposed by Levine (*239*) that assumes a reconstruction of the surface according to which the surface atoms rotate themselves to form adjacent rows of "pedestals" and surface "caves." Adsorbed Cs atoms reside on top of the pedestals, while oxygen adsorption takes place primarily in the cave sites. Thus a Cs–O–Si dipole is formed, which lowers the surface barrier. The surface dielectric constant deduced from AES experiments (*201*) yields a dipole length of 2.8 Å, which is in good agreement with the structural model. Additional support for this model comes from LEED, AES, desorption (*240*), and other experiments (*241–243*).

For activation with Cs the clean surface of Si is slightly overcesiated, that is, the activation with Cs is continued past the photoemission peak. Oxidation of this surface only once gives the minimum work function and maximum sensitivity. The work function is reduced to a value of about 0.85 to 0.93 eV (*244, 245*). The thickness of the Cs–O layer on the surface is $\sim 3$ Å and there is no evidence of an interfacial barrier similar to that observed in III–V compounds. This is probably because of the thinness of the barrier, which results in a tunnelling probability near unity (*185*). Activation of the Si surface has been successfully done with Rb + O (*14*). However, while Cs–F gave good NEA with GaAs (*193*), the Si(100) face could not be activated to NEA with Cs–F. In general, the Si surfaces are less tolerant of any rough treatment regarding the cleanliness or vacuum than is GaAs.

In the case of GaP the basic investigations on the surface show that, like GaAs(110), the GaP(110) surface undergoes similar reconstruction in which the P atoms are displaced outward and Ga atoms inward (246). LEED studies (247) indicate that Cs is adsorbed amorphously on the surface but heating of the Cs–GaP surface can produce several ordered structures. In contrast to GaAs, however, thermal desorption studies show three distinct desorption peaks, at 160, 360, and 550°C (247). Photoemission and ELS studies (248) on the MBE GaP(111) B surface have identified the empty and filled surface states, in general agreement with the model for GaAs (Fig. 30). Activation of the GaP surface is easier because of its large band gap and hence larger degree of NEA. The $GaAs_xP_{1-x}$ compounds have also been activated to NEA. The band gap being composition dependent, the threshold energy can be varied. These compounds have higher band gap and so shorter threshold wavelength than GaAs. Figure 37 shows the quantum efficiency at 2.34 eV (5300 Å) versus GaAsP band gap (249). At higher phosphorous concentration the band gap becomes indirect, like that in GaP, and the quantum efficiency at near-gap regions decreases.

The $In_xGa_{1-x}As$ alloys have smaller band gap and hence lower threshold wavelengths than GaAs, and the band gap decreases with increase of the

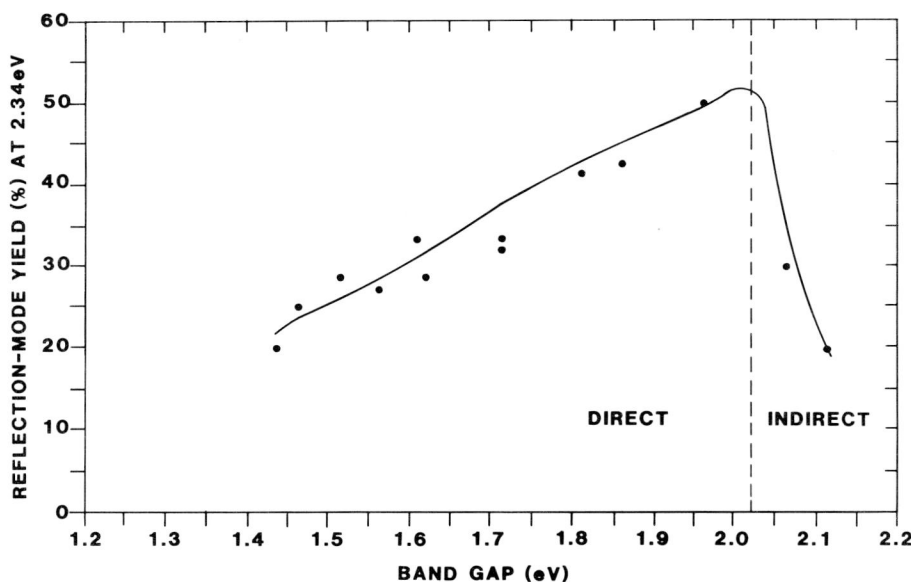

FIG. 37. Experimental reflection mode quantum efficiency at 2.34 eV (5300 Å) versus $GaAs_{1-x}P_x$ band gap.

IN:Ga ratio. The situation with $InAs_xP_{1-x}$ alloys is similar. In $InAs_xP_{1-x}$ the band gap decreases with increase of As concentration. With smaller band gaps the degree of NEA is less, hence a thicker Cs–O layer is needed for surface activation. With decrease of band gap a situation would arise in which the top of the interfacial barrier becomes higher than the bottom of the conduction band in the bulk (250) (Fig. 38A). This requires that all the thermalized electrons tunnel through the interfacial barrier, and quantum efficiency drops abruptly. Figure 38B shows the fall of quantum efficiency with decreasing band gap. If the band gap is further reduced, NEA will not be achieved, which is why such materials as Ge cannot be activated to NEA.

Other quaternary systems such as InGaAsP have been successfully activated to NEA (251). Quantum efficiencies of about 0.01 electrons photon at 1.06 μm have been achieved with these cathodes.

## 2. Carrier Transport

The transport of photogenerated carriers can be described by the simple diffusion equation (249). The electrons quickly thermalize at the bottom of the conduction band or other minimum, depending on the band structure, and from there they can either recombine or reach the surface and be emitted.

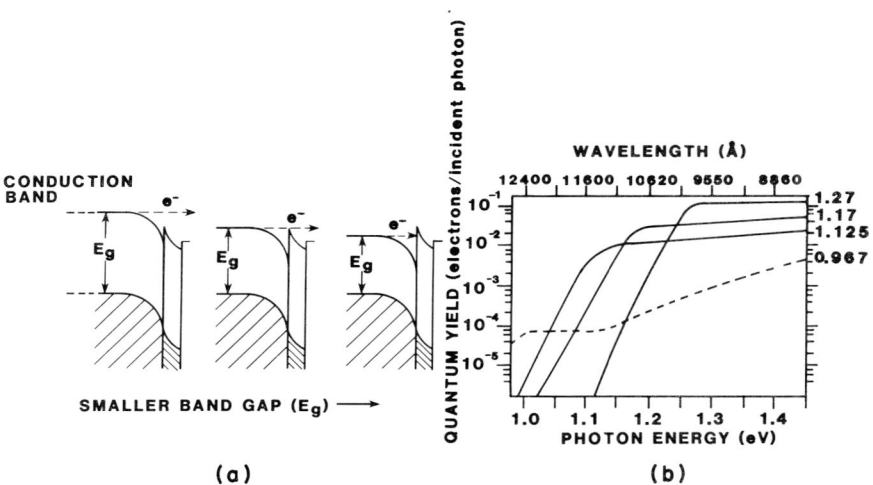

Fig. 38. (a) Schematic band diagram for decreasing III–V band gap. (b) Change in yield of $InAs_xP_{1-x}$ alloys for various band gaps. The band gaps are given at the extreme right for each curve. The dashed curve illustrates the drop and change in the shape of the yield curve when the conduction band minimum drops below the interfacial barrier. From Spicer (250).

The diffusion equation is

$$D\left(\frac{\partial^2 n}{\partial x^2} + \frac{\partial^2 n}{\partial y^2}\right) - \frac{n}{\tau} + G(x, y) = 0 \qquad (28)$$

where $n(x, y)$ is the density of minority carriers (electrons), $D$ is the diffusion constant for the electrons, $\tau$ is the electron lifetime, and $G(x, y)$ is the generation function for the electrons. The electron diffusion length is given by $L = (D\tau)^{1/2}$. The simplest case for analysis is the reflection mode for thick photocathode. The generation function can be given by $(1 - R)\exp(-\alpha_c x)$. The boundary conditions becomes $n(x) = 0$ at $x = \infty$ and $n(x) = 0$ at $x = 0$, because at the surface the electrons would be subjected to the field and would be swept away. The solution for $n(x)$ can be given by

$$n(x) = A[\exp(-\alpha_c x) - \exp(-x/L)] \qquad (29)$$

where $A = [\alpha_c I_0(1 - R)\tau]/(1 - \alpha_c^2 L^2)$, $R$ is the reflectivity, and $I_0$ is the incident intensity. The collection efficiency for electrons in the bent-band region is given by the ratio of the electron current flowing into the bent-band region $qD\, dn/dx|_{x=0}$ and the maximum possible photogenerated current. This can be expressed as

$$Y(\text{int}) = \alpha_c L(1 - R)/(1 + \alpha_c L) \qquad (30)$$

Equation (30) describes the internal collection efficiency of the emitting surface region. Some of the electrons will be lost in this region because they may not be able to tunnel through the surface barrier or may suffer scattering such that their energy is not above the vacuum level at the surface. If we introduce an escape probability factor $P$, then the photoemission quantum yield $Y(\text{PE})$ is given by

$$Y(\text{PE}) = PY(\text{int}) \qquad (31)$$

For the purpose of analysis of experimental quantum yield data:

$$(1 - R)/Y(\text{PE}) = P^{-1} + (PL\alpha_c)^{-1} \qquad (32)$$

The slope and intercept of a plot of the left-hand side versus $1/\alpha$ give $P$ and $L$.

In the case of semitransparent cathodes, one additional important parameter is the surface recombination velocity of the minority carriers at the substrate–cathode interface. This is because more of the photoelectrons will be generated near the substrate–surface interface. There have been some calculations of transmission mode (TM) photocathode quantum efficiency (*252–254*). These calculations show that for optimal response the important conditions are (*185*):

(1)   That the thickness should be such that the cathode will be thick

enough to absorb most of the incident light, i.e., $t$ should not be much less than $1/\alpha$ or the absorption length when $\alpha$ is the absorption coefficient. At the same time it should not be such that $t \gg L$, the diffusion length, in which case the electrons generated near the substrate face cannot be emitted. The values of $\alpha$ and $L$ (in the case of GaAs at 0.85 $\mu$m) are about 1 $\mu$m$^{-1}$ and 2 $\mu$m, respectively, and the optimal value of $t$ should be about 1 $\mu$m.

(2) When $S \gg 1.0$, then the increase of value of $L$ beyond 1 $\mu$m does not produce any increase of quantum efficiency (q · E). If $S \lesssim 1.0$ the quantum efficiency increases with increase of $L$ up to 3 $\mu$m. The value of $L$ between 1 and 3 $\mu$m is realistic, and hence the $S$ value should be $\lesssim 1.0$ for utilizing the increase of q · E due to an increase of $L$. These requirements are met with AlGaAs/GaAs, InGaP/GaAs, InGaAsP/InP, and passivated Si structures (255–257).

The modulation transfer function (MTF) or spatial frequency response calculation, using the diffusion equation, has been made by Bell (258) and Fisher and Martinelli (259). The requirements of higher q · E and higher MTF are in conflict because the lateral movement of the photogenerated electrons in the material reduces the MTF. However, the MTF obtainable is much higher than other components in the case of night vision devices, where MTF is quite important.

### 3. Diffusion Length and Doping

The diffusion length $L$ that characterizes transport of the electrons from the depth of the material to the surface is given by

$$L = (D\tau)^{1/2} = (\mu k T \tau / e)^{1/2} \tag{33}$$

where $\mu$ is the electron mobility, $e$ is the electronic charge, $k$ is Boltzmann's constant, and $T$ is the temperature. For a given material, $L$ is influenced by impurities and crystalline defects. The diffusion length decreases with increase of impurity concentration. For example, in the case of LPE-grown Zn-doped GaAs the diffusion length decreases from 10 to 5.3 $\mu$m when impurity concentration increases from $9 \times 10^{17}$ to $3 \times 10^{19}$ cm$^{-3}$ (185). Diffusion length is a very important parameter, particularly for materials with lower absorption coefficient, as it signifies the depth from which the electrons can emerge to the surface for photoemission.

For good photoemission the width of the bent-band region should not be larger than 100 Å, i.e., the optical phonon–electron scattering mean free path. As can be seen in Fig. 35, the actual electron affinity at the surface is positive, and if the electrons cannot be accelerated at the bent-band region by the field then they cannot be emitted from the surface. If the bent-band region is very large, the electrons will lose the energy gained from the field by electron–phonon scattering and will not be able to escape. Thus the width of the bent-band region should be less than the electron–phonon scattering mean free path. This implies a bulk doping of about $10^{19}$ cm$^{-3}$ (*260, 261*). At higher doping concentrations the decrease in diffusion length more than offsets the increase of surface escape probability. The doping levels that are used for practical cathodes are between $5 \times 10^{18}$ and $1 \times 10^{19}$ cm$^{-3}$.

## 4. Surface Escape Probability

The surface escape probability $P$ is defined as the probability of an electron's being emitted when it reaches the bent-band region. The amount of band-bending relative to the band gap or the degree of electron affinity, along with the component of the electron's momentum perpendicular to the surface, which in turn is related to the band structure and the depth of the band-bending region have a major influence on the escape probability of the electrons from the surface.

The probability of emission from the surface is strongly dependent on the bent-band profile. If the depth of band-bending is much greater than the electron–phonon scattering mean free path, the acceleration by the built-in field will be insufficient to overcome phonon losses, and most electrons will be thermalized when they arrive at the surface. The probability of emission will thus be low. Fisher *et al.* (*229*) devised a model that assumes an initial Boltzmann-like distribution of thermalized electrons when they arrive at the bent-band region, are transported through it, and escape over the surface barrier. This model was found to fit well to InGaAs alloy cathodes in the band gap range of 1.38–0.74 eV. The calculated value of the work function is 0.97 eV and the height of the barrier is 1.24 eV with respect to the Fermi level or 0.28 eV with respect to the vacuum level. The calculated value of the barrier width is $8 \pm 2$ Å, which is consistent with the experimental width of the Cs–O layer (*228*). Figure 39 shows the correlation of the experimental results with the theoretical model. This model has been found to give good correlation with the experimental work function versus surface escape probability of GaAs (*249, 262*) and InGaAsP (*263*).

Escape probability depends on surface crystallographic orientation. In the case of GaAs(Cs–O) cathode the surface escape probability values for the (111) A, (100), (110), and (111) B faces are 0.212, 0.317, 0.307, and 0.489, respectively (*264*). In the case of Si only the (100) face can be activated to NEA. The requirement of conservation of electron energy and momentum parallel to the surface demands that only the electrons with momenta primarily normal to the surface be emitted (*265*). Using this as a basis, the calculated transmission coefficients in GaAs for the (100), (110), (111) A, and (111) B faces have been calculated for $\Gamma$ and X electrons (*265, 266*).

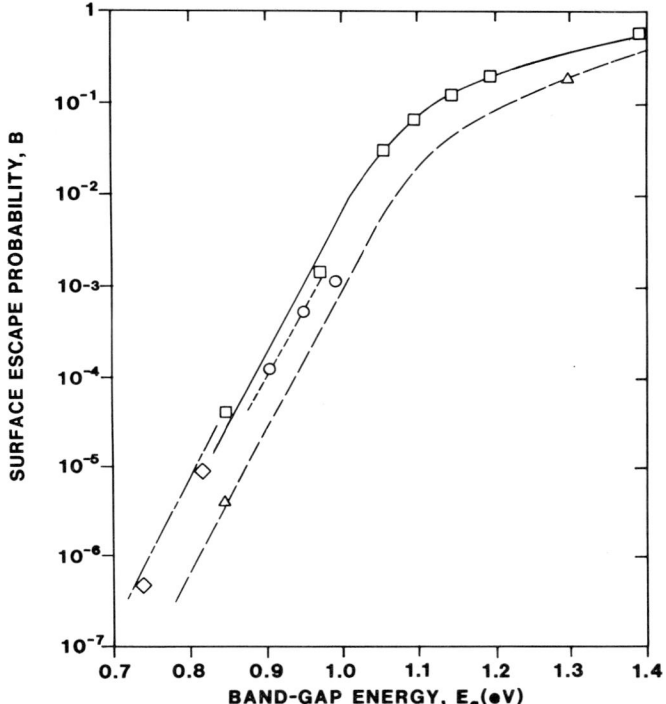

FIG. 39. Experimental and theoretical surface escape probability versus band gap for (In, Ga)As:Cs–O cathodes of various doping concentrations. Locations of the work function $\phi$ and the barrier height $E_b$ are shown [from Fisher et al. (229)]:

| $n_A$ (cm$^{-3}$) | $\phi$ Data | $E_b$ model |
|---|---|---|
| $3 \times 10^{19}$ | ◇ | —·— |
| $1 \times 10^{19}$ | □ | — |
| $5 \times 10^{18}$ | ○ | --- |
| $2 \times 10^{18}$ | △ | --- |

These calculations show that the escape probability is a sensitive function of the amount of band-bending, independent of the phonon scattering consideration. The strong band-bending dependence derives from the fact that during transport across the band-bending region electrons gain energy and may move into a higher valley, e.g., from $\Gamma$ to L in GaAs, which may have higher or lower probability of emission depending on the electrons' momenta relative to the surface. In general agreement with experiment

(*264*) the (111) B face is predicted to have a higher escape probability then the (111) A face, due to the heteropolar nature of GaAs. The escape probability for $\Gamma$ or X electrons from the (110) face, however, is calculated to be significantly lower than for other low-index faces. Experimentally this is not the case, and it has been suggested that multiple electron reflections within the band-bending region may be important (*266*). James and Moll (*267*) arrived at the same conclusion from their energy distribution measurements for vacuum-cleaved GaAs (110) surfaces. Not all the electrons that are reflected back from the surface are lost to recombination; some can be reaccelerated towards the surface and be emitted.

Application of conservation of energy and momentum in the case of Si surfaces (*256*, *265*) shows that there should be effective blockage of the thermalized x-electron emission from (111) and (110) surfaces. This is consistent with the failure to achieve NEA on these surfaces. This model gives a good estimate of the escape probability (0.33) from the (100) face, which is also consistent with experimental observation (*8*, *261*).

The surface escape probability is a very sensitive function of the surface preparation, cleanliness, and crystal quality. Different values of surface escape probability have been reported for the same face by following different methods of crystal growth and heat treatment (*185*).

## 5. Activation Techniques

For activating the materials to NEA the wafers are put in an UHV system capable of giving $10^{-10}$ Torr after chemical cleaning of the surface. Ultrahigh vacuum is very important to avoid adsorption of undesired gas atoms on the surface. The surface is cleaned prior to activation by means of heat cleaning, cleaving, or argon-sputter annealing. For device applications heat cleaning is preferred, because the other practical method (argon sputtering) leaves too many defects on the surface and does not produce high sensitivity (*268*, *269*). In the heat-cleaning procedure the material is heated to a temperature near the decomposition point, which is about 670°C for GaAs. Even at lower temperatures, for example, 640°C, there is some loss of the group V element. The removal of carbon from the surface is very important in the cleaning procedure, which is not considered to be complete until the AES spectrum shows complete removal of carbon. Activation is achieved by admitting Cs, usually by heating a metallic cesium reservoir or a channel containing the generating mixtures, which have been properly outgassed. Initially the surface is cesiated to the maximum photosensitivity (in the case of Si beyond the point of maximum photosensitivity), after which oxygen is admitted through a leak valve, when sensitivity further increasing sensitivity. For Si, admitting oxygen only once is enough, whereas for GaAs,

Cs, and oxygen has to be added alternately a number of times. Simultaneous activation with Cs + O after the initial Cs is often found to give somewhat better near-threshold yields than alternate addition (*249*) usually called the yo-yo technique. For optimum photoemission the activation should be carried out at room temperature (*270*). Activation at slightly elevated temperature helps to distribute the Cs all over the enclosure, which improves the stability of the cathode, but at the cost of a loss of sensitivity. Usually the activated cathode is heat cleaned once again at a temperature about 100°C lower than the initial heat-cleaning temperature and reactivated. This so-called hi-lo technique (*269–272*) may increase the yield near threshold from 10 to 500% over that with only first activation (*249*). The mechanism of the increase is not well understood; however, it could be due to rearrangement of surface atoms or to the change of stoichiometry of the Cs–O layer (*269*).

## 6. Fabrication of the Material for NEA Activation

The first NEA photocathodes were fabricated from vacuum-cleaved single crystals (*7*), but it soon became apparent that epitaxial layers would be preferable because the doping is easier to control and the crystalline quality and surface would be better. Two methods for epitaxial deposition are commonly employed, liquid phase epitaxy (LPE) and chemical vapor deposition (VPE). Each method is best suited for particular materials. For example, growth of (Ga, Al)As alloys is relatively straightforward by LPE but difficult by VPE. Similarly, (In, Ga)P alloys are more easily grown by VPE (*185*). It is commonly observed that VPE surfaces are more free of gross defects than are LPE surfaces. However, LPE materials have better electron transport properties, such as electron diffusion length and mobility. For example, the diffusion length for LPE GaAs is about two to three times greater than that of comparably doped VPE GaAs (*185*). At present, however both these techniques are used with good results for fabricating NEA layers.

Another method for depositing good epitaxial layers has also been described, namely the organometallic method (*273, 274*). In this method the reaction of metal alkyl vapors and the group V hydrides produces various III–V compounds. For example,

$$Ga(CH_3)_3 + AsH_3 \xrightarrow[\text{heat}]{H_2 \text{ carrier gas}} GaAs + 3CH_4 \qquad (34)$$

which is similar to the silane reaction used to produce epitaxial silicon. For n-type doping of the film, hydrogen sulfide may be added to the gas stream, and for p-type doping, dimethyl zinc may be mixed into the gas stream. The alkyl vapor deposition system has a number of advantages over the other methods of epitaxy. As all the reactants and the dopants are in the vapor phase and only the substrate is heated, the equipment (Fig. 40) is relatively simple and is capable of fast turn-around time. High gas flow rates can be used, as there is no source material with which equilibrium has to be reached. This, plus the absence of halides, limits autodoping. Bass (275) obtained a high sensitivity of 1150 $\mu$A lm$^{-1}$ using p-type layers doped at $6 \times 10^{18}$ cm$^{-3}$. Escher et al. (276) also obtained high sensitivity using organometallic-grown layers. Because of the simplicity and other advantages, the organometallic growth method is likely to become very useful for NEA material preparation.

The fabrication of reflection-mode cathodes is relatively straightforward because the thickness of the substrate and the cathode layer could be quite large. However, in the case of transmission-mode cathodes the active layer thickness should be in the range of 1-2 $\mu$m, making fabrication more difficult. The most straightforward approach is to polish or etch a single-crystal material to the 1–2 $\mu$m thickness. Self-supporting layers of this thickness can be made, but they have a tendency to cleave and hence are very fragile. This problem can be circumvented by first bonding the cathode material to a transparent substrate and then thinning it (277). When thinned to 1–2 $\mu$m the layer is supported by the transparent substrate, and the composite structure is not so fragile.

The other technique is to grow epitaxial layers on transparent materials such as sapphire ($Al_2O_3$) and spinel ($MgAl_2O_4$). However, the sensitivity obtained is not very high (278–280). This is because of the poor lattice match at the substrate–active layer interface, so that there is a large number of interface states, giving rise to a high surface recombination velocity. The other higher-band gap substrate materials such as GaP can be tried, depending on the wavelength of operation. GaP, being transparent beyond 0.55 $\mu$m, is suitable material for most applications. Because of the large lattice mismatch of GaP with GaAs, a buffer layer of GaAlAs that has the

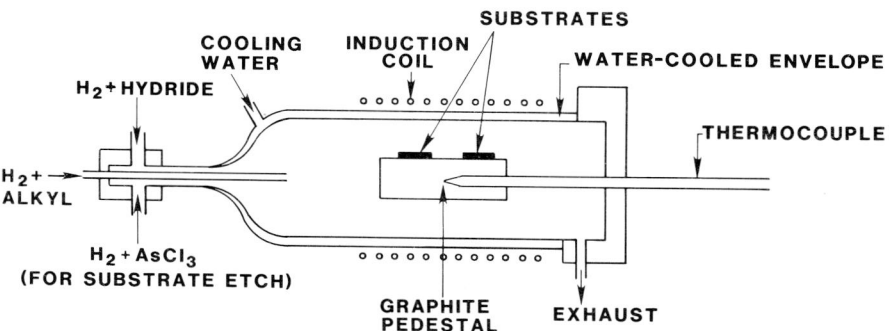

FIG. 40. Reactor for the epitaxial growth of GaAs by the alkyl process.

same lattice constant as GaAs is introduced (*185*). Figure 41 shows a plot of lattice constant versus band gap for these compounds. A number of possible combinations of buffer layers are possible for GaAs, such as (GaAl)As, (InGa)P, (In, Ga, As, P), (In, Al, As, P), (Ga, Al, Sb, P), (In, Al, Sb)P, and Ga(Sb, P). Of these, the (GaAl)As system is apparently the most advantageous, because in this system all the compositions match with GaAs. As the As:Ga ratio increases the spectral window broadens, and the lattice match is guaranteed independent of composition. The GaAs/InGaP/GaP system (*225*) also has been found to produce very good lattice match and large diffusion length with small surface recombination velocity. Good lattice match has been obtained with the composition $In_xGa_{1-x}P$ at $x = 0.504$. Since lattice constant changes with composition, strict control of composition is essential. Nearly ideal semitransparent response can be obtained from the two structures (GaAl)As/GaAs and GaAs/InGaP, with quantum efficiencies generally higher than reflection with an antireflection coating on the back of the cathode (*249*). With InGaP/GaP the higher-energy cutoff is at ~6500 Å and with GaAlAs it can be changed with composition. By suitable growth and etching techniques the GaAs substrate can be removed and the GaAlAs backing can be fused to the glass. These glass-bonded

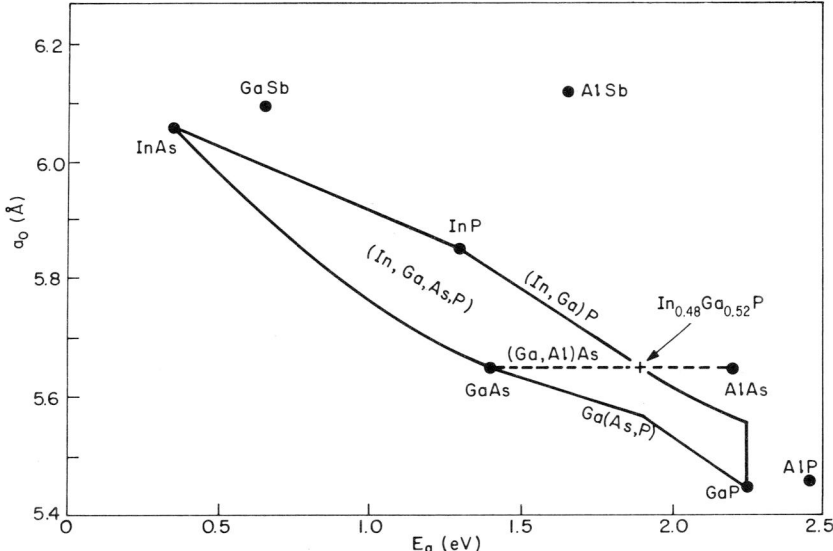

Fig. 41. Lattice constant versus band gap for several III–V compounds. Alloys for which a lattice match to GaAs occurs are shown by the dashed line. [From Martinelli and Fisher (*185*). Copyright © 1974, IEEE.]

cathodes have been found to give good photosensitivity along with the desired mechanical stability for device applications (258). Antypas et al. (281) first reported a photomultiplier (PM) tube with glass-bonded GaAs cathode. It has a broad-band response (1.4–3.0 eV), achieved with an optically thin (~0.3 μm) AlGaAs layer and is available commercially (Varian Associates, Inc., Palo Alto). Figure 42 shows the q · E of such a cathode.

Development of TM Si photocathode has been relatively straightforward, as the technology for producing large-area, uniformly thin (~10 μm), mechanically stable Si wafers was well developed for Si-vidicon applications. Techniques have been developed to reduce effectively the back surface recombination velocity by means of a thin $p^{2+}$ layer at the back surface (282), and semitransparent sensitivities as high as 1000 μA lm$^{-1}$ have been reported from optimized cathodes (261). Dark current emission is high from these cathodes and is similar to that in S-1 cathodes ($10^{-11}$–$10^{-12}$ A cm$^{-2}$).

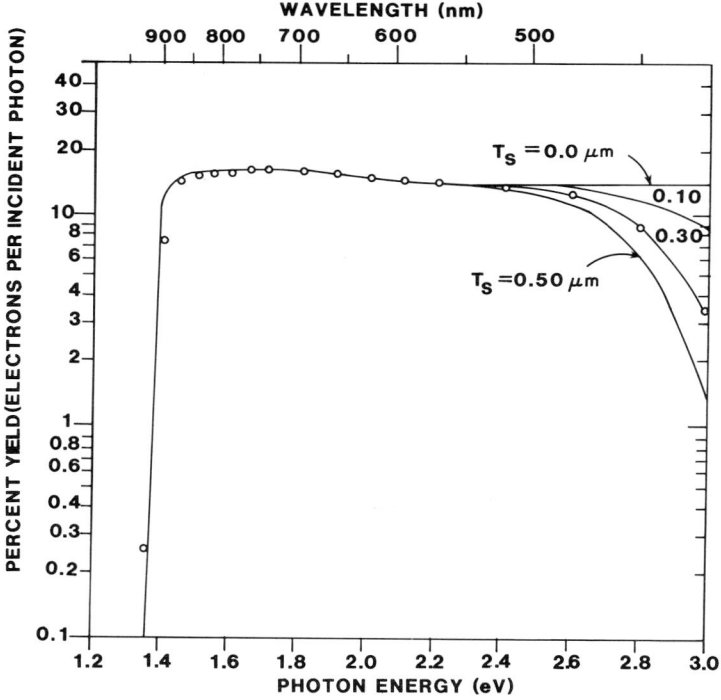

FIG. 42. Semitransparent (TM) quantum yield from a GaAs–AlGaAs 7056 glass PMT tube: (—) calculated; (○) experiment. The AlGaAs is grown relatively thin so that it becomes optically semitransparent but still provides a low recombination interface to GaAs. From Antypas et al. (281).

For low light-level applications some cooling is needed during operation. A TM Si photocathode coupled to an isocon low light-level TV system has been reported (283).

### 7. Secondary Emission

Since the mechanism for electron emission is the same for photoemission and secondary emission, the only difference being the mode of excitation of the electrons, good photoemitters should be good secondary emitters also. Therefore the NEA materials are expected to be good secondary emitters. Like photoemission, secondary emission also takes place in reflection (RM) and transmission (TM) modes, which means in the first case that the secondary electrons are emitted from the same side as the primary electron incidence and that in the second case they are emitted from the other side. Very high secondary electron gain has been achieved in such NEA materials as GaP (284), GaN (285), and Si (286), compared to conventional emitters such as MgO (287) (Fig. 43). The secondary emission yield becomes a maximum when the range of the primaries becomes equal to the escape depth. The conventional emitters have a small escape depth of a few hundred angstroms, so the maximum in secondary yield is obtained in a few hundred volts of primary energy. In the case of NEA emitters the escape depth is equal to the diffusion length. In the case of GaP the diffusion length is about 0.2 $\mu$m, so the peak in secondary emission is achieved in about 5 keV and the maximum gain is about 140, compared to a gain of 4–10 for practical conventional secondary emitters such as Cu–Be at 400–600 V primary energy. In the case of Si the diffusion length is 4–10 $\mu$m so the gain does not saturate even at

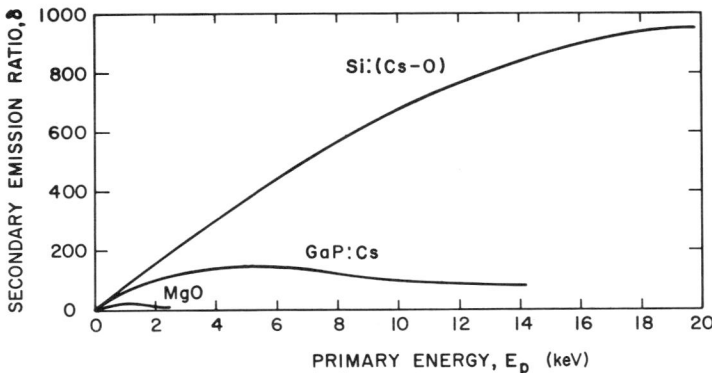

Fig. 43. Comparison of reflection mode secondary emission gain curves for a conventional emitter, for MgO, and for NEA emitters Si:Cs–O and GaP:Cs–O. From Martinelli and Fisher (185).

20 kV and the maximum gain obtained at 20 kV is about 1000. At 20 kV the range is about 2 μm, and thus an increase in gain will be achieved at still higher energies until the range of the primaries equals the escape depth.

The time resolution of the secondary emission process in the case of NEA emitters is likely to be worse than that of conventional emitters. In conventional emitters the hot electrons escape before they thermalize, that is, in $\sim 10^{-13}$ sec. In the case of NEA, however, since the lifetime of thermalized carriers is about $10^{-7}$ sec in the case of Si and about $10^{-10}$ sec in the case of GaAs, a much longer response time is expected. In the case of transmission secondary emitters (TSE), this time is smaller because of the thinness of the emitter layer. It has been shown that the impulse time response or the FWHM for the secondary pulse with excitation of very short duration can be given by (185)

$$t_p = 0.6T^2/D \tag{35}$$

where $T$ is the thickness and $D$ is the diffusion constant $(L^2/\tau)$. Calculated values of $t_p$ for 5 μm thick dynode are about 10 nsec for Si and 150 psec for GaAs. Si response is slower because of the longer lifetime of the minority carriers than in GaAs.

The statistics of the emission process from the NEA emitters is also another important parameter for secondary electron emission. The fluctuation of gain in the multiplication process is generated due to statistical variation in (1) the number of internal secondaries produced, (2) the number of secondaries transported to the NEA surface, and (3) the number of secondaries escaping to vacuum. The fluctuation in the number of internal secondaries for primary electron is given by $\Delta n_s^2 = Fn_s$, where $F$ is the Fano factor (288). For Ge and Si, $F \gg 0.1$, which means that $\Delta n_s^2 < n_s$ and the generation process has a smaller variance than a Poisson process. In GaAs and Si almost all the excited electrons are emitted because the escape depth is much larger than the range of the primaries, and the transport to surface therefore introduces little noise to the emission process. Variation in escape probability for the electrons can result from nonuniform activation over the surface. If the surface is uniformly activated, the variance in the surface escape process is $B(1 - B)$, where $B$ is the surface escape probability. In general, the secondary emission from NEA emitters has extremely good statistics, making it very useful for use in photomultipliers.

## 8. Cold Cathodes

In NEA materials if the minority carriers can be injected not far from the surface, they have a good probability of escape. The cold cathodes are based on this idea, in which there is a p–n junction with the surface of the p-type material activated to NEA. The thickness of the p layer should be less than the diffusion length; otherwise the injected electrons may not escape. Cold cathode emission was first observed from forward-biased p–n junctions of GaAs by Williams and Simon (*289*), and for Si by Kohn (*290*). A variety of cold cathode structures have been reported, employing such materials as GaAsP (*291, 292*), GaP (*293*), heterojunction GaAs–AlGaAs (*294*), and GaP–GaAlP (*295*). Peak efficiencies (emitted electrons/injected current) on the order of 1% with emission current density of about 2 A cm$^{-2}$ under dc conditions and 225 A cm$^{-2}$ under pulsed conditions have been reported (*290*).

## 9. Dark Current

The dark current is an important parameter of photoemissive materials for low-level signal detection. NEA materials such as Si have high dark currents (*296*) of about $10^{-9}$–$10^{-12}$ A cm$^{-2}$. This has been the most serious limitation of Si, even though it is otherwise a very good quantum efficiency photoemitter. The dark current in GaAs is in the range of $10^{-14}$–$10^{-16}$ A cm$^{-2}$, in InGaAsP it is $\sim 10^{-11}$–$10^{-13}$ A cm$^{-2}$ and in GaAsP it is less than $10^{-14}$ A cm$^{-2}$. On cooling to about $-20°C$ from room temperature the thermionic current falls by a factor of about $10^{-3}$. The decrease with temperature is consistent with thermionic emission over the work function rather than band gap generation. It is believed that thermionic emission from the surface activation layer and surface states contributes substantially to the total thermionic emission from these materials (*245, 249, 297*).

## 10. Stability

The NEA GaP dynodes were easily incorporated in PM tubes and were almost as stable as other conventional dynode materials. However, the stability of GaP is due to its large band gap and hence the large degree of NEA. Moreover, in GaP dynodes most of the electrons are not thermalized (*298*) and therefore their emission would not be much affected by slight changes in work function that might occur due to prolonged storage or operation. This also contributes to the higher stability of GaP dynodes.

Other III–V materials, such as GaAsP and GaAs, also have shown good storage life and equally good operating life if operated at low current levels, e.g., $10^{-9}$–$10^{-11}$ A (*249*). At higher current levels the ion desorption from the surfaces that are bombarded by photoelectrons causes the degradation of the cathodes when the ions are accelerated toward the photocathodes. The effect would be a continuous rise of work function (*294*).

NEA Si apparently has good shelf life (*245, 296, 299*), which may be due to the well-positioned Cs and O atoms in the cave and pedestal sites.

However, other similar low band-gap III–V materials are generally not stable at room temperature. They have to be stored at a low temperature of about $-25°C$. Stability of the cathodes is enhanced if all parts of the tube are saturated with Cs vapor and there is no imbalance in the tube. The Cs apparently gets desorbed in large UHV systems under dynamical vacuum, but in small closed tubes the Cs vapor present stabilizes, and is in turn helpful in stabilizing the Cs–O layer.

## 11. Photoemitters beyond 1.1 μm

The negative electron emitters are limited to a band gap corresponding to a cutoff of ~1.1 μm. Materials with band gaps below this cannot be activated to NEA because of the interfacial barrier. Only S-1 cathodes have a cutoff up to 1.2 μm, but the quantum efficiency is poor beyond 1.1 μm. There have been some efforts to extend the photoemission beyond 1.1 μm. One attempt was by means of a field-emission array-type photoemitter, which gave good photoemission up to 1.1 μm for Si (*300–303*) and was considered to be capable of going to longer wavelengths in smaller band gap materials such as Ge.

Another approach is by means of transferred electron photoemission. Good photoemission has been obtained up to a cutoff wavelength of 2.1 μm (*9*), and the method looks quite promising.

In the case of field-emission arrays (*300*) an array of spikes 12 μm high, with tip diameters of less than 0.5 μm and tips separated from each other by 25 μm, is placed on a silicon substrate by etching. When this array is subjected to a high field in proximity configuration the electric field lowers the surface barrier and good photoemission is obtained. Both reflection and transmission mode operation yield good quantum efficiency (Fig. 44). The electron affinity becomes nearly zero, so the band gap determines the threshold energy for photoemission. It may be possible to use Ge to go to a threshold at 1.6 μm. However, the dark emission is too high from these cathodes. At room temperature the dark current is $10^{-6}$ A cm$^{-2}$, which is predominantly the result of surface generation in the depleted space-charge region at the emitting surface. Even at a temperature of 90 K the dark current is $\sim 10^{-12}$ A cm$^{-2}$. In addition, the spatial frequency response of these emitters is poor due to the discrete nature of the photoemission surface. However, MTF values of about 0.4 have been obtained at about 20 line pairs (lp)/mm. Because of these problems there is not much interest in these types of emitters at present.

The transferred-electron type of photoemitter depends on transferring the photogenerated electrons, by means of electric field (as in the Gunn effect), from a lower to upper valley, from which they can be emitted. A

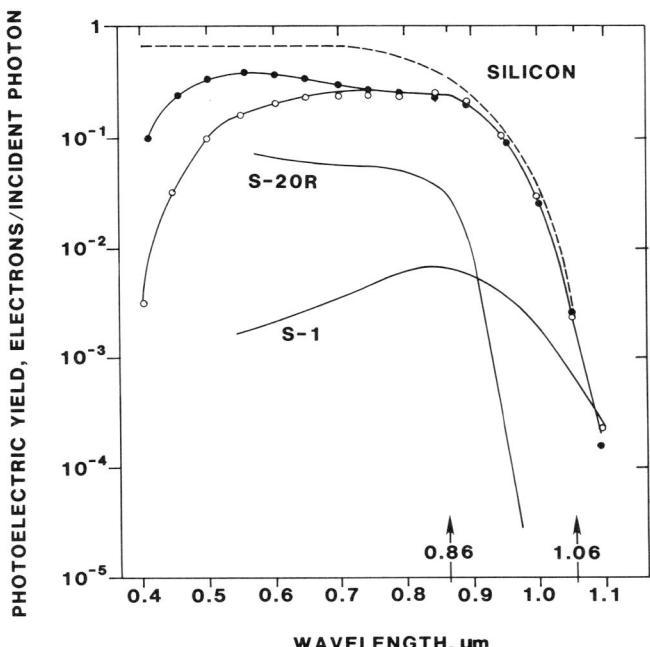

FIG. 44. Transmission (○) and reflective (●) photoelectric yields of a silicon array photoemitter (10 Ω cm, p-Si, 30 μm thick) compared to conventional red sensitive S·S0 and S-1 photocathodes. Experimental measurements were made at 90 K and 4000 V. Calculated absorption in the Si layer is indicated by the broken line. From Thomas et al. (302).

number of such external field-assisted photoemission geometries have been proposed and studied (304–311). Bell et al. (308) studied a structure made of p-type InP on which a thin layer of Ag forms a Schottky barrier. If a reverse bias is applied to the Schottky barrier, the photogenerated electrons are accelerated towards the surface. The accelerated electrons may be placed in the L or the X valley from which they can penetrate through the metal (Ag), which is activated with Cs–O to lower its electron affinity to 1 eV. The L and X valleys are at energies above the vacuum level. The improvement of quantum efficiency on application of bias can be seen in Fig. 45.

Sahai et al. (302) used a double heterojunction type of structure. This structure was meant for operation in the transmission mode for 1.06 μm radiation. GaAsSb serves the purpose of an absorber material having an energy gap of about 1.1 eV. The middle layer of GaAlAs acts as a hole barrier and improves the lattice matching to GaAs. The entire lattice mismatch is between the absorber layer and the hole barrier layer. An internal

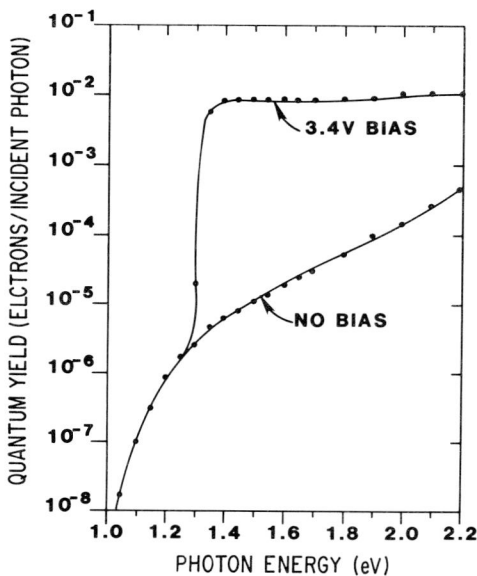

FIG. 45. Experimental reflection mode quantum yield from an Ag/p-InP transferred electron photocathode. From Bell et al. (308).

transfer efficiency as high as 70% has been obtained, which should produce a good quantum efficiency for 1.06 μm operation. However, work on this kind of structure has been discontinued.

Escher and Sankaran (140) used smaller band gap materials matched to InP, which can give lower threshold energy, with an Ag layer that was activated with Cs–O. Since there is no band gap limitation as in NEA, the cutoff wavelength can be increased with the reduction of band gap. Escher and Sankaran obtained a cutoff near 1.5 μm using InGaAsP with a band gap of 0.85 eV. Gregory et al. (9) used a TM structure that gave photoemission up to 2.1 μm (Fig. 46). This was achieved by using an $In_{0.77}Ga_{0.23}As$ layer with band gap of 0.52 eV as the emitter layer. Since this is not lattice matched to InP, a layer of $InAs_xP_{1-x}$ was used in between. In this configuration the photons are incident on the back surface (transmission mode), but only those for which $0.83 > hv > 0.52$ eV are absorbed in the InGaAs layer. Photons for which $1.35 > hv > 0.83$ eV are absorbed in the InAsP layer, which has a graded composition meant to produce a field that accelerates the photoelectron towards the surface. Photons with energy greater than 1.35 eV would be absorbed in InP. The silver film thickness was about 50 Å, and the film was activated to Cs and $O_2$.

The limit of band gap for this system should give a maximum threshold of

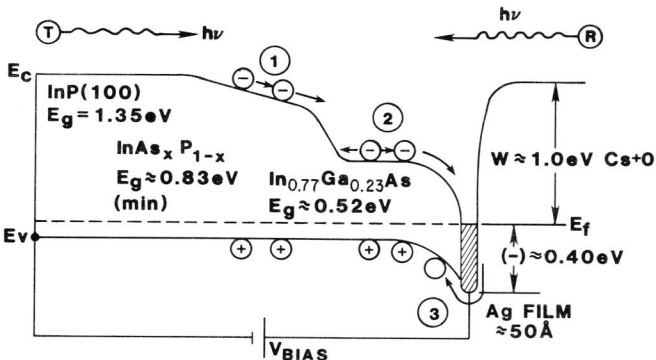

FIG. 46. Energy band diagram for bias-assisted photocathode. From Gregory et al. (9).

up to 3.54 μm for InAs. But the main problem of operation of these cathodes is the dark current. The dark current for Ag/p-In–GaAsP-type structures is shown in Fig. 47. Even with cooling to −100°C the dark current is very high and rises very sharply with increase of bias voltage. The rapid rise is considered to be due to impact ionization. These cathodes should become

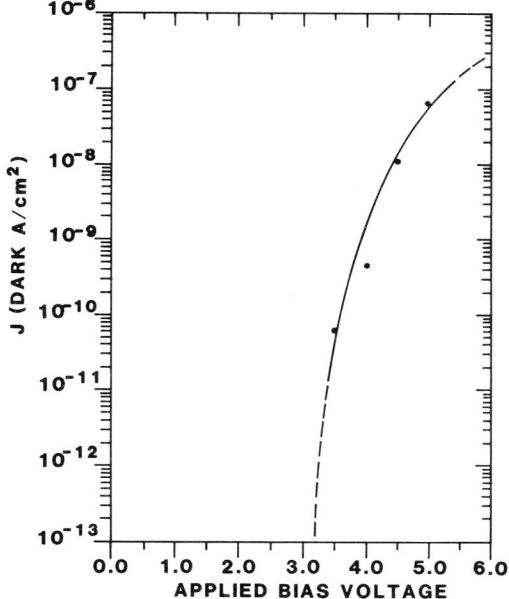

FIG. 47. Experimental dark current emission versus applied bias from an Ag/p-InGaAsP direct emitter cathode ($T = 100°C$). The rapid increase in dark current emission with bias is thought to derive from impact ionization. From Escher and Sankaran (310).

quite useful for certain types of applications if the dark current can be reduced.

## VI. Applications of Photoemissive Materials

Historically, the first applications of the photoemissive materials were for the detection of radiation. Subsequently, they have been used for another important application, namely imaging.

Two types of devices are used for the detection of radiation. One is the simple photodiode, and the other is the photomultiplier. Photodiodes, which are the simplest photoemissive devices, contain only two components, a photoemissive cathode and a collecting anode. Photodiodes are primarily used in applications in which high light fluxes are involved. The large resistance of the alkali antimonide type of photocathodes makes it difficult to draw large currents from these materials, because of the problems of electrolytic decomposition of the cathode and nonlinearity (66). To overcome these limitations the photocathodes are processed on thin metal films having high transparency and good conductivity, or deposited on transparent tin oxide coating. Though it is difficult to obtain high sensitivity from photocathodes on tin oxide, because the alkalis tend to react with the substrate, reasonably high sensitivity cathodes have nevertheless been produced in many laboratories.

### 1. Photomultipliers

The photomultiplier is the most important type of photodetector. In a photomultiplier, photoelectrons from the photocathode are multiplied by the process of secondary emission from a number of electrodes, which are called dynodes. The potential of the dynodes is made progressively more positive so that the electrons generated out of secondary multiplication in one dynode are accelerated towards the next. If $\delta$ is the multiplication per stage, then the total multiplication is

$$M = (\delta)^n \tag{36}$$

where $n$ is the number of dynode stages. Generally the multiplication produced per stage is between 3 and 6, and a total multiplication of $10^5$ to $10^7$ is obtained using 10–16 multiplying stages. Since the secondary emission yield is a function of the primary electron energy, the gain of the system can be chosen by varying the voltage across the dynodes.

There are many different types of dynode configurations designed to focus the electrons properly from one stage to the next higher stage. Figure 48a,b,c shows the three types of focusing geometries that are most commonly used by the commercial manufacturers. (a) and (b) are linear and circular focusing designs and (c) is the Venetian-blind type of dynode system. The choice of a particular dynode system and the number of dynodes are determined by many considerations, such as the minimum or maximum total gain, dark current, and speed of response.

The dark current in photomultipliers is an important consideration for many applications. The dark current of different tubes usually varies from 1 $\mu$A to 1 nA. It is finally limited by the thermionic emission from the cathode. By using low thermionic emission cathodes such as $K_2CsSb$ or by reducing the area of the cathode in low dark current photomultipliers, the dark current has been reduced to $1 \times ^{-10}$ A or in some cases to $1 \times 10^{-11}$ A. Cooling of the photocathode reduces the dark current even further.

The maximum current that can be drawn from the photomultipliers is about 1 mA. This limit is imposed because of the heating of the final stages of the dynodes and consequent degeneration of the dynode material. Of course, much higher currents can be sustained in short pulses. Operation at higher current levels (1 mA) gives rise to fatigue effects in which the multiplication gradually falls. The safe level of operation is with the anode current at about 1 $\mu$A or lower.

The choice of the photocathode also depends on the photomultiplier's intended use. In the earlier days most of the general-purpose photomultipliers used $Cs_3Sb$ photocathodes, as the multialkali (S-20) photocathode was more expensive. However, with the advent of computerized processing the cost has been reduced, nowadays most general-purpose PM tubes are composed of S-20 photocathodes. Special purpose tubes requiring low dark emission often use $K_2CsSb$ as the cathode material. For the high temperatures (up to 150°C) required for certain applications, such as oil exploration, the $Na_2KSb$ photocathode is used. The use of NEA photocathodes is still restricted to certain applications requiring high speed or good counting efficiency. Improvement of stability and reduction of cost would make NEA cathodes available for commonplace applications.

The conventional dynode materials are the Ag–Mg and Cu–Be alloys and $Cs_3Sb$, which is also a good photoemitter. With the development of NEA materials, NEA GaP–Cs has been found to be a good dynode material. GaP can be used for dynode applications as a polycrystalline material and can easily be activated to NEA. GaP has a very high secondary emission field, the maximum being about 140 at 5 keV (296). In photomultipliers the primary energy is fixed at about 0.6 keV, an energy near which the gain is linear with

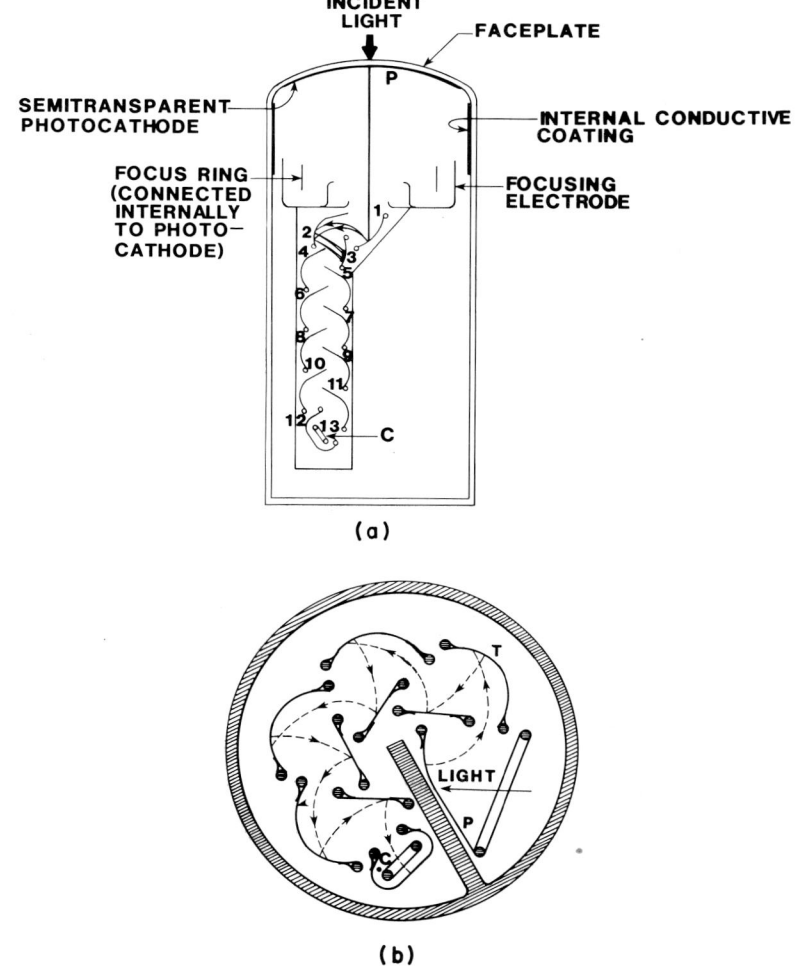

Fig. 48. Focusing designs for the dynode systems; (a) linear design; (b) circular design; (c) venetian-blind design; P represents the photocathode and C represents the collecting electrode.

primary energy. These dynodes are rugged and stable and have been available commercially for the last 10 years. Their high gain makes it possible to achieve the same multiplication factor in fewer stages. A photomultiplier using five stages of GaP dynodes has been reported by Persyk and Crawshaw (*312*). They use a circular focusing geometry (Fig. 48b), and the dynode configuration is very carefully designed so that the time broadening of the

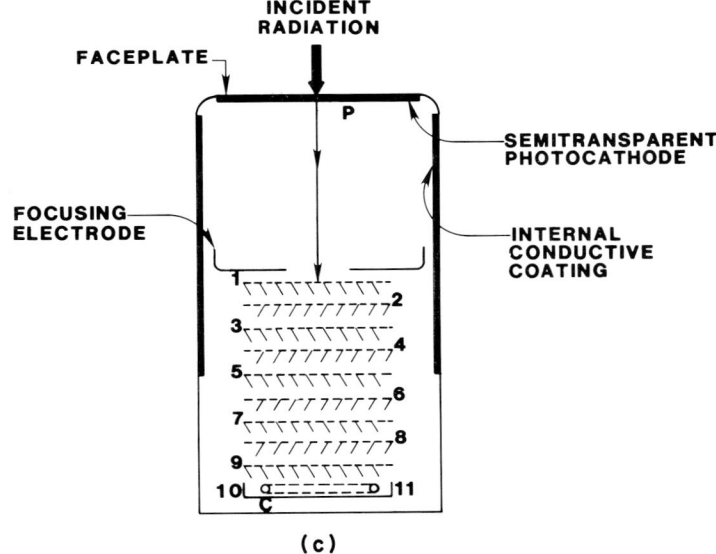

Fig. 48 (*Continued*)

pulses due to electron path-length difference and field difference is minimized. The smaller number of stages is very useful in reducing the time-broadening effect. Another advantage of GaP dynodes is the essentially thermalized secondaries, compared to the broader emission energies of conventional dynode materials, which add to the time-broadening effect. With five GaP–Cs dynodes a rise time of 800 psec and single-electron time resolution of about 570 psec have been obtained. Photomultipliers with only three GaP dynodes (RCA) have been found to give a rise time of only 300 psec. However, with fewer dynodes there is less total gain and the tubes cannot be used for observation of single-photoelectron pulses. Wilcox *et al.* (*313*) have reported a tube with crossed electric and magnetic fields using four stages, in which they obtained a rise time of 120 psec and a gain of $5 \times 10^3$.

GaP dynodes have good statistics of electron multiplication along with high gain. It has been observed (*314*) that in a photomultiplier in which only the first dynode is GaP the photoelectron statistics are degraded by less than 5%, as compared with 20% for conventional multipliers. This makes it possible to obtain good pulse height resolution because the pulse height resolution expressed in terms of fractional FWHM is a function of the dynode gain of the PMT tube. If the photoelectrons leave the photocathode as the result of a series of scintillations, the FWHM is given by (*314*)

$$\mathrm{FWHM}(Ne) = 2.35(Ne)^{-1/2}\{1 + \Delta^{-1}[\epsilon\delta/(\delta - 1)]\}^{1/2} \qquad (37)$$

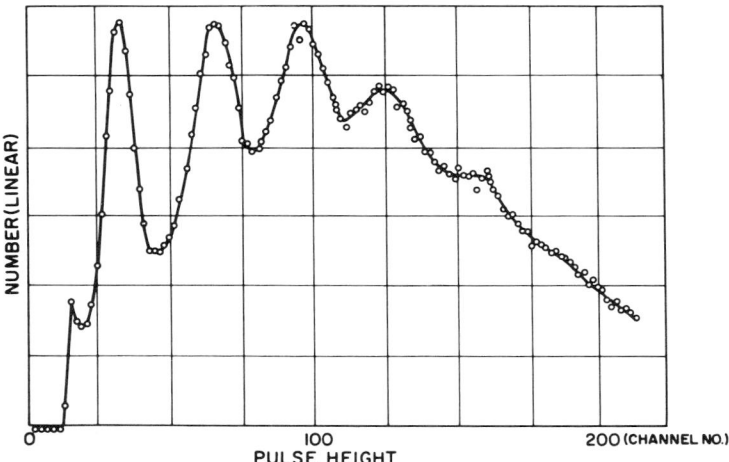

FIG. 49. Pulse height distribution for a photomultiplier with a GaP first dynode. Integral photoelectron peaks are clearly distinguishable. From Morton et al. (*314*).

where $\Delta$ is the gain of the first dynode, $\delta$ is the gain of the other dynodes, and $\epsilon$ is an experimental coefficient approximately equal to 1.6. In the photomultiplier with a first dynode of GaP ($\Delta = 20$) and other dynodes of conventional materials ($\delta = 5$), an excellent pulse height resolution was obtained, with which it was possible to resolve the peaks corresponding to emission of 1, 2 to 5 electrons in a group (Fig. 49). The dark emission mostly contains single-electron pulses; thus it is possible to adjust the discriminator level so as to eliminate the single-electron pulses, thereby reducing the dark counts by an order of magnitude. The excellent pulse-resolution properties of these photomultipliers make them very suitable for photon counting, which has become the standard technique for many applications. These high-gain GaP–Cs dynodes, along with NEA photocathodes, are available commercially. The use of NEA photocathodes solves the problem of having to use two photomultipliers from the visible to the near IR for many applications, because now a single photomultiplier can be used from short wavelengths to the near IR with almost uniform quantum efficiency. Even the semitransparent glass-bonded cathodes (AlGaAs/GaAs cathodes) are available commercially with an active diameter of 18 mm and a gain of $10^6$ (VPM-192 series, Varian Associates, Inc.). However, the shelf life of photomultipliers using NEA materials with $E_g \lesssim 1.3$ eV is not very good, and they have to be cooled to lower temperatures to extend their lives.

## 2. Image Tubes

The image tubes using photoemissive materials can be divided into two categories. One is the direct-view type and the other is the signal-generating type. In the direct-view devices, such as image converters and intensifiers, the photoemitter converts an optical image into an electron image, which is focused by suitable electron optics on a phosphor screen to produce an image that can be viewed better. In the case of signal-generating tubes, such as the image orthicon and secondary-emission conduction (SEC) camera tubes, the photoelectrons help to generate a charge image of the scene from which the video signal is produced, usually by scanning with an electron beam.

*a. Image Intensifiers.* Image intensifiers are devices that produce an output image of greater intensity than is available at the input. In such devices the photoemitter converts an optical image into an electron image. The electrons are focused and accelerated towards a phosphor screen, where they convert their energy to optical photons. There are various mechanisms by which the electron image is focused on the phosphor.

*Proximity focusing.* The simplest and earliest developed electron optics for an imaging tube is that of proximity focusing, in which the object plane of the emitting electrons is placed in close proximity to the image plane (Fig. 50) and an appropriate voltage is applied between them. The advantages of this are basic simplicity of design and fabrication, small size, freedom from distortion, and uniformity of resolution (*315*). The major limitation of this kind of focusing is that the resolution and MTF are lower than for other kinds of focusing. However, for many applications other resolution-degrading mechanisms dominate the system performance. The resolution and MTF of this type of electron optics are determined by (*316*) the displacement of the image plane given by

$$r = 2d(\phi_{r0}/\phi)^{1/2}\{(1 + \phi_{z0}/\phi)^{1/2} - (\phi_{z0}/\phi)^{1/2}\} \tag{38}$$

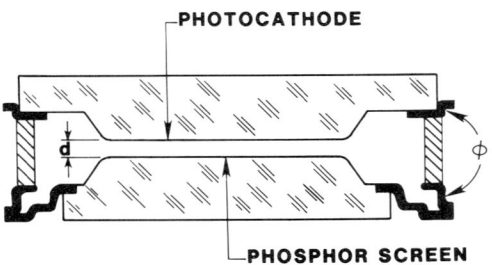

FIG. 50. Proximity-focused image tube.

where $d$ is the spacing between the object and image planes, $\phi$ is the applied voltage, $\phi_{r0}$ is the initial lateral velocity of the electrons in eV, and $\phi_{z0}$ is the longitudinal electron velocity in eV. Since $\phi_{z0}$ is much less than $\phi$, in most cases

$$r \approx 2d(\phi_{r0}/\phi)^{1/2} \qquad (39)$$

Usually the maximum field that can be applied is $\sim 10^4$ V mm$^{-1}$. Since $\phi_{r0}$ is generally below 1 eV the values of $r$ that can usually be achieved are less than 0.02 mm.

*Magnetic focusing.* Of all types of focusing, magnetic focusing provides the best resolution and MTF performance. In magnetic focusing an axial magnetic fileld is applied along with the electric field between parallel object and image planes (Fig. 51). The condition for optimum focusing is

$$dB \simeq 10.6n(\phi)^{1/2} \qquad (40)$$

where $n$ is the number of nodes, $d$ is the spacing between object and image planes in cm, $B$ is the magnetic flux density in Gauss, and $\phi$ is the applied potential between two planes in volts. The resolution is related to $r$ by

$$r \cong 4d(\phi_{r0}/\phi)^{1/2}(\phi_{z0}/\phi)^{1/2} \qquad (41)$$

where $r$ is the displacemeut in the image plane for perfect focusing. Since both $r_{r0}$ and $\phi_{z0}$ are small, $r$ in this case is much smaller than in proximity focusing, resulting in much higher resolution. In general, the advantages of magnetic focus image intensifiers (*317*) are their high resolution, low distortion, freedom from fiber optic windows, and large apertures with short tubes. The fiber optic windows restrict the choice of suitable windows for imaging in the UV, but in a magnetically focused tube the input window can

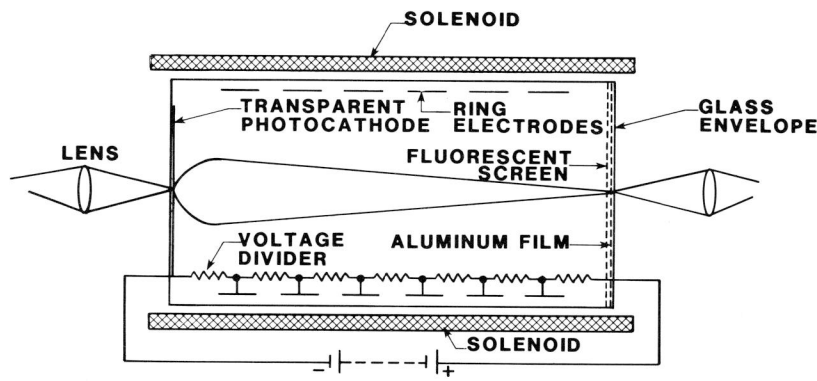

FIG. 51. Single-stage magnetically focused image intensifier.

be of almost any material. Magnetically focused tubes of diameter as large as 146 mm with a limiting resolution of 90 lp mm$^{-1}$ on the axis and better than 80 lp mm$^{-1}$ at 70 mm off axis have been constructed (*317*). Such high resolution is unmatched by any other focusing techniques. The limitations of magnetic focusing are that it requires a very stable magnetic field and thus needs highly regulated current supplies, and that the solenoids are bulky. Outdoor application is limited for these reasons. However, these tubes are very suitable for astronomical observations, such as imaging of faint stars.

Since the brightness gain of a single-stage tube can be a maximum of about 200 (*317*), it is desirable for many purposes to increase the gain of the intensifier. This can be done by cascading two or more single-stage intensifiers by using the output of one as the input of the other. McGee *et al.* (*318*) developed a cascaded tube in which all the stages were in the same envelope and the photocathode and phosphor were deposited on the opposite sides of a mica sheet about 3.5 $\mu$m thick. The use of the thin mica sheet reduces the spread of light in the coupling between the phosphor and photocathode. Varma (*319*) developed a tube with three-stage cascading that gave an overall gain of about $10^6$ and with limiting resolution at the center $\sim$ 50 lp mm$^{-1}$ and better than 40 lp mm$^{-1}$ over a diameter of 30 mm. Increasing the number of stages beyond three is not profitable because the advantage of increase in gain is offset by the degradation in MTF due to addition of one more stage, and overall picture quality becomes poorer. Mica sheets have in some cases been replaced by thin fiber-optic plates for coupling between two stages. For a two-stage tube a limiting resolution of 63 lp mm$^{-1}$ at the center and 56 lp mm$^{-1}$ at 50 mm off the axis have been reported with that type of coupling (*317*).

*Electrostatic focusing.* A large number of image intensifiers utilize an electrostatic lens as the focusing mechanism in the tube. The electron optics consists of a curved cathode and a cone-shaped anode, with the top of the cone at the center of curvature of the cathode (Fig. 5). The emitted electrons converge to a point near the top of the cone and cross over to produce an inverted image on the phosphor screen. The image surface is also curved. This poses serious problems for interfacing one tube with another and also for coupling with the lens system. However, this problem has been solved with the development of fiber optics. The major disadvantages of the electrostatic lens are that the resolution drops off markedly from the center to the edges, and that there is pin-cushion distortion associated with the overall image quality, due to the crossover. The electrostatic lens can be designed so that the center resolution is as high as that of a magnetic focus lens. However, the resolution falls rapidly toward the edges. In practical tubes the high resolution in the center is usually compromised to prove a reasonable resolution over a larger diameter (*320*). For cascading, three independent tubes are usually brought together; three-stage cascaded tubes have a gain of $0.4-1 \times 10^5$ and center resolution of $\sim 25-30$ lp mm$^{-1}$, falling to 15–20 lp mm$^{-1}$ at a distance of 10 mm from the center.

The advantage of an electrostatic lens compared to a magnetic lens is that it does not require a heavy solenoid or magnet or a large power supply. The operation of the tube takes very little power, and the power supply need not be stabilized. This has led to the development of a very lightweight power supply, and the system is easily portable.

*Microchannel plate intensifier.* The development of microchannel plates has resulted in a single-stage tube of gain comparable to that of three-stage cascaded tubes. The microchannel plate (MCP) is an array of hollow glass tubes with high electrical resistivity and secondary emission gain of the glass greater than unity. With the application of a high field along the length of the tube any electron entering the channel produces secondaries when it strikes the wall. These secondaries are accelerated down the tube and again strike the wall to generate more secondaries. In this way a large electron gain is obtained. Typical gain is about 1000–3000 by application of a potential in the range of 600–1000 V across a channel plate a few mm wide. Figure 52 shows a MCP intensifier of the inverter type. In this device the electron image is focused on the MCP by an electrostatic lens. Since the MCP cannot be curved to match the image plane of the simple electrostatic lens, extra focusing electrodes are added to flatten the image surface. The output image from the MCP is focused on the phosphor screen by means of proximity

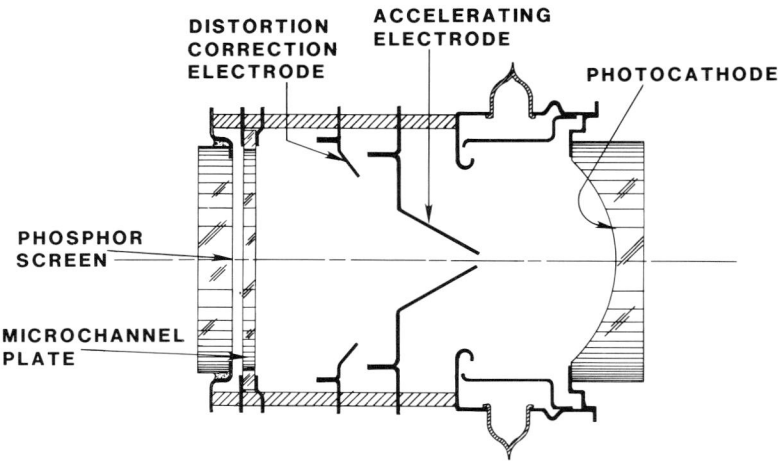

FIG. 52. Microchannel plate (MCP) type of image intensifier.

focusing. Proximity focusing has to be used because for electrostatic focusing the electron-emitting surface should be curved. Moreover, the electrons at the output of the MCP have a large spread in energy, which makes them almost impossible to focus with an electrostatic lens. Typical gains obtainable from these tubes are about 40,000, with center resolution of about 25 lp mm$^{-1}$ and edge resolution of about 18 lp mm$^{-1}$. There is another kind of tube, called a wafer tube, in which both photocathode–MCP and MCP–phosphor focusings are done by proximity. In this the inversion of the image is achieved by fiber optic twisters. The greatest advantage of this tube is the extremely small size; its biggest problem has been the reduction of tube life by outgassing of the MCP. However, this problem has been overcome and in certain conditions, such as high light, the cathode-to-MCP voltage can remain low and can still give a resolvable image at the output, compared to that of an electrostatic inverter.

*b. Gating Tubes.* These tubes are used for high-speed photography and for recording of high-speed transients. Image converter tubes have been used as photographic shutters since 1949 (*321*). However, an ordinary image converter or intensifier is not very suitable for gating, as substantial amounts of gating power are required, and such devices cannot be used for very high-speed shuttering. In 1957, Stoudenheimer and Moor (*322*) developed an image tube specifically for shuttering applications in which there is a gating grid close to the photocathode and a pair of deflecting electrodes inside the anode cone (Fig. 53). Application of a negative voltage of about 90 V with respect to the photocathode cuts off the photoelectrons, in the manner of the grid in a vacuum triode. During exposure a positive voltage of 175 V is applied, so that the grid has almost no effect on the transmission of the electrons from the photocathode towards the anode. A positive potential is applied to the focusing grid (grid No. 2) and the anodes to focus the image on the phosphor. Since the gating voltage is low, the power requirement is

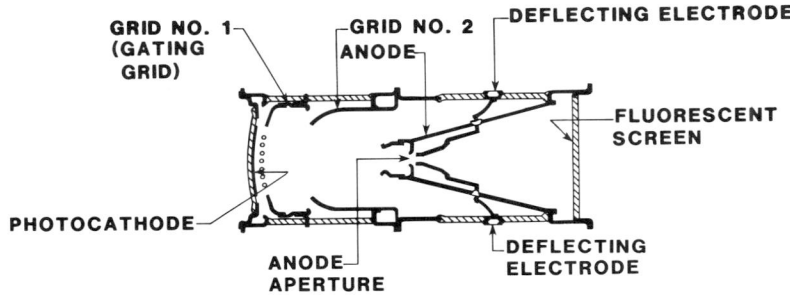

Fig. 53. Gating type of image tube.

small and the rise time is shorter. By use of deflecting electrodes, four sequentially displaced images can be formed on the photocathode using the whole photocathode area. If a smaller area on the photocathode is used, more pictures can be formed on the phosphor. Images can be formed with an exposure time of $10^{-8}$ sec without distortion due to transit-time effects.

In the streak mode the gating grid is made positive and the deflection electrodes are pulsed with a linear ramp voltage. Very high sweep speeds and consequently very short resolution times are obtained with this mode. The output on the phosphor consists of a line trace of the incoming event in which the intensity of the trace is a measure of the intensity of the input signal. Higher gating and streak mode speed have been obtained by applying higher positive voltage up to 2500 V to the gating grid, by which means the transit time of the photoelectrons is reduced and higher speed is obtained (*322*).

*c. Image Converters.* Image converters are tubes that convert a near-IR or an UV image to a visible image. IR image converters for night vision are generally active devices which are used by illuminating the object by with an IR floodlight invisible to the human eye. IR image converters normally use S-1 photocathodes, and the electron image is accelerated and focused on the phosphor screen in the same way as in image intensifiers. Cascading of the tubes is not necessary because sufficient scene illumination can be achieved in the active mode.

*d. Characteristics of Image Intensifiers.* The specifications of image intensifier tubes include a number of items, of which the more important ones are brightness gain, photocathode sensitivity, equivalent screen background input, limiting resolution, and modulation transfer function. Since image intensifiers usually deal with visual systems, the brightness gain is usually defined as the ratio of the output brightness in foot-lamberts divided by the input illumination on the photocathode in foot-candles. In order to define gain it is also necessary to specify the nature of the input light. Normally this is a tungsten source operating at 2854 K.

*Equivalent screen background input.* Equivalent screen background input (ESBI) is a measure of the residual illumination on the phosphor screen when there is no input. It is normally due to field emission, thermionic emission from the photocathode, ion noise feedback, and light feedback. Again, since most image intensifiers are used for visual purposes, the ESBI is defined in terms of lumens per square centimeter. There are two ways of measuring ESBI. One is to measure the gain and then to measure the screen brightness with no light input, after which dividing the second by the first should give ESBI. An alternative is to measure the output brightness with no input and then apply an input such that the output brightness is just doubled. The input brightness would then be a measure of ESBI. For most applications a typical value of ESBI is $2 \times 10^{-11}$ lm cm$^{-2}$ or less.

*Modulation transfer function and limiting resolution.* The modulation transfer function describes the capability of an image tube to reproduce details of a picture without much loss of contrast. In the past, this capability has been approximated by measurement of the limiting resolution, which is defined as the maximum number of black and white line pairs (cycles mm$^{-1}$) that the observer's eye can recognize at the output with a high-contrast (almost 100%) input. Usually, the human eye can recognize pairs up to a contrast level of 3–5%, though this ability is somewhat subjective. Thus, the limiting resolution can be described as the frequency at which the output to input contrast ratio or contrast transfer function is about 0.03. However, the limiting resolution does not give information about the performance of the tube at other frequencies, although tubes with higher limiting resolution often have better contrast transfer performance at lower frequencies. Two tubes with the same limiting resolution may have a substantial difference in contrast transfer capability at lower frequencies. The contrast available in most natural scenes is seldom higher than 20–30%, and since the eye does not usually respond well below a contrast level of 5%, the frequencies at which the contrast transfer factor for the tube is about 0.03 have little use for recognition of natural scenes (*323*). In fact, if the input scene contrast is 25%, then the spatial frequencies at which the contrast transfer function is below 0.2 are of little consequence. Thus the limiting resolution does not give much information about the actual field performance of the tubes, so the MTF is measured at a number of frequencies. The MTF is referred to as the ratio of the output contrast at a particular frequency to the contrast at zero frequency with a high-contrast input.

There are a number of methods for measurement of MTF. One of the direct methods is to project an image of a high-contrast bar pattern like the Westinghouse resolution chart (*323*) at the input of the tube and scan the output image after magnification by a very fine slit. A photomultiplier measures the intensity of light through the slit. Usually the chart pattern is moved and the output image patterns for black and white bars at different frequencies are plotted on the recorder. This method of measurement gives the MTF for a square-wave input. It is possible to convert square-wave MTF data to sine-wave data by simple conversion schemes (*324*).

There are other methods for determining MTF much more quickly. The Optics Technology modulation transfer analyser uses the image of a narrow slit at the input of the photocathode, and the output light is imaged on a movable test pattern consisting of bars of varying spatial frequency whose optical transmission varies as the square of the cosine of the angular position in each cycle. Each bar is parallel to the image of the input slit (Fig. 54).

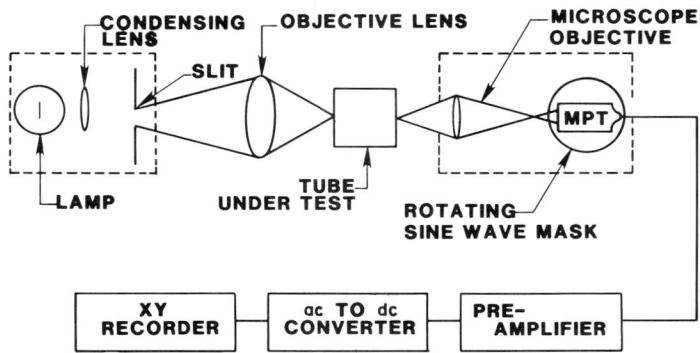

FIG. 54. Schematic of optics technology modulation transfer function analyzer. From Hall (323).

Behind the test pattern, which is actually formed on a drum rotating on a vertical axis, is a PMT whose signal current is proportional to the convolution integral of the image of the slit and of a sine function whose spatial frequency is made to vary slowly through the range of interest by rotating the drum about its axis. This integral is the Fourier transform of the image of the slit formed by the tube under test. In actual practice the drum is rotated and the PMT output and spatial frequency are fed to an $x$–$y$ recorder which gives a direct plot of the MTF of the system. In this way the MTF can be plotted in a matter of seconds. However, extreme care should be taken about the proper focusing of the images and the alignment of the slit with the direction of the pattern. Figure 55 shows the MTF of three image intensifiers (316). The one with highest MTF at all frequencies corresponds to the first-generation cascaded image intensifier. The other two correspond to the second-generation wafer, in which both sides of the MCP are proximity focused, and the second-generation inverter, in which the photocathode side of the MCP is electrostatically focused and the MCP to phosphor is proximity focused. It can be seen that while the second-generation wafer has higher MTF at higher frequencies, at low frequencies its performance is worse than that of the second-generation inverter. By further decreasing the phosphor–MCP separation and the channel-to-channel spacing in the MCP, the MTF of the second-generation tubes can be improved to approach the MTF of first-generation cascaded tubes.

*e. Signal-Generating Tubes.* The earliest television camera tube using photoemissive cathodes was the image dissector tube (Fig. 56) (325). This tube uses a photocathode as the image sensor, from which the electron image of the scene is accelerated toward and focused upon a disk with a small

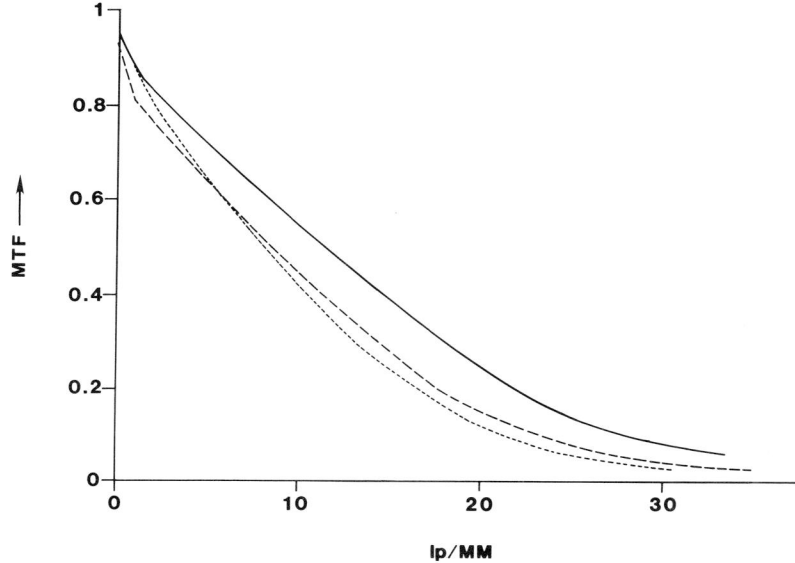

Fig. 55. MTF of image intensifiers. (—) First generation; (----) second-generation inverter; (---) second-generation wafer. From Freeman (*316*).

sampling or dissecting aperture. Those electrons that pass through this aperture are directed into a multistage electron multiplier that provides a gain of about $10^6$. The electron image from the photocathode is scanned across the dissecting aperture by the magnetic fields from two sets of deflecting coils that provide orthogonal deflections. The instantaneous current

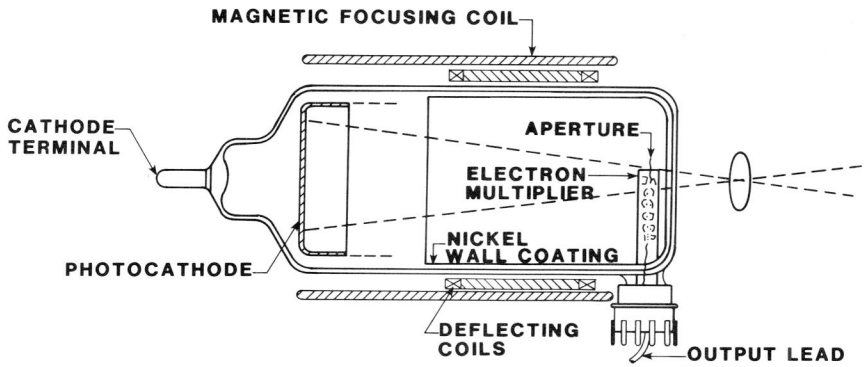

Fig. 56. Image dissector tube.

through the aperture is proportional to the instantaneous irradiation on that element of the photocathode that corresponds to the electronic image of the aperture. The signal current is thus extremely small and has to be amplified by an electron multiplier. The signal-to-noise ratio is very poor in these tubes at normal illumination. A high scene illumination such as that at broad daylight ($\sim 10^3$ fL) would produce a good quality picture. With this illumination, use of an $f/2$ lens and a photocathode quantum efficiency of 10% should provide the eye with a picture whose signal-to-noise ratio at 400 lines and 0.2 sec integration time is approximately 100 (326). This can be considered to be a good quality picture.

With the development of tubes of higher sensitivity which use storage targets to integrate information the image dissector tube ceased to be competitive for most applications. However, because of its inherent simplicity and greater reliability, and its suitability for fast-moving scenes in which the use of storage targets is much more complicated and for some other applications in which other considerations are more important than sensitivity, the image dissector tube is still used (327).

The development of the iconoscope (328) was the next major step in the television camera tube technology using photoemissive cathodes. This used a storage target made by depositing the photocathode on an insulating substrate or target which continuously sensed the radiation. An electron beam was used for scanning the targets.

The simple addition of a storage target to the nonstorage dissector should have increased the sensitivity $10^5$ times and produced the ideal television camera tube. The iconoscope, however, realized an improvement in sensitivity over the dissector tube of only a factor of 10. The major reason for not achieving the $10^5$-fold increase was that the video signal was amplified by a conventional vacuum triode rather than by an electron multiplier. The vacuum triode introduced its own noise level, which was 100–1000 times larger than the noise associated with the signal current in a multiplier tube.

Some improvement of the iconoscope was made in the image iconoscope (329), which increased the sensitivity by about a factor of 10. However, the potential of high sensitivity for storage targets was better realized with the development of the orthicon (330) and image orthicon (331).

*Image orthicon.* Figure 57 is a schematic diagram of an image orthicon. The scene to be transmitted is focused on a semitransparent conducting photocathode. The electron image is focused onto one side of a two-sided target. The two-sided target was a very thin glass sheet a few microns thick. The ionic conductivity of glass was sufficient to allow charges on opposite faces to neutralize each other in less than 0.1 sec and thin enough that very little spread of the charge pattern takes place before the charges from opposite sides neutralize each other. The electron image from the photocathode

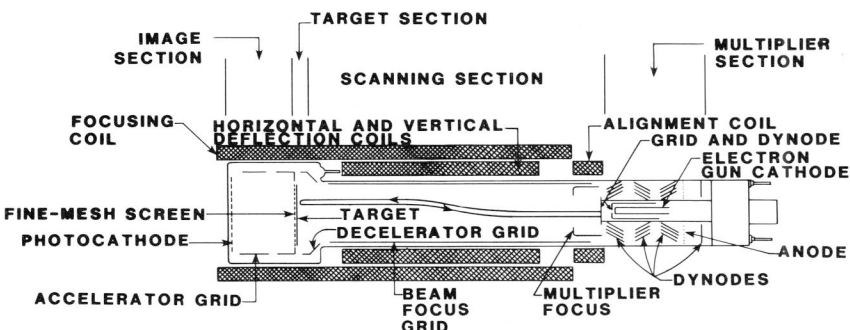

FIG. 57. Image orthicon tube.

is accelerated towards the target with a voltage such that it produces secondary emission from the target, the secondary electrons being collected by the mesh in the front of the target. The gain is greater than unity, which gives an amplified potential pattern on the target. This pattern is stored and is then sampled by the electron beam scanning the other side. This charge pattern is positive. The scanning beam is a low-velocity beam such that the secondary emission coefficient is less than unity and thus a negative charge is built on the other side. Actually the scanning beam raises the potential of the scanned surface to cathode potential, which is depleted according to the charge pattern on the other side. Thus the scanning beam has to deposit different numbers of electrons at different spots, and the rest of the charges in the beam are reflected. In the image orthicon this return beam constitutes the video signal, because it is the beam current minus the current that has leaked across the target during the scan period. The return beam is amplified by an electron multiplier of five stages that gives a gain of about 1000. With the low noise gain associated with the electron multiplier the shot noise of the scanning beam can be raised to the level where it is larger than the noise of the rest of the camera chain. Thus the shot noise in the scanning beam is made the limiting noise.

*Target characteristics.* In the operation of an image orthicon, electrons are removed from one face of the target by secondary emission and replaced on the other face by the electron beam. The target must satisfy two conditions. (1) It has to be sufficiently conductive that there is no charge buildup from frame to frame, and (2) the lateral resistance of the target must be very high so that the definition of point charges is not excessively reduced due to lateral leakage.

Most image orthicon tubes use conventional glass targets having ionic conductivity $\sim 10^{11}\ \Omega^{-1}\ \mathrm{cm}^{-1}$ sufficient to transport charges across the thickness quite quickly. The charge is conducted primarily by sodium atoms and their depletion after only a few hundred hours of operation causes image retention or "stickiness" of the tube. The use of thin-film MgO targets has solved this problem because in MgO the condution is electronic, and the electrons are always supplied by the scanning beam, so there is no depletion process and very little image retention. The secondary emission of MgO is also very high and thus it can give a gain of about 15 by secondary emission at the operating voltage, which is about five times that of a glass target. The principal disadvantage of the thin-film MgO target is that it lacks the strength to be used with the close target mesh spacing used in conventional image orthicons. However, for low light-level tubes, a wide-spaced mesh target structure is ordinarily used, and MgO is quite satisfactory with these larger spacings. The larger spacing is used to decrease the target capacitance. With a smaller capacitance a given charge pattern produces a larger potential pattern and improves the beam landing characteristics (*332*).

Although the wide-spaced tube is better at low light levels, it cannot produce as good a picture at high light levels as a close-spaced tube. Electron conducting glass targets have been developed to produce nonsticky, close-spaced targets. These glasses, called brown glasses, usually contain transition metal oxides in both divalent and trivalent states. The mixture of doubly and triply ionized sites throughout the material permits electron conduction by hopping.

*Resolution.* The lateral leakage of charge often limits resolution, particularly for low light-level applications, in which a longer integration time is used. It is found, for example, that the improvement in resolution becomes much less when the integration time is increased to 1/4 to 2 sec from 1/30 to 1/4 sec.

Image orthicons cannot perform well over a wide range of illumination. About 100:1 illumination ratio is the maximum range, because at high light levels, target saturation sets the upper limit to the signal-to-noise ratio. At low illuminations, beam modulation produces a very low signal-to-noise ratio. For high signal levels, distortion of the stored charge pattern at the target is a serious problem with the image orthicon. This is because the high flux emission from the very bright portion of the image is not completely collected by the target mesh and it migrates to surrounding areas of the target. Some method of eliminating this type of distortion must be achieved before the image orthicon can provide a completely reliable relationship between the signal amplitude and the absolute brightness of the object in the scene. This is especially true if the scene contains a broad range of brightness levels.

*Other television camera tubes.* A number of other tubes with still more improved capability have been developed. For example, the image isocon (*333*) reduced the noise characteristics in the dark areas by a factor of 2–3, making a significant improvement in the sensitivity, noise properties, and low light-level performance of the image orthicon. Subsequently, tubes based on electron- bombardment-induced conductivity (334) have been found to give very good performance. Secondary emission conductivity tubes (335), as they were called, used porous alkali halide material as the target. The high-energy electrons from the photocathode produce charge multiplication, which is scanned by an electron beam. The target in this case produces a gain of about 100. The most recent and successful of this class of tubes is the silicon intensifier tube (SIT) (336), in which the target is a slab of silicon 15 mm thick. The scanned surface has an array of back-biased p–n junctions. The image side of the target is field-free n-type Si. The electron–hole pairs generated by the electrons from the photocathode diffuse from the field-free image surface to the p–n junctions at the scanned surface, where they are split by the junction field so that the holes reach the scanned surface. With 10 kV on the photocathode the primary photocurrent generates a thousandfold-enhanced current in the target. A further 30-fold enhancement is achieved by I-SIT (intensifier-silicon intensifier target) camera tubes, in which an SIT tube is coupled by fiber optics to a single-stage image intensifier tube. With these two types of tubes it has been possible to extend the operation of TV camera tubes to very low light levels down to $10^{-7}$ fL.

## VII. Conclusions

The development of high-efficiency photoemissive materials, starting with Ag–O–Cs, has been followed by their extensive application in device technology. The NEA photoemitters are the latest addition to this class of materials. Even though the discovery of the first NEA photocathode dates back to the year 1965, the technology for fabrication and application of these cathodes has progressed rather slowly, and as of today even though the fabrication-related technical problems have been mostly solved, they remain expensive. With better mass production methods their cost will probably decrease, and their use become as commonplace as that of the $S \cdot 20$ today.

The highest quantum efficiency available with NEA cathodes is about 0.4 electrons/photon and is almost constant for a wide range of frequency. The maximum quantum efficiency that could be achieved with normal semiconductors is about 0.5 electrons/photon if we consider the generation of photoelectrons in the bulk, where they would have almost equal probability of coming to the surface as of going in the opposite direction. Thus the quantum efficiency of the NEA emitter approaches the theoretical quantum efficiency. To bring the rest of the photoelectrons to the surface, an electric field can be applied. Application of an electric field can also give rise to multiplication of the carriers by the photoelectrons. In addition, the generation of more than one electron by a single photon by scattering from the valence band (39) presents the possibility of exceeding the barrier of 0.5 electrons/photon. As a matter of fact, with the generation of more than one electron by application of an electric field or by scattering by valence band electrons it might even be possible to exceed the limit of 1 electron/photon. But such photocathodes should either have a built-in electric field or should somehow create the necessary conditions by using very low band-gap materials, so that electron multiplication can take place more easily in the solid. Bias-assisted photocathodes giving rise to photoemission with a cutoff at 2.1 μm (9) may be an indication of the future trend of development of photoemissive materials.

#### Acknowledgments

I wish to thank G. K. Bhide and B. P. Varma for their suggestions during preparation of the manuscript. Thanks are also due to my wife Malathi Ghosh for her help in preparation and careful reading of the manuscript. Careful typing of the manuscript by Valerie K. Fisher is greatly appreciated.

## References

1. H. Hertz, *Ann. Phys.* **31**, 983 (1887).
2. L. R. Koller, *Phys. Rev.* **36**, 1639 (1930).
3. N. R. Campbell, *Philos. Mag.* **12**, 173 (1931).
4. P. Gorlich, *Z. Phys.* **101**, 335 (1936).
5. A. H. Sommer, U.S. Parent 2,285,062; Brit. Patent 532259 (1939).
6. A. H. Sommer, *Rev. Sci. Instrum.* **26**, 725 (1955).
7. J. J. Scheer and J. Van Laar, *Solid State Commn.* **3**, 189 (1965).
8. R. U. Martinelli, *Appl. Phys. Lett.* **16**, 261 (1970).
9. P. E. Gregory, J. S. Escher, R. R. Saxena, and S. B. Hyder, *Appl. Phys. Lett.* **36**, 639 (1980).
10. R. H. Fowler, *Phys. Rev.* **38**, 45 (1931).
11. K. Mitchell, *Proc. R. Soc. Ser. A* **146**, 442 (1934).
12. K. Mitchell, *Proc. Cambridge Philos. Soc.* **31**, 416 (1935).
13. K. Mitchell, *Proc. R. Soc. Ser. A* **153**, 513 (1936).
14. H. Frohlich and R. A. Sack, *Proc. Phys. Soc.* **59**, 30 (1947).
15. A. H. Sommer and W. E. Spicer, " Photoemissive Materials and Devices" (S. Larach, ed.), p. 175. Van Nostrand-Reinhold, Princeton, New Jersey, 1965.
16. W. E. Spicer, *Phys. Rev.* **112**, 114 (1958).
17. G. Chiarotti, "Electronic Materials" (B. Hannay, ed.), p. 199. Plenum, New York, 1973.

18. L. Van Hove, *Phys. Rev.* **89,** 1189 (1963).
19. J. C. Phillips, *Phys. Rev.* **104,** 1263 (1956).
20. D. Brust, J. C. Phillips, and F. Bassani, *Phys. Rev. Lett.* **9,** 94 (1962).
21. W. E. Spicer, *Phys. Rev. Lett.* **11,** 243 (1963).
22. R. C. Eden, PhD. Thesis, Stanford University, 1967. (Published by University Microfilms, Ann Arbor, Mich. p. 71.)
23. W. E. Spicer, *Phys. Lett.* **20,** 325 (1966).
24. W. E. Spicer, J. P. Hernandez, and F. Wooten, *Bull. Am. Phys. Soc.* **8,** 614 (1963).
25. C. Ghosh, Unpublished.
26. J. Tauc, "Amorphous and Liquid Semiconductors" (J. Tauc, ed.), p. 159. Plenum, New York, 1974.
27. K. G. McKay and K. B. McAfee, *Phys. Rev.* **91,** 1079 (1953).
28. K. G. McKay, *Phys. Rev.* **94,** 877 (1954).
29. A. G. Chynoweth and K. G. McKay, *Phys. Rev.* **102,** 369 (1959); **106,** 418 (1957), *J. Appl. Phys.* **30,** 811 (1959).
30. B. Senitzsky, *Phys. Rev.* **116,** 874 (1959).
31. A. G. Chynoweth, *J. Appl. Phys.* **31,** 1161 (1960).
32. J. L. Moll, N. I. Meyer, and D. J. Bartelink, *Phys. Rev. Lett.* **7,** 87 (1961).
33. W. E. Spicer and R. E. Simon, *Bull. Am. Phys. Soc.* **7,** 537 (1962); *Phys. Rev. Lett.* **9,** 385 (1962).
34. L. Apker, E. Taft, and J. Dickey, *J. Opt. Soc. Am.* **43,** 78 (1953).
35. E. Taft and H. Philipp, *Phys. Rev.* **115,** 1583 (1959).
36. P. A. Wolff, *Phys. Rev.* **95,** 1415 (1954).
37. V. Vavilov, *J. Phys. and Chem. Solids* **8,** 223 (1959).
38. W. E. Spicer, *J. Phys. Chem. Solids* **22,** 365 (1961).
39. C. Ghosh, *Phys. Rev.* **22,** 1972 (1980).
40. J. J. Quinn, *Phys. Rev.* **126,** 1453 (1962).
41. N. V. Smith and G. B. Fisher, *Phys. Rev. B* **3,** 3662 (1971).
42. J. Monin, *Acta Electron.* **16,** 139 (1973).
43. W. F. Krolikowski and W. E. Spicer, *Phys. Rev.* **185,** 882 (1969).
44. E. J. Ryder and W. Shockley, *Phys. Rev.* **81,** 139 (1951).
45. W. Shockley, *Bell Syst. Tech. J.* **30,** 990 (1951).
46. F. Seitz, *Phys. Rev.* **73,** 550 (1948).
47. C. Ghosh, PhD. Thesis, Bombay University, 1976.
48. H. M. Hebb, *Phys. Rev.* **81,** 707 (1951).
49. R. G. Lye and A. J. Dekker, *Phys. Rev.* **107,** 977 (1957).
50. D. J. Bartelink, J. L. Moll, and N. I. Meyer, *Phys. Rev.* **130,** 972 (1963).
51. R. A. Smith, "Semiconductors," p. 219. Cambridge Univ. Press, London and New York, 1978.
52. C. Ghosh and B. P. Varma, *J. Appl. Phys.* **49,** 4549 (1978).
53. E. O. Kane, *Phys. Rev.* **127,** 131 (1962).
54. C. Ghosh and B. P. Varma, *Thin Solid Films* **46,** 151 (1977).
55. K. R. Crowe and J. L. Gumnick, *Appl. Phys. Lett.* **11,** 249 (1967).
56. J. R. Howorth, A. L. Harmer, E. W. L. Trawny, R. Holtom, and C. J. R. Sheppard, *Appl. Phys. Lett.* **23,** 123 (1973).
57. K. H. Kingdon, *Phys. Rev.* **24,** 510 (1924).
58. A. Asao and M. Suzuki, *Proc. Phys. Math. Soc. Jpn.* **12,** 247 (1930).
59. C. H. Prescott and M. J. Kelly, *Bell Syst. Tech. J.* **11,** 334 (1932).
60. P. G. Borzyak, V. F. Bibik, and G. S. Kramerenko, *Bull. Acad. Sci. USSR Ser. Phys.* **20,** 939 (1956).

61. N. A. Soboleva, A. S. Shefov, and V. N. Tolmasova, *Bull. Acad. Sci. USSR Ser. Phys.* **26**, 1393 (1962).
62. A. H. Sommer and W. E. Spicer, Contract DA-44-009-ENG 3642 (1960).
63. N. S. Zeitsov and N. S. Kelbanikov, AERDL-1689 (1963).
64. M. Srinivasan, B. M. Bhat, and N. Govindarajan, *J. Phys. E* **7**, 859 (1974).
65. E. Bovey and P. Unger, *J. Sci. Instrum.* **26**, 68 (1951).
66. A. H. Sommer, "Photoemissive Materials," p. 134. Wiley, New York, 1968.
67. K. L. Chopra, "Thin Film Phenomenon," p. 160. McGraw-Hill, New York, 1969.
68. R. S. Sennett and G. D. Scott, *J. Opt. Soc. Am.* **40**, 203 (1950).
69. R. C. Faust, *Philos. Mag.* **41**, 1238 (1950).
70. R. Philip, *J. Phys. Radium* **21**, 165 (1960).
71. G. Rasigni and P. Rourd, *J. Opt. Soc. Am.* **53**, 604 (1963).
72. J. C. Maxwell Garnett, *Philos. Trans. R. Soc.* **203**, 385 (1904); **205**, 237 (1906).
73. R. W. Cohen, G. D. Cody, M. D. Coutts, and B. Abeles, *Phys. Rev. B* **8**, 3689 (1973).
74. R. H. Doremus, *J. Chem. Phys.* **40**, 2389 (1964); *J. Appl. Phys.* **37**, 2775 (1966).
75. E. L. Hoene and B. Saggan, *Z. Agnew Phys.* **20**, 502 (1966).
76. V. V. Tjapkina and P. D. Dankov, *Dokl. Akad. Nauk SSSR* **54**, 419 (1946).
77. H. Wilman, *J. Chem. Phys.* **53**, 607 (1956).
78. T. Suzuki, *Z. Naturforsch.* **12A**, 497 (1957).
79. F. K. McTaggart, "Plasma Chemistry in Electrical Discharges." Elsevier, Amsterdam, 1967.
80. J. P. Cougulin, "Heat and Free Energies of Formation of Inorganic Oxides." Bull. 542, Bur. of Mines, Washington, D.C., 1954.
81. W. Heimann, E. L. Hoene, S. Jeric, and E. Kansky, *Exp. Tech. Phys.* **21**, 193 (1973); **21**, 325 (1973); **21**, 431 (1973).
82. C. Ghosh and B. P. Varma, *J. Phys. D* **7**, 1773 (1974).
83. Y. Sayama, *J. Phys. Soc. Jpn.* **1**, 13 (1946).
84. C. R. Helms and W. E. Spicer, *Phys. Rev. Lett.* **28**, 565 (1972); **31**, 1307 (1973); **32**, 228 (1974); *Appl. Phys. Lett.* **21**, 237 (1972).
85. P. E. Gregory, P. Chye, H. Sunami, and W. E. Spicer, *J. Appl. Phys.* **46**, 3525 (1975).
86. A. H. Sommer, *J. Appl. Phys.* **51**, 1254 (1980).
87. S. J. Yang and C. W. Bates, Jr., *Appl. Phys. Lett.* **36**, 675 (1980).
88. S. Asao, *Proc. Phys. Math. Soc. Jpn.* **22**, 448 (1940).
89. N. Govindarajan and M. Srinivasan, Personal communications.
90. W. J. Harper and W. J. Choyke, *J. Appl. Phys.* **27**, 1358 (1956).
91. J. E. Davey, *J. Appl. Phys.* **28**, 1031 (1957).
92. J. J. Uebbing and L. W. James, *J. Appl. Phys.* **41**, 4505 (1970).
93. H. Ehrenreich and H. R. Philipp, *Phys. Rev.* **128**, 1622 (1962).
94. A. H. Sommer, *RCA Rev.* **28**, 543 (1967).
95. C. N. Berglund and W. E. Spicer, *Phys. Rev.* **136**, A 1030 (1964); **136**, A 1044 (1964).
96. K. S. Neil and C. H. B. Mee, *Phys. Status Solidi A* **2**, 43 (1970).
97. F. Eckart, *Ann. Phys.* **16**, 322 (1955).
98. S. M. Sze, J. L. Moll, and T. Sugano, *Solid State Electron.* **7**, 509 (1964).
99. C. R. Crowell, W. G. Spitzer, L. E. Howarth, and E. E. LaBate, *Phys. Rev.* **127**, 2006 (1962).
100. A. H. Sommer and W. E. Spicer, *J. Appl. Phys.* **32**, 1036 (1961).
101. G. K. Bhide, L. M. Rangarajan, and B. M. Bhat, *J. Phys. D* **4**, 568 (1971).
102. G. A. Condas, *Rev. Sci. Instrum.* **33**, 987 (1962).
103. M. Rowe, *J. Appl. Phys.* **26**, 166 (1955).
104. C. Ghosh and B. P. Varma, *Indian J. Pure Appl. Phys.* **13**, 785 (1975).

105. A. H. Sommèr, *J. Appl. Phys.* **37,** 2789 (1966).
106. A. H. Sommer, *Proc. Phys. Soc. (London)* **55,** 145 (1943).
107. C. Kunze, *Ann. Phys.* **6,** 89 (1960).
108. J. J. Polkosky, U.S. Patent 2,676,282.
109. W. H. McCarroll, *J. Appl. Phys.* **39,** 3414 (1968).
110. F. Wooten, *J. Appl. Phys.* **37,** 2965 (1966).
111. A. H. Sommer and W. H. McCarroll, *J. Appl. Phys.* **37,** 174 (1966).
112. B. P. Varma and C. Ghosh, India Patent 291274 (1974).
113. T. Ninomiya, K. Taketoshi, and H. Tachiya, *Adv. Electron. Electron Phys.* **28A,** 337 (1969).
114. H. H. Hofman and K. Deutscher, *Z. Phys.* **236,** 288 (1970).
115. B. P. Varma and C. Ghosh, *J. Phys. D* **6,** 628 (1973).
116. B. R. C. Garfield, *Adv. Electron. Electron Phys.* **33A,** 339 (1972).
117. A. H. Sommer, *Appl. Phys. Lett.* **3,** 62 (1963).
118. V. Kanev and K. Nanev, *C. R. Acad. Bulg. Sci.* **15,** 123 (1962).
119. V. Kanev, K. Nanev, and R. Petrova, *Radio Eng. Electron* **10,** 338 (1965).
120. F. W. Dorn and W. Klemm, *Z. Anorg. Allgem. Chem.* **309,** 189 (1961).
121. B. R. C. Garfield and R. F. Thumwood, *Br. J. Appl. Phys.* **17,** 1005 (1966).
122. M. Hagino and T. Takahashi, *J. Appl. Phys.* **37,** 3741 (1966).
123. K. H. Jack and M. M. Wachtel, *Proc. R. Soc. London A* **239,** 46 (1957).
124. G. Guntzmann, F. W. Dorn, and W. Klemm, *Z. Anorg. Allgem. Chem.* **309,** 210 (1961).
125. J. J. Scheer and P. Zalm, *Philips Res. Rep.* **14,** 143 (1959).
126. W. H. McCarroll, *J. Appl. Phys.* **32,** 2051 (1961).
127. J. C. Robbie and A. H. Beck, *J. Phys. D* **6,** 1381 (1973).
128. A. A. Dowman, T. H. Jones and A. H. Beck, *J. Phys. D* **8,** 69 (1975).
129. G. Guntzmann, PhD. Dissertation, Munich, 1953.
130. N. N. Zhuravlev, V. A. Smirnov, and T. A. Mingazin, *Sov. Phys. Crystallogr.* **5,** 124 (1960).
131. J. Chikawa, S. Imamura, K. Tanaka, and M. Shiojiri, *J. Phys. Soc. Jpn.* **16,** 1175 (1961).
132. G. Brauer and E. Zintl, *Z. Phys. Chem. B* **37,** 327 (1937).
133. A. H. Sommer, *J. Appl. Phys.* **29,** 1588 (1958).
134. W. H. McCarroll, *J. Phys. Chem. Solids* **26,** 191 (1965).
135. W. H. McCarroll and R. E. Simon, *Rev. Sci. Instrum.* **35,** 508 (1964).
136. W. H. McCarroll, *J. Phys. Chem. Solids* **16,** 30 (1960).
137. E. L. Hoene, *Adv. Electron. Electron Phys.* **33A,** 369 (1972).
138. W. E. Spicer and F. Wooten, *Proc. IEEE* **51,** 1127 (1963).
139. M. Garbuny, T. P. Vogel, and J. R. Hansen, *J. Opt. Soc. Am.* **51,** 261 (1961).
140. A. R. Boileau and F. D. Miller, *Appl. Opt.* **6,** 1179 (1967).
141. H. Miyazawa and S. Fukuhara, *J. Phys. Soc. Jpn.* **7,** 645 (1952).
142. A. H. Sommer, *Appl. Opt.* **12,** 90 (1973).
143. A. H. Sommer and W. E. Spicer, "Methods of Experiments in Physic," Vol. 6B, p. 384. Academic Press, New York, 1959.
144. A. S. Shefov and G. A. Lisina, *Bull. Acad. Sci. USSR* **25,** 1415 (1962).
145. A. H. Sommer, *Nature (London)* **148,** 468 (1941).
146. P. G. Borzyak, *C. R. Acsd. Sci. URSS* **31,** 546 (1941).
147. K. Miyake, *J. Appl. Phys.* **31,** 76 (1960).
148. H. Miyazawa, K. Noga, S. Chikazumi, and A. Kobayashi, *J. Phys. Soc. Jpn.* **7,** 647 (1952).
149. G. Oertel, *Phys. Status Solidi* **3,** 314 (1963).
150. G. Wallis, *Ann. Phys.* **17,** 401 (1956).
151. T. Sakata, *J. Phys. Soc. Jpn.* **8,** 723 (1953).
152. T. Sakata, *J. Phys. Soc. Jpn.* **9,** 1030 (1954).

153. T. Sakata, *J. Phys. Soc. Jpn.* **9**, 1031 (1954).
154. S. Imamura, *J. Phys. Soc. Jpn.* **14**, 1491 (1959).
155. S. Imamura, *J. Phys. Soc. Jpn.* **16**, 1036 (1961).
156. M. Hagino, T. Takahashi, and M. Wada, *J. Appl. Phys.* **35**, 2112 (1964).
157. P. G. Borzyak, *J. Tech. Phys. USSR* **20**, 923 (1950).
158. W. E. Spicer, *J. Appl. Phys.* **31**, 2077 (1960).
159. R. Suhrmann and A. Kangro, *Naturwissenschaften* **40**, 137 (1953).
160. R. Gobrecht, *Phys. Status Solidi* **13**, 429 (1966).
161. C. Ghosh and B. P. Varma, *Indian J. Pure Appl. Phys.* **13**, 15 (1975).
162. W. H. McCarroll, R. J. Paff, and A. H. Sommer, *J. Appl. Phys.* **45**, 487 (1974).
163. M. B. Oliver, PhD. Thesis, London University, 1970.
164. E. Kansky, *Adv. Electron. Electron Phys.* **33A**, 357 (1973).
165. R. Nathan and C. H. B. Mee, *Int. J. Electron.* **23**, 349 (1967).
166. D. G. Fisher, A. F. McDonie, and A. H. Sommer, *J. Appl. Phys.* **45**, 487 (1974).
167. A. H. Sommer, *J. Appl. Phys.* **43**, 2479 (1972).
168. A. A. Mostovskii, O. B. Vorobeva, and G. B. Struchinskii, *Sov. Phys. Solid State* **5**, 2436 (1964).
169. C. Ghosh and B. P. Varma, *J. Appl. Phys.* **49**, 4554 (1978).
170. A. J. Dekker and R. G. Lye, *Phys. Rev.* **107**, 977 (1957).
171. J. A. Burton, *Phys. Rev.* **72**, 531 (A) (1947); Burton's data are given by V. K. Zworykin and E. Ramberg, *in* "Photoelectricity," p. 59. Wiley, New York, 1949.
172. N. D. Morgulis, P. G. Borzyak, and B. I. Dyatlovitskaya, *Bull. Acad. Sci. USSR Fiz. Ser.* **12**, 126 (1948).
173. P. G. Borzyak, *Trud. Inst. Fiz. Akad. Nauk Ukr. SSR* **4**, 28 (1953).
174. A. Ebina and T. Takahashi, *Phys. Rev. B* **7**, 4712 (1973).
175. C. Ghosh and B. P. Varma, *Proc. Symp. Photoelectron. Dev.*, 7th, Imp. College, London, *1978* p. 221.
176. A. A. Mostovskii, V. A. Chaldyshev, G. F. Karavaev, A. I. Klimin, and I. N. Ponomarenko, *Izv. Akad. Nank SSSR Ser. Fiz.* **38**, 195 (1974).
177. A. A. Mostovskii, V. A. Chaldyshev, V. P. Kiseler, and A. I. Klimin, *Izv. Akad. Nauk SSSR Ser. Fiz.* **40**, 2490 (1976).
178. V. Heine and I. Abarenkov, *Philos. Mag.* **9**, 451 (1964).
179. F. Wooten, J. P. Hernandez, and W. E. Spicer, *J. Appl. Phys.* **44**, 1112 (1973).
180. C. Ghosh and B. P. Varma, *Indian J. Phys.* **53A**, 14 (1979).
181. H. Philipp, E. A. Taft, and L. Apker, *Phys. Rev.* **120**, 49 (1960).
182. C. Ghosh, Unpublished.
183. R. Nathan and C. H. B. Mee, *Phys. Status Solidi A* **2**, 67 (1970).
184. A. A. Turnbull and G. B. Evans, *J. Phys. D* **1**, 155 (1968).
185. R. U. Martinelli and D. G. Fisher, *Proc. IEEE* **62**, 1339 (1974).
186. B. F. Williams and R. E. Simon, *Phys. Rev. Lett.* **18**, 485 (1967).
187. R. L. Bell and J. J. Uebbing, *Appl. Phys. Lett.* **12**, 76 (1968).
188. R. E. Simon, A. H. Sommer, J. J. Tietjan, and B. F. Williams, *Appl. Phys. Lett.* **15**, 43 (1969).
189. H. Sonnenberg, *Appl. Phys. Lett.* **16**, 245 (1970).
190. R. U. Martinelli and M. Ettenberg, *J. Appl. Phys.* **45**, 3896 (1974).
191. G. A. Antypas, R. L. Moon, L. W. James, J. Edgecumbe, and R. L. Bell, *in* "GaAs and Related Compounds," Ch. 1. London, Inst. Phys., 1973.
192. G. A. Antypas and L. W. James, *J. Appl. Phys.* **41**, 2165 (1970).
193. S. Garbe, *Phys. Status Solidi A* **2**, 497 (1970).

194. R. L. Bell, L. W. James, G. A. Antypas, J. Edgecumbe, and R. L. Moon, *Appl. Phys. Lett.* **19,** 513 (1971).
195. R. U. Martinelli, *J. Appl. Phys.* **44,** 2566 (1973).
196. D. E. Eastman, F. J. Himpsel, J. A. Knapp, and J. A. Van Vechten, *Bull. Am. Phys. Soc.* **24,** 403 (1979).
197. K. H. Kingdon and I. Langmuir, *Phys. Rev.* **21,** 380 (1923).
198. J. B. Taylor and I. Langmuir, *Phys. Rev.* **44,** 423 (1933).
199. I. Langmuir, *J. Am. Chem. Soc.* **54,** 2798 (1932).
200. J. J. Uebbing, *J. Appl. Phys.* **41,** 802 (1970).
201. B. Goldstein, *Surf. Sci.* **35,** 227 (1973).
202. R. Holtom and P. M. Gundry, *Surf. Sci.* **63,** 263 (1977).
203. J. D. Joannopoulos and M. L. Cohen, *Phys. Rev. B* **10,** 5075 (1974).
204. J. Van Laar and J. J. Scheer, *Surf. Sci.* **8,** 342 (1967).
205. A. Huijser and J. Van Laar, *Surf. Sci.* **52,** 202 (1975).
206. I. Lindau, P. Pianetta, C. M. Garner, P. W. Chye, P. E. Gregory, and W. E. Spicer, *Surf. Sci.* **63,** 45 (1977).
207. P. W. Chye, I. A. Babalola, T. Sukegawa, and W. E. Spicer, *Phys. Rev. Lett.* **23,** 1602 (1975).
208. A. Huijser, J. Van Laar, and T. L. Van Rooy, *Surf. Sci.* **62,** 472 (1977).
209. R. H. Williams, R. R. Varma, and A. McKinley, *J. Phys. C* **10,** 4545 (1977).
210. W. Monch and H. J. Clemens, *J. Vac. Sci. Tech.* **16,** 1238 (1979).
211. G. M. Guichar, C. A. Sebenne, and C. D. Thuault, *J. Vac. Sci. Tech.* **16,** 1212 (1979).
212. W. E. Spicer, P. W. Chye, P. R. Skeath, C. Y. Su, and I. Lindau, *J. Vac. Sci. Tech.* **16,** 1422 (1979).
213. R. Ludeke and L. Esaki, *Phys. Rev. Lett.* **33,** 653 (1974).
214. G. J. Lapeyre and J. Anderson, *Phys. Rev. Lett.* **35,** 117 (1975).
215. D. J. Chadi, *J. Vac. Sci. Tech.* **15,** 1244 (1978).
216. J. R. Chelikowsky and M. L. Cohen, *J. Vac. Sci. Tech.* **16,** 1307(A) (1979).
217. W. E. Spicer, I. Lindau, P. E. Gregory, C. M. Garner, P. Pianetta, and P. W. Chye, *J. Vac. Sci. Tech.* **13,** 780 (1976).
218. A. R. Lubinsky, C. B. Duke, B. W. Lee, and P. Mark, *Phys. Rev. Lett.* **36,** 1058 (1976).
219. W. Ranke and K. Jacobi, *Surf. Sci.* **63,** 33 (1977).
220. P. Pianetta, I. Lindau, and W. E. Spicer, in "Quantitative Surface Analysis of Materials" (N. S. McIntyre, ed.), ASTM STP 643, p. 105. Amer. Society for Testing of Materials, Philadelphia, Pennsylvania.
221. P. E. Gregory and W. E. Spicer, *Phys. Rev. B* **12,** 2370 (1975).
222. I. Lindau, P. W. Chye, C. M. Garner, P. Pianetta, C. Y. Su, and W. E. Spicer, *J. Vac. Sci. Tech.* **15,** 1332 (1978).
223. J. Derrien and F. Arnaud D'Avitoya, *Surf. Sci.* **65,** 668 (1977).
224. J. O. McCaldin, T. C. McGill, and C. A. Mead, *J. Vac. Sci. Tech.* **13,** 802 (1976).
225. D. L. Smith and D. A. Huchital, *J. Appl. Phys.* **43,** 2624 (1972).
226. B. Goldstein and D. Szostak, *Appl. Phys. Lett.* **26,** 111 (1975).
227. J. Derrien and F. Arnaud d'Avitoya, *Rev. Phys. Appl.* **11,** 377 (1976).
228. A. H. Sommer, H. H. Whitaker, and B. F. Williams, *Appl. Phys. Lett.* **17,** 273 (1970).
229. D. G. Fisher, R. E. Enstrom, J. S. Escher, and B. F. Williams, *J. Appl. Phys.* **43,** 3815 (1972).
230. P. E. Gregory and W. E. Spicer, *J. Appl. Phys.* **47,** 510 (1976).
231. L. W. James and J. J. Uebbing, *Appl. Phys. Lett.* **16,** 370 (1970).
232. H. Sonnenberg, *J. Appl. Phys.* **40,** 3414 (1969).
233. H. Sonnenberg, *Appl. Phys. Lett.* **14,** 289 (1969).

234. W. Heiman, E. L. Hoene, and E. Kansky, *Exp. Technol. Phys.* **21**, 193 (1973).
235. P. E. Gregory and W. E. Spicer, *Phys. Rev. B* **12**, 2170 (1975).
236. H. Rougeot and C. Baud, *Adv. Electron. Electron Phys.* **49**, 1 (1979).
237. L. F. Wagner and W. E. Spicer, *Phys. Rev.* **9**, 1512 (1974).
238. J. E. Rowe, G. Margaritondo, and S. B. Christman, *Phys. Rev. B* **15**, 2195 (1977).
239. J. D. Levine, *Surf. Sci.* **34**, 90 (1973).
240. B. Goldstein, *Surf. Sci.* **47**, 143 (1975).
241. P. M. Gundry, R. Holtom, and V. Leverett, *Surf. Sci.* **43**, 647 (1974).
242. R. Holtom and P. M. Gundry, *Surf. Sci.* **63**, 263 (1977).
243. I. F. Koval, P. V. Melnik, N. G. Nakhodkin, and S. N. Goisa, *Sov. Phys. Solid State* **20**, 1769 (1978).
244. D. Edwards and W. T. Peria, *Appl. Surf. Sci.* **1**, 419 (1978).
245. J. R. Howorth, C. J. R. Sheppard, R. Holtom, and A. L. Harmer, *J. Appl. Phys.* **46**, 151 (1975).
246. D. J. Miller and D. Haneman, *Surf. Sci.* **82**, 102 (1979).
247. A. J. Van Bommel and J. E. Crombeen, *Surf. Sci.* **76**, 499 (1978).
248. K. Jacobi, *Surf. Sci.* **51**, 29 (1975).
249. J. Escher, "Semiconductors and Semimetals." (In press).
250. W. E. Spicer, *Appl. Phys.* **12**, 115 (1977).
251. J. S. Escher, L. W. James, and R. Sankaran, G. A. Antypas, R. L. Moon, and R. L. Bell, *J. Vac. Sci. Tech.* **13**, 874 (1976).
252. G. A. Antypas, L. W. James, and J. J. Uebbing, *J. Appl. Phys.* **41**, 2888 (1970).
253. G. A. Allen, *J. Phys. D* **4**, 208 (1971).
254. G. Frank and S. Garbe, *Acta Electron.* **16**, 237 (1973).
255. D. G. Fisher and G. H. Olsen, *J. Appl. Phys.* **50**, 2930 (1979).
256. G. A. Antypas and J. Edgecumbe, *Appl. Phys. Lett.* **26**, 371 (1975).
257. L. W. James, *J. Appl. Phys.* **45**, 1326 (1974).
258. R. L. Bell, "Negative Electron Affinity Devices," p. 127. Oxford Univ. Press (Clarendon), London and New York, 1973.
259. D. G. Fisher and R. U. Martinelli, *in* "Advances in Image Pickup and Display" (B. Kazan, ed.), Vol. 1, p. 71. Academic Press, New York, 1973.
260. S. Garbe, *Solid State Commun.* **12**, 893 (1969).
261. J. R. Howorth, J. R. Folks, I. C. Palmer, R. Holtom, C. J. R. Sheppard, and E. W. L. Trawny, *J. Phys. D* **9**, 785 (1976).
262. V. L. Korotkikh, A. D. Koriufskii, A. A. Matyash, Al. L. Musatov, S. S. Strl'chenko, and V. A. Titov, *Sov. Phys. Solid State* **19**, 1681 (1977).
263. L. W. James, G. A. Antypas, R. L. Moon, J. J. Edgecumbe, and R. L. Bell, *Appl. Phys. Lett.* **22**, 270 (1973).
264. L. W. James, G. A. Antypas, J. Edgecumbe, R. L. Moon, and R. L. Bell, *J. Appl. Phys.* **42**, 4976 (1971).
265. M. G. Burt and J. C. Inkson, *J. Phys. D* **8**, L3 (1975).
266. M. G. Burt and J. C. Inkson, *J. Phys. D* **9**, L5 (1976).
267. L. W. James and J. L. Moll, *Phys. Rev.* **183**, 740 (1969).
268. P. Skeath, W. A. Saperstein, P. Pianetta, I. Lindau, and W. E. Spicer, *J. Vac. Sci. Tech.* **15**, 1219 (1978).
269. B. J. Stocker, *Surf. Sci.* **47**, 501 (1975).
270. D. G. Fisher, *IEEE Trans. Electron Dev.* **21**, 541 (1974).
271. D. G. Fisher and G. W. Fowler, *Conf. Photoelectr. Second. Emission, Univ. Minnesota Minneapolis* (*1973*).

272. G. H. Olsen, D. J. Szostak, T. J. Zamerowski, and M. Ettenberg, *J. Appl. Phys.* **48**, 1007 (1977).
273. H. M. Manasevit and W. I. Simpson, *J. Electrochem. Soc.* **116**, 1725 (1969).
274. P. Raichoudhury, *J. Electrochem. Soc.* **116**, 1745 (1969).
275. S. J. Bass, *J. Cryst. Growth* **31**, 172 (1975).
276. J. S. Escher, P. E. Gregory, S. B. Hyder, and R. Sankaran, *J. Appl. Phys.* **49**, 2591 (1978).
277. L. W. James, US Patent 3,769,536 (1973).
278. C. H. A. Syms, *Adv. Electron. Electron Phys.* **28A**, 399 (1969).
279. Y. Z. Liu, J. L. Moll, and W. E. Spicer, *Appl. Phys. Lett.* **17**, 60 (1970).
280. S. B. Hyder, *J. Vac. Sci. Tech.* **8**, 228 (1971).
281. G. A. Antypas, J. S. Escher, J. Edgecumbe, and R. S. Enck, *J. Appl. Phys.* **49**, 4301 (1978).
282. J. R. Howorth and P. J' Pool, US Patent no. 4,099,198 (1978).
283. G. P. Hopkins, J. R. Howorth, I. C. Palmer, and H. J. Pettas, *Int. Conf. Low Light Level Therm. Imaging*, 2nd p. 22 (1979).
284. R. U. Martinelli, *J. Appl. Phys.* **45**, 3203 (1974).
285. R. U. Martinelli and J. I. Pankove, *Appl. Phys. Lett.* **25**, 549 (1974).
286. R. U. Martinelli, *Appl. Phys. Lett.* **17**, 313 (1970).
287. N. R. Whetten and A. B. Laponsky, *J. Appl. Phys.* **30**, 432 (1959).
288. U. Fano, *Phys. Rev.* **72**, 26 (1947).
289. B. F. Williams and R. E. Simon, *Appl. Phys. Lett.* **14**, 214 (1969).
290. E. S. Kohn, *IEEE Trans. Electron. Dev.* **ED20**, 321 (1973).
291. K. R. Faulkner, R. A. Astridge, J. R. Howorth, and R. K. Surridge, *Appl. Phys. Lett.* **23**, 298 (1973).
292. J. R. Howorth, R. Holtom, C. J. R. Sheppard, and E. W. L. Trawny, *Adv. Electron. Electron Phys.* **40A**, 387 (1976).
293. E. Stupp, A. Pelissier, M. Kidder, and A. Mitch, *J. Appl. Phys.* **48**, 4741 (1977).
294. H. Schade, H. Nelson, and H. Kressel, *Appl. Phys. Lett.* **20**, 385 (1972).
295. H. Kan, T. Nakamura, H. Katsumo, M. Hagino, and T. Sukegawa, *Appl. Phys. Lett.* **34**, 545 (1979).
296. R. U. Martinelli, *J. Appl. Phys.* **45**, 1183 (1974).
297. R. L. Bell, *Solid State Electron.* **12**, 475 (1969).
298. C. Piaget, P. Saget, and J. Vannimenns, *J. Appl. Phys.* **48**, 3907 (1977).
299. E. M. Yee and D. A. Jackson, *Solid State Electron.* **15**, 245 (1972).
300. R. M. Thomas and H. C. Nathanson, *Appl. Phys. Lett.* **21**, 384 (1972).
301. R. N. Thomas and H. C. Nathanson, *Appl. Phys. Lett.* **21**, 387 (1972).
302. R. N. Thomas, R. A. Wickstrom, D. K. Schroeder, and H. C. Nathanson, *Solid State Electron.* **17**, 155 (1974).
303. D. K. Schroder, R. N. Thomas, J. Vine, and H. C. Nathanson, *IEEE Trans. Electron. Dev.* **ED21**, 785 (1974).
304. J. A. Burton, *Phys. Rev.* **108**, 1342 (1957).
305. R. E. Simon and W. E. Spicer, *Phys. Rev.* **119**, 521 (1960); *J. Appl. Phys.* **31**, 1505 (1960).
306. P. Schagen and A. A. Turnbull, *Adv. Electron. Electron Phys.* **28A**, 393 (1969).
307. V. Dalal, *J. Appl. Phys.* **43**, 1160 (1972).
308. R. L. Bell, L. W. James, and R. L. Moon, *Appl. Phys. Lett.* **25**, 645 (1974).
309. R. Sahai, J. S. Harris, R. C. Eden, L. O. Bubulac, and J. C. Chu, *CRC Crit. Rev. Solid State Sci.* **5**, 565 (1975).
310. J. Escher and R. Sankaran, *Appl. Phys. Lett.* **29**, 87 (1976).
311. J. S. Excher, P. E. Gregory, and T. J. Maloney, *J. Vac. Sci. Tech.* **16**, 1394 (1979).
312. D. E. Persyk and D. D. Crawshaw, *RCA Rev.* **34**, 344 (1973).

313. D. A. Wilcox, W. G. Abraham, D. Bardas, G. F. Gwilliams, and R. S. Enck, *Electro Opt. Syst. Design*, March 41 (1979).
314. G. A. Morton, H. M. Smith, and H. R. Krall, *Appl. Phys. Lett.* **13,** 356 (1968).
315. B. P. Varma and C. Ghosh, *Proc. Symp. Quantum Optoelectron., Bombay* (*1974*) p. 484.
316. C. F. Freeman, *Proc. SPIE* **42,** 3 (1973).
317. D. H. Chckowski, *Proc. SPIE* **42,** 25 (1973).
318. J. D. McGee, R. W. Airey, M. Aslam, J. R. Powell, and C. E. Catchpole, *Adv. Electron. Electron Phys.* **22A,** 571 (1966).
319. J. D. McGee, R. W. Airey, and B. P. Varma, *Adv. Electron. Electron Phys.* **28A,** 89 (1969).
320. T. B. Bhatia, G. K. Bhide, C. Ghosh, G. N. Kelkar, M. Srinivasan, B. P. Varma, and R. L. Verma, *Adv. Electron. Electron Phys.* **40A,** 409 (1976).
321. J. S. Courtney Pratt, *Research* (*London*) **2,** 287 (1949).
322. R. G. Stoudenheimer and J. C. Moor, *Image Intensifier Symp.*, Ft. Belvoir, p. 1 (1958).
323. J. A. Hall, *in* "Photoelectronic Imaging Devices" (L. M. Biberman and S. Nudelman, eds.), p. 53. Plenum, New York, 1971.
324. I. Limansky, *Electron. Eng.* (June 1968).
325. P. T. Farnsworth, *J. Franklin Inst.* **218,** 411 (1934).
326. A. Rose, "Vision, Human and Electronic," p. 55. Plenum, New York, 1973.
327. J. A. Hall, "Photoelectronic Imaging Devices" (L. M. Biberman and S. Nudelman, eds.), p. 483. Plenum, New York, 1971.
328. V. K. Zworykin, *Proc. IRE* **22,** 16 (1934).
329. H. A. Iams, G. A. Morton, and V. K. Zworykin, *Proc. IRE* **27,** 541 (1939).
330. A. Rose and H. A. Iams, *RCA Rev.* **4,** 186 (1939).
331. A. Rose, P. K. Weimer, and H. B. Law, *Proc. IRE* **34,** 424 (1946).
332. R. W. Reddington, "Photoelectronic Imaging Devices" (L. M. Biberman and S. Nudelman, eds.), p. 193. Plenum, New York, 1971.
333. P. K. Weimer, *RCA Rev.* **10,** 366 (1949).
334. L. Pensak, *Phys. Rev.* **75,** 472 (1949).
335. G. W. Goetz and A. H. Boerio, *Proc. IEEE* **52,** 1007 (1964).
336. R. L. Rogers, III, G. S. Briggs, W. M. Henry, P. W. Kaseman, R. E. Simon, and R. L. Van Asselt, *Int. Conf. Solid State Circuits*, Feb. 18–20 (*1970*).

# Chemical Solution Deposition of Inorganic Films

K. L. CHOPRA, R. C. KAINTHLA, D. K. PANDYA, AND A. P. THAKOOR

*Thin Film Laboratory*
*Indian Institute of Technology, Delhi*
*New Delhi, India*

| | |
|---|---|
| I. Introduction | 168 |
| II. Spray Pyrolytic Process | 169 |
|    1. Physical Aspects | 169 |
|    2. Spray Setup | 169 |
|    3. Atomization Process | 171 |
|    4. Kinetics of Growth | 172 |
|    5. Chemical Aspects | 174 |
| III. Characteristic Features of the Spray Pyrolytic Process | 178 |
|    1. Growth Rate | 178 |
|    2. Substrate Effects | 178 |
|    3. Film Composition | 179 |
| IV. Multicomponent Doping and Alloying Effects | 181 |
| V. Structural Properties | 187 |
| VI. Electrical and Optical Properties | 192 |
| VII. Solution Growth Process | 201 |
|    1. Solution Growth Chemistry | 202 |
|    2. Film Growth and Deposition Parameters | 206 |
| VIII. Impurity and Dopant Effects | 211 |
| IX. Multicomponent Films | 212 |
| X. Oxide Films | 213 |
| XI. Structure | 214 |
| XII. Transport Properties | 217 |
|    1. Optical Properties | 218 |
|    2. Electrical Properties | 219 |
|    3. Photoconductivity | 223 |
| XIII. Some Large-Area Applications | 223 |
|    1. Thin-Film Solar Cells | 224 |
|    2. Photoelectrochemical Solar Cells | 226 |
|    3. Selective Coatings | 226 |
|    4. Transparent Conducting Coatings | 228 |
|    5. Heat Mirrors | 228 |
|    6. Photon Detectors (Visible and IR) | 228 |
| XIV. Concluding Remarks | 230 |
|    References | 232 |

## I. Introduction

Thin-film processes, technology, and devices have undergone phenomenal development in the last decade, primarily as a result of the ever-increasing demand for thin films of a variety of conventional and exotic materials for applications in a host of physics-based industries. Yet another era of large-area thin-film technology, for photothermal and photovoltaic conversion of solar energy, has already been ushered in in this decade. Considerations of simplicity, economics, and input energy dictate that large-area thin films be deposited by chemical solution techniques. As Table I shows, many such techniques are found in the literature. Of these, spray pyrolysis and solution growth are presently attracting considerable attention. As with other thin-film techniques, both of these have undergone experimentation for a long time. For example, spray pyrolysis of $SnCl_4$ solution for obtaining conducting transparent oxides was achieved as early as 1910, as reported by Foex (1) in his review. The extension of the technique to sulfide and selenide films by Chamberlin (2–4) took place during the 1960s. The solution growth of PbS films for infrared (IR) applications was carried out in 1946 (5). Both techniques have been used extensively ever since. However, it is only recently that large-area and large-scale applications of these techniques to obtain doped and undoped multicomponent semiconductor films of usual, unusual, and metastable structures have necessitated the understanding of the physics and chemistry of the two processes. Thus two full-fledged techniques of great promise have emerged and are the subject of this review.

Spray pyrolysis involves a thermally stimulated chemical reaction between clusters of liquid/vapor atoms of different chemical species. Solution growth, on the other hand, involves a recombination of ions of different chemical species. Accepting the strict definition of thin-film growth as an atom/molecule/ion-by-atom/molecule/ion condensation process, solution growth is a thin-film technique, whereas spray pyrolysis lies somewhere between

TABLE I

CHEMICAL SOLUTION DEPOSITION TECHNIQUES
FOR INORGANIC FILMS

| | |
|---|---|
| Electroplating | Electroless conversion |
| Anodization | (chemiplating) |
| Electroconversion | Spray pyrolysis |
| Electrophoresis | Controlled precipitation |
| Electroless deposition | (solution growth) |
| | Dip growth (wetting) |

the thin-film and thick-film techniques, depending on the atom cluster size. The two techniques use similar chemicals and reactions in the preparation of sulfides, selenides, and oxides, and yield films of similar microstructure and properties. Therefore, the two techniques, though apparently different, are covered in the same review.

Limited and old reviews (*6–8*) of the two techniques have been published in the literature. Spray pyrolysis has been developed extensively by Chamberlin (*2–4, 9–11*) and co-workers (now at Photon Power, Inc., El Paso), Bube *et al.* (*12–18*) at Stanford University, Savelli *et al.* (*19–22*) at Montpellier University, and Chopra and co-workers (*23–30*) at the Indian Institute of Technology, New Delhi. The solution growth technique has been pioneered by the works of Bode and co-workers (*31–33*) at Santa Barbara Research Center, G. A. Kitaev and co-workers (*34–39*) at Ural Polytechnic, USSR, and Chopra and co-workers (*40–43*) at the Indian Institute of Technology, New Delhi.

## II. Spray Pyrolytic Process

### 1. Physical Aspects

Spray pyrolysis involves spraying a solution, usually aqueous, containing soluble salts of the constituent atoms of the desired compound onto a heated substrate. Every sprayed droplet reaching the hot substrate surface undergoes pyrolytic (endothermic) decomposition and forms a single crystallite, or a cluster of crystallites of the product. The other volatile by-products and the excess solvent escape in the vapor phase. The substrate provides the thermal energy for the thermal decomposition and subsequent recombination of the constituent species, followed by sintering and recrystallization of the clusters of crystallites. The result is a coherent film. The chemical solution is atomized into a spray of fine droplets by a spray nozzle (described below) with the help of a carrier gas, which may (as in the case of $SnO_x$ films) or may not (as for CdS films) play an active role in the pyrolytic reaction involved. The solvent liquid serves to carry the reactants and distribute them uniformly over the substrate area during the spray process. In most cases the carrier liquid takes part in the pyrolytic reaction.

### 2. Spray Setup

A block diagram of a typical spray pyrolysis set-up is shown in Fig. 1. The filtered carrier gas and solution are fed into a spray nozzle at a pre-

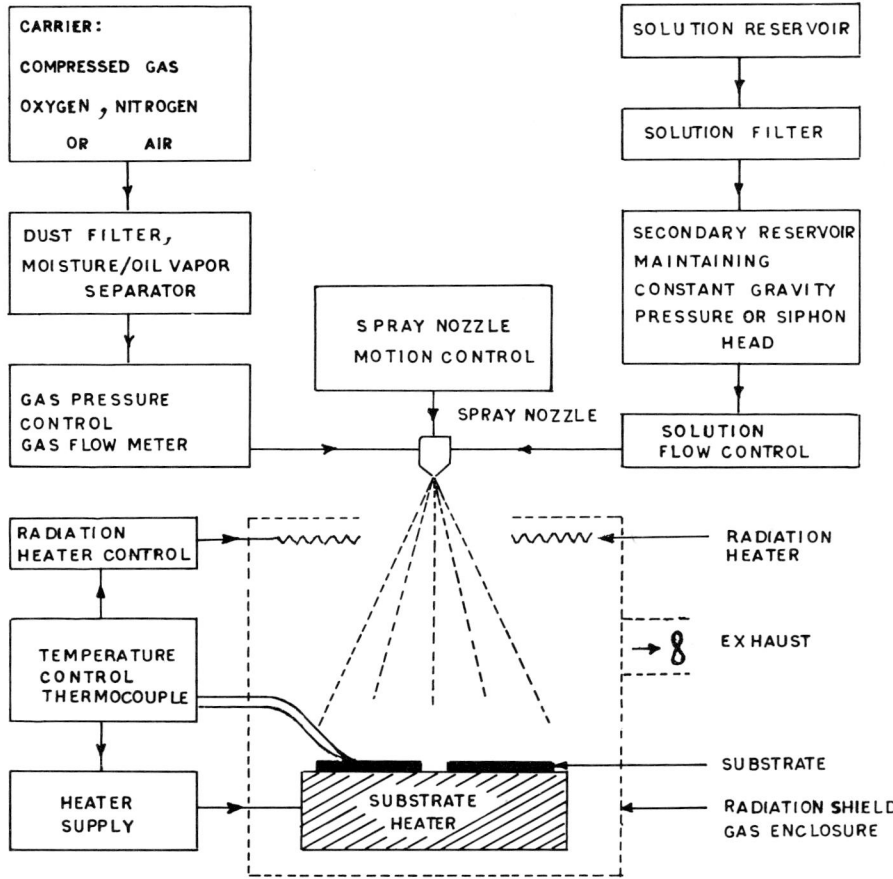

Fig. 1. Block diagram of a typical spray setup (27).

determined constant pressure and flow rate. The substrate temperature is maintained by a feedback circuit that controls a primary and auxiliary heater power supply. Large-area uniform coverage of the substrate may be obtained by employing several mechanical or electromechanical arrangements for scanning either the spray head or the substrate, or both. The setup is generally enclosed in a chamber provided with an exhaust duct to remove the vaporized constituents and to provide a stable flow pattern.

The spray pattern, size distribution of droplets, and spray rate depend sensitively on the geometry of the gas and liquid nozzles. A variety of nozzles

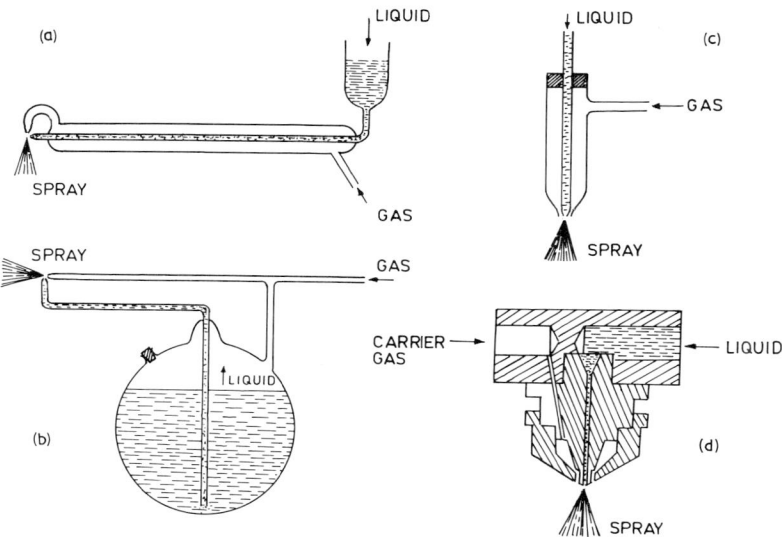

FIG. 2. (a, b, c) Different types of air-atomizing spray nozzles made of glass. (d) Cross section of a commercial spray nozzle made of stainless steel lucite.

have been employed for spray pyrolysis on stationary and moving substrates ranging from a few square centimeters to a few square meters. Some nozzles and a cross-sectional view of a typical nozzle are shown in Fig. 2. Spray heads made entirely of glass, Teflon, rubber, Lucite, quartz, or stainless steel are commercially available from the Spraying Systems Co. (*44*).

## 3. Atomization Process

Figure 3 shows the basic mode of droplet formation in a typical air-atomizing nozzle. There exist three regions in the spray. Region A is in front of the nozzle, where the liquid is being lifted off the tip and accelerated into the main air cone turbulent-vortex region. The length of this region is proportional to the liquid flow rate. In region B, the droplets are formed by the turbulence of the air stream. The vortex of the expanding helical shape of the rest of the aerosol envelope is seen at the confluence of regions A and B. This process leads to the erosion of the tip of the nozzle with time, as shown by the dotted region in Fig. 3. Region C is the region outside the main aerosol cone where low-velocity droplets are seen when the nozzle design is not mechanically optimized.

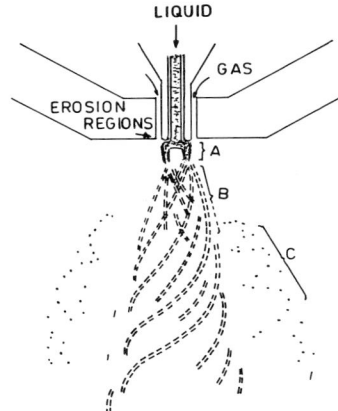

Fig. 3. Basic mode of droplet formation in a typical air-atomizing nozzle (45).

## 4. Kinetics of Growth

The aerodynamics of the atomization and droplet impact processes has been studied by Lampkin (45) by means of high-speed photography. He has further correlated the dynamic features of the process with the kinetics of film growth and surface topographic features obtained in the films. According to Lampkin, good quality (optical) and smooth films of CdS are obtained when both the size and momentum of the spray droplets are uniform. Applying high voltage between the nozzle and the substrate, Lampkin observed that the spray pattern becomes better defined in the presence of an electric field, presumably because of the presence of electrostatic charges on the droplets. How the field affects the velocity and shape distribution of variously charged droplets has not been established. In any case, the field-induced higher droplet velocities and the coalescence kinetics of the droplets on the substrate surface are expected to have a considerable effect on the microstructural details of the films so obtained.

Banerjee et al. (25, 27) have studied the growth kinetics of spray-deposited CdS films by optical and electron microscopy. According to these studies, the liquid droplet tends to flatten out at impact on the substrate surface, due to its momentum. The radially outward forces are balanced by the surface tension, and the drop thus spreads into a disk-shaped structure having more mass at its edges. The disk geometry depends on the momentum and volume of the droplet and the substrate temperature. The disk undergoes a complicated oscillatory behavior, presumably due to the thermal evaporation

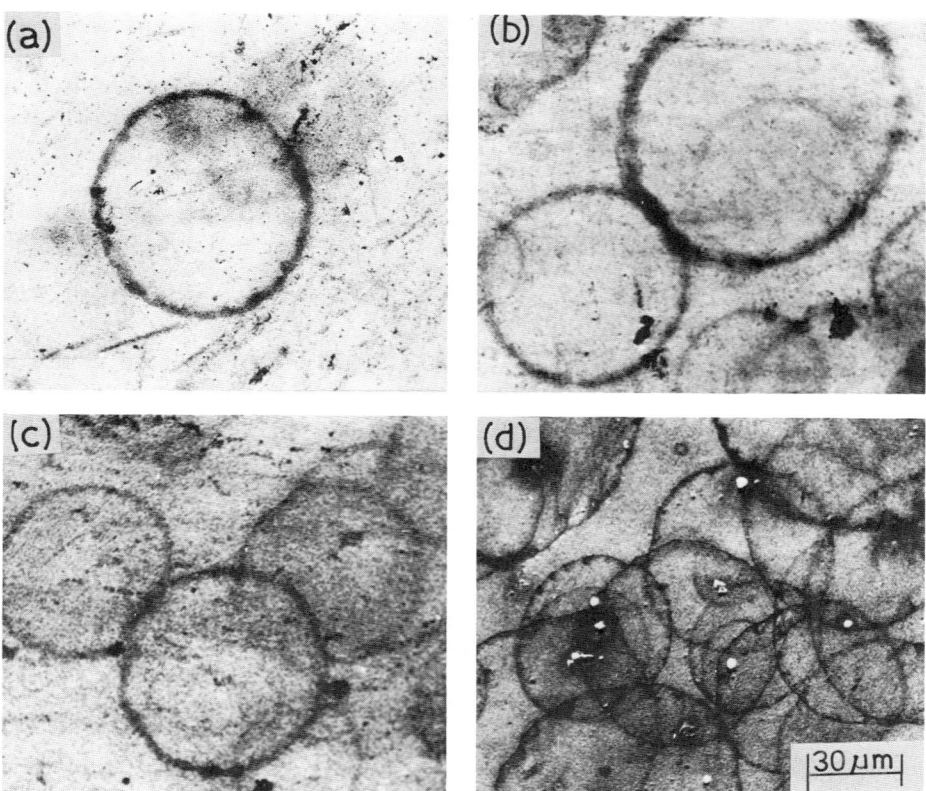

FIG. 4. SEM micrographs showing the initial stages of growth of spray-deposited CdS films (27).

processes. Thus, the deposition process is the net result of (a) spreading of a drop into a disk, (b) pyrolytic reaction between the decomposed reactants, (c) evaporation of the solvent, and (d) repetition of the preceding processes with succeeding droplets. Consequently, the film generally contains overlapping disks, as shown in Fig. 4. Each disk corresponds to a single droplet, and the details of its shape are determined by the balance of dynamics, surface energy, and thermal processes. In contrast to the adatom-mobility- and coalescence-dominated processes prevalent in an atom-by-atom condensation process for a typical thin film, the lateral mobility of the droplets and the coalescence and sintering kinetics of the overlapping disk crystallites determine the growth kinetics and microstructural features of the spray-deposited films. This mode of growth, like that of a thin film, is expected to have the following important characteristics:

(1) The random disk-by-disk growth exposed to a continuous flow of pressurized liquid droplets eliminates microscopic and macroscopic voids and cavities in the growing film. Spray-deposited films are thus pinhole-free even at thicknesses as low as 1000 Å, provided the substrate temperature is sufficiently high (see Section III,2) for complete pyrolytic reaction.

(2) The microstructure of the films should depend very sensitively on a host of deposition conditions, such as spray head geometry, carrier gas and liquid, flow pattern and rate, droplet velocities, sizes and geometries, nature and temperature of substrate, kinetics and thermodynamics of the pyrolytic reactions, and temperature profile during deposition process. This aspect is discussed in Section III.

(3) By incorporating other soluble and reactive cationic and/or anionic complexes in the sprayed liquid it should be possible to dope the films, to extend the solubility limits of two or more component films, and to prepare several-component homogeneous and single-phase, single- or multilayer films with spatial control of the composition and hence of the properties (see Section IV).

### 5. Chemical Aspects

Chemicals used for spray pyrolysis must satisfy a variety of conditions. On thermal decomposition, the chemicals in a solution form should provide the species/complexes necessary to undergo a thermally activated chemical reaction to yield the desired thin-film material. The remainder of the constituents of the chemicals, including the carrier liquid, should be volatile at the spray temperature. For a given material, these conditions are met by a number of combinations of chemicals. Clearly, each combination has its own thermodynamic and kinetic considerations under the prevailing spray conditions so that different deposition parameters are required to obtain films of comparable (structural) quality. These points are best illustrated by the familiar examples of sulfide, selenide, and oxide films discussed below.

*a. Sulfides and Selenides.* The chemistry of spray pyrolysis to obtain CdS (and ZnS) films has been studied (22, 27, 46, 47) extensively. A dilute (0.001–0.1 $M$) aqueous solution of a water-soluble cadmium salt and a sulfoorganic salt is commonly used. The commonly used chemicals $CdCl_2$ and thiourea yield CdS films by the reaction

$$CdCl_2 + (NH_2)_2CS + 2H_2O \rightarrow CdS\downarrow + 2NH_4Cl\uparrow + CO_2\uparrow$$

Similar reactions are obtained by using other cadmium salts, such as $Cd(NO_3)_2$, $Cd(SO_4)$, $Cd(CH_3COO)_2$, $Cd(CHO_2)_2$, and $Cd(C_3H_5O_2)$. Similarly, thiourea can be replaced by N,N-dimethylthiourea ($N_2(CH_3)_2H_2CS$), allylthiourea ($H_2NCSNHCH_2CH:CH_2$), thiolocetic acid ($CH_3COSH$), and ammonium thiocyanate ($NH_4CNS$).

The decomposition process and the various intermediate chemical reactions and products are quite complex, as is illustrated for the case of $CdCl_2 \cdot SCN_2H_4$ complex in Scheme I. In most other cases of thin-film materials very little is known. This area should be investigated, because the quality of the films and the nature of the residual trapped impurities are dependent on these processes.

$$\boxed{3\langle CdCl_2SCN_2H_4\rangle}$$

$$\downarrow \sim 240°C$$

$$\boxed{\langle CdS\rangle + 2\langle CdCl_2SCN_2H_4\rangle + [CH_2N_2 2HCl]}$$

$$\downarrow$$

$$\boxed{\langle CdS\rangle + \langle Cd_2Cl_6CN_2H_4\rangle} + \tfrac{1}{4}(N_2H_8CS_3) + \tfrac{7}{6}(NH_4SCN) + \tfrac{1}{6}(NH_3) + \boxed{\tfrac{1}{12}\langle C_6H_9N_{11}HSCN\rangle}$$

$$\downarrow \sim 290°C \qquad\qquad\qquad\qquad\qquad\qquad\qquad\qquad\qquad\qquad \downarrow$$

$$\boxed{\langle CdS\rangle + 2\langle CdCl_2\rangle + \tfrac{1}{2}\langle NH_4Cl\rangle} + \tfrac{3}{2}(HCl) + \boxed{\tfrac{1}{6}\langle C_6H_9N_{11}\rangle N\tfrac{1}{12}\langle C_6H_9N_{11}\rangle} + \tfrac{1}{12}(HSCN)$$

$$\downarrow \sim 350°C \qquad\qquad\qquad\qquad\qquad\qquad \downarrow$$

$$\boxed{\langle CdS\rangle + 2\langle CdCl_2\rangle} + \tfrac{1}{2}(NH_4Cl) \qquad\qquad \boxed{\tfrac{1}{4}\langle C_6H_6N_{10}\rangle} + \tfrac{1}{4}(NH_3)$$

$$\downarrow$$

$$\boxed{\tfrac{1}{4}\langle C_6H_3N_9\rangle} + \tfrac{1}{4}(NH_3)$$

SCHEME 1. Pyrolytic decomposition scheme of the complex $CdCl_2 \cdot SCN_2H_4$ (46): $\langle\ \rangle$, solid; [ ], liquid; ( ), gas.

Sulfide and selenide films of a number of other elements (Table II), such as Zn, Cu, In, Ag, Ga, Sb, Pb, and Sn, have been obtained by using similar pyrolytic reactions. For selenide films, thiourea is replaced by selenourea

TABLE II

SPRAY PYROLYTICALLY DEPOSITED BINARY COMPOUNDS (4)

| IB VIA | IIB VIA | IIIA VIA | IIIB VIA | IVA VIA | VA VIA | VIII VIA |
|---|---|---|---|---|---|---|
| $Cu_2S$ | CdS | $In_2S_3$ | GdSe | PbS | $Sb_2S_3$ | CoSe |
| CuS | CdSe | $In_2Se_3$ | — | — | — | — |
| $Ag_2S$ | ZnS | $Ga_2S_3$ | SmS | PbSe | — | — |
| — | ZnSe | $Ga_2Se_3$ | — | — | — | — |

or other suitable selenium compounds, such as $N,N$-dimethylselenourea. The corresponding reaction is

$$CdCl_2 + (NH_2)_2CSe + 2H_2O \rightarrow CdSe\downarrow + 2NH_4Cl\uparrow + CO_2\uparrow$$

In principle, it should be possible to form telluride films using telluroorganic salts. But, these salts are extremely unstable and difficult to synthesize. However, stable inorganic tellurium compounds with tellurium in a positive oxidation state could be used to generate elemental tellurium at the pyrolysis surface (48). This tellurium will subsequently undergo a solid-state reaction with the elemental metal made available also at the pyrolysis surface through the dissociation of the inorganic metal salt. Films of CdTe have been deposited (48a) by spraying acidic solution of $(NH_4)_2TeO_4$, an inorganic tellurate, and a cadmium salt by using nitrogen as the carrier gas in a nitrogen-purged enclosure.

We have established (49) the possibility of utilizing liquid media other than water. For example, solutions of $CdCl_2$ and thiourea in different alcohols have been used to spray deposit CdS films. For this purpose, the pre-synthesized $CdCl_2 \cdot (NH_2)_2CS$ complex is precipitated (4) from its aqueous solution and then redissolved in an alcohol (ethanol, methanol, or propanol). Since the latent heat of vaporization of alcohols is considerably smaller than that of water, CdS films can be deposited at relatively much lower temperatures with alcohol solutions.

b. *Oxide Films.* An aqueous metal salt solution is sprayed onto a hot substrate in air to obtain the corresponding metal oxide films. Generally, metal chlorides such as $SnCl_4$ for $SnO_2$ (26, 49a), $InCl_3$ for $In_2O_3$ (50), $AlCl_3$ for $Al_2O_3$ (51), $FeCl_3$ for $Fe_2O_3$ (52), $CoCl_3$ for $Co_2O_3$ (53), and $ZnCl_2$ for ZnO (54) have been used. Other salts, such as $Co(CH_3COO)_3$ for $Co_2O_3$ (55); $Fe(NO_3)_3$ (52), $FeBr_3$ (52), and $Fe(CH_3COO)_3$ (52) for $Fe_2O_3$; $Mo(CO)_6$ (56) for $MoO_2$; tetraisopropyl titanate for $TiO_2$ (57); and nitrates, carbonates, and acetates of zinc for ZnO (54), have also been used.

Normally 0.01–0.1 $M$ aqueous chloride solutions are used for good optical quality $SnO_2$ films, although concentrations as high as 2.85 $M$ (58) have been used. The optimum concentration is dictated by the desired optical and electrical quality of the film, deposition rate, and chemistry of the reaction. A typical chemical reaction for $SnO_2$ films is

$$SnCl_4 + 2H_2O \rightarrow SnO_2 + 4HCl$$

The choice of the anion in the metal salt depends on the thermodynamic driving force. For example, in the case of ZnO films, nitrates, carbonates, lower aliphatic acids, and halides are in decreasing order of thermodynamic driving force (54). The heat of reaction for the anion A in the reaction

$$(ZnA)_{(aq)} + H_2O \rightarrow ZnO + 2HA$$

is 30 kcal $mol^{-1}$ for chlorides, $-0.1$ kcal $mol^{-1}$ for acetate, and $-10$ kcal $mol^{-1}$ for nitrate at room temperature. The unfavorable thermodynamics of the zinc halides suggests that even if the nonequilibrium reaction proceeds towards oxide formation, an accumulation of the anion species is expected to be incorporated in the film. This has indeed been observed and, further, the concentration has been found to decrease with increasing substrate temperature resulting in better electrical and optical quality films (54). Incorporation of unreacted anion species could also be reduced by the addition of corresponding acid in the spray solution. For example, addition of acetic acid (few drops) to zinc acetate solution clears the solution of any turbidity/precipitate occurring due to a slight lowering of pH on addition of alcohol and thus helps to obtain better quality ZnO films (54a).

Organometallic compounds have also been used for obtaining oxide films. Although relatively more expensive, these compounds offer the advantages of low decomposition temperature, so that substrate interaction with the vapor and growing film is reduced. Their high vapor pressure allows the use of a vapor-transport rather than a liquid-transport technique for pyrolytic deposition. Some of the compounds used are dibutyl tin diacetate $[(C_4H_9)_2Sn(CH_3COO)_2]$ for $SnO_2$ (59), and indium acetyl acetonate $[In(C_5H_2O_2)]$ (60, 61) and an indium chelate derived from dipivaloylmethane (62) for $In_2O_3$.

*c. Carbide and Nitride Films.* Like oxides, carbide and nitride films can be prepared by spray pyrolysis. For example, vanadium carbide ($VC_{0.84}$–$VC_{0.89}$) films on carbon steel have been obtained (63) by vapor pyrolysis of $VCl_2$ on steel at 1000°C in an atmosphere of hydrogen and argon. Carbon for the reaction is obtained on the steel surface by diffusion. We have obtained vanadium carbide films by spray pyrolysis in a controlled atmosphere.

## III. Characteristic Features of the Spray Pyrolytic Process

### 1. Growth Rate

The growth rate in this process is expected to depend on the chemical and topographical nature and temperature of the substrate, the chemical nature and concentration of the spray solution and its additives, and the spray parameters. Since commercial interest in this process has primarily been to obtain films of the desired qualities for a particular application on an empirical basis, not much systematic work has been reported in the literature. In the case of $SnO_x$ (TO) and $In_xO_y$:Sn (ITO) films, the thickness increases nearly linearly with time of spray (that is, with the amount of sprayed solution). The growth rates can be as large as 1000 Å min$^{-1}$ for oxide films and ~500 Å min$^{-1}$ for sulfide films. The marked substrate temperature dependence of the deposition rate for CdS and ZnO films under various spray conditions is shown in Fig. 5. It should be emphasized that the microstructure and hence the properties of the films are also very strongly dependent on the substrate temperature.

### 2. Substrate Effects

In general, the spray pyrolysis process affects the substrate surface. If it is not desirable for the substrate to take part in the pyrolytic reactions, the

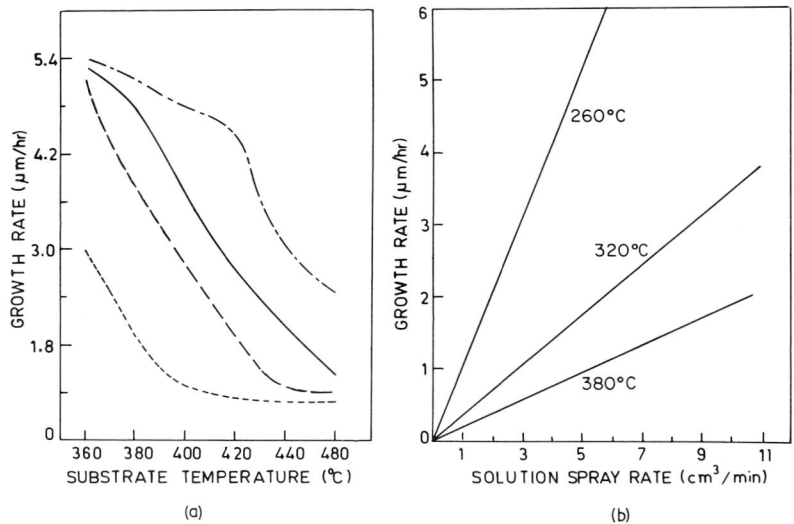

FIG. 5. Deposition rate as a function of substrate temperature and solution spray rate for (a) ZnO and (b) CdS films (54, 27).

choice of neutral substrates is limited to glass, quartz, ceramics, appropriate oxide/nitride/carbide coated substrates, Ge, Si, CdS, CdTe, etc. In case of certain oxide films on silicon, some desirable etching takes place (64) during deposition. It is generally difficult to deposit a coherent film on any metallic substrate (27).

Even if the substrate is chemically inactive, it may contain mobile alkali and other rare earth ions, such as $Li^+$, $Na^+$, $Ca^{2+}$, $Sr^{2+}$, and $Mg^{2+}$, which will be readily incorporated (50) in the film. For example, $Na^+$ and $Li^+$ ions will be incorporated substitutionally and $Ca^{2+}$ and $Sr^{2+}$ ions interstitially in TO and ITO films with consequent effects on their electrical properties. Clearly, the extent of inclusion of the ions would increase with the substrate temperature.

Since the dynamics of evaporation and pyrolytic reaction are strongly temperature-dependent processes, the substrate temperature has the most significant effect on the quality of the films. The minimum temperature at which a film can be deposited by spray is determined by the decomposition and pyrolytic reaction temperatures of the reacting salts under the existing spray conditions. Generally, slow reaction at lower temperatures would yield foggy and diffusely scattering films. Because of insufficient time for proper spreading of the droplets, fast evaporation of the solvent, rapid onset of the reactions, and partial reevaporation of the constituents, high substrate temperatures yield thinner, continuous, and hard films. Such films may have drastically different stoichiometry and impurity contents than films deposited at lower temperatures. At still higher temperatures, reevaporation of anionic species may take place, resulting in metal-rich deposits.

## 3. Film Composition

The composition of the film is expected to depend on the kinetics of the spray process and the thermodynamics of the pyrolytic processes. Under appropriate conditions, stoichiometric sulfide and selenide films and nearly stoichiometric oxide films are obtained. Although some increase of sulfur does take place with increasing concentration of sulfur ions in the spray solution, the stoichiometry of sulfide and selenide films, as determined by Auger analysis in our laboratory (65), is found to be within $\sim 1$ at. % and is independent of the metal to sulfur ion ratio between 1:1 and 1:5. Note, however, that the microstructure of the films is dependent on this ratio (see Section V).

If the pyrolytic reactions have not been completed, some by-products or intermediate compounds will be trapped as impurities in the film. In the case of chloride salts, residual chlorine is often obtained (54) in films. Its concen-

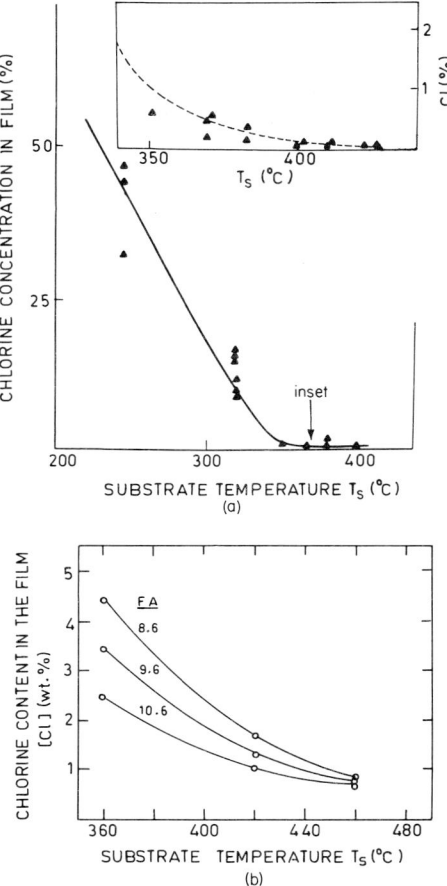

FIG. 6. Chlorine concentration in spray-deposited (a) CdS (22) and (b) ZnO (54) films.

tration in CdS and ZnO films has been shown (Fig. 6) to decrease with increasing substrate temperature during pyrolysis. Due to the cooling effect at the growing film surface (a consequence of the technique), a higher concentration of chlorine exists at the surface. Chlorine concentration has been found (22, 46) to depend sensitively on the chloride to sulfide salt ratio in the spray solution. Also, the addition of HCl to the solution has been found (65) to reduce the chlorine content in the film.

For oxide films the stoichiometry is dependent on relatively more complex reactions (50, 66–68). For example, the presence of oxygen ion vacancies ($V_0^{2+}$) in TO films is associated with the conversion of $Sn^{4+}$ to $Sn^{2+}$ species (both of which are present to form $Sn^{4+}_{(1-\delta)}Sn^{2+}_{\delta}O^{2-}_{(2-\delta)}$ according to the relations

$$O_0 \rightleftharpoons \tfrac{1}{2}O_2 + 2e^- + V_0^{2+}, \quad Sn^{4+} + 2e^- \rightleftharpoons Sn^{2+}$$

Thus, the number of oxygen-ion vacancies ($V_0^{2+}$) is equal to the number of $Sn^{4+}$ species reduced to $Sn^{2+}$ ions which form a donor band in $SnO_2$. We can substitute $\delta$ moles of oxide ions by chloride ions to give rise to $Sn^{4+}e_\delta^- O^{2-}_{(2-\delta)}Cl^-_\delta$ which is a controlled-valence semiconductor. Since chlorine has one less 2p orbital to fill than oxygen, the tin atom retains an extra 5s electron, which enters the conduction band. Presence of $Sn^{2+}$ and $Cl^-$ ions has indeed been observed in pure TO films. The conductivity of TO films is attributed to the presence of multivalency tin ions (thus oxygen deficiency) and/or chlorine ions. The deviation from stoichiometry (or oxygen vacancies) is controlled by the water and alcohol content in the spray solution. Water molecules provide oxygen and alcohol acts as a reducing agent. Use of a carrier gas (oxygen or nitrogen) does not seem to affect the concentration of incorporated oxygen vacancies. The oxygen content in films is also affected by the rate of cooling of the films after the spray is over, due principally to the adsorption of oxygen.

In general, the addition of a controlled quantity of oxygen in a reducing atmosphere, where electrons for reduction of oxygen are provided from within the system, is a necessary condition for forming oxygen-deficient metal oxide films.

## IV. Multicomponent Doping and Alloying Effects

If simultaneous pyrolysis of different salts can be effected, one should in principle obtain microscopically homogeneous multiple-component mixtures, alloys, or compounds, depending on the reaction products and their interactions. Simultaneous pyrolysis (or copyrolysis) is made possible by choosing appropriate salts and by spraying the common solution from one nozzle, or by using multiple nozzles spraying different solutions and, if necessary, with different carrier gases (as used by Deshotals *et al.* (*69, 70*) to dope CdS films with In and Ga). One may sequentially spray different solutions and thus employ sequential pyrolysis to obtain multilayer films of different materials, or films of spatially gradient composition. Since mixing of multicomponent materials takes place in micron or submicron clusters under nonequilibrium conditions, one expects the relaxation of solubility criteria for alloy and compound formation, in a way somewhat similar to vapor deposition processes.

## TABLE III
### Stannite, Adamantine, and Chalcopyrite Compounds and Alloys Prepared by Spray Pyrolysis (72)

| Formula | Sphalerite lattice constant (Å) | | Resistivity ($\Omega$ cm) | Type | Energy gap (eV) |
|---|---|---|---|---|---|
| | Measured | Theoretical | | | |
| **Stannite compounds** | | | | | |
| $Cu_2ZnSnS_4$ | 5.36 | 5.42 | 0.02 | p | — |
| $Cu_2ZnSnSe_4$ | 5.69 | 5.65 | 0.05 | p | — |
| $Cu_2CdSnS_4$ | 5.65 | 5.56 | 0.2 | p | — |
| $Cu_2CdSnSe_4$ | 5.72 | 5.78 | 1.0 | p | ~1.5 |
| **I III IV $\Box$ $VI_4$ materials** | | | | | |
| $CuGaSnS_4$ | 5.35 | — | High | — | — |
| $CuGaSnSe_4$ | 5.62 | 5.60 | 20 | p | — |
| $CuInSnS_4$ | 5.53 | Spinel | 200 | p | 1.1 |
| $CuInSnSe_4$ | 5.69 | 5.67 | High | p | 0.9 |
| **Solid solutions based on I III $VI_2$–$III_2$ $\Box$ $VI_3$** | | | | | |
| $CuIn_5Se_8$ | 5.8 | — | 200 | p | 1.3 |
| $CuIn_5S_4Se_4$ | 5.7 | — | High | — | 1.3 |
| $CuGa_3S_5$ | Amorphous | — | 0.5 | p | 1.8 |
| $CuGa_5S_8$ | Amorphous | — | 0.3 | p | 2.0 |
| $CuGa_3Se_5$ | 5.58 | — | 10 | p | — |
| $CuGa_5Se_8$ | 5.60 | — | High | — | 1.2 |
| $CuGa_5S_4Se_4$ | 5.58 | — | High | — | 1.6 |
| $CuGa_{2.5}In_{2.5}Se_8$ | 5.76 | — | High | — | 1.0 |
| $CuGa_{2.5}In_{2.5}S_4Se_4$ | 5.74 | — | High | — | 1.4 |
| **Chalcopyrite compounds with alloys** | | | | | |
| $CuInS_2$ | 5.51 | 5.51 | 0.01 to very high | n or p | 1.3 |
| $CuInSe_2$ | 5.79 | 5.79 | 0.01 to very high | n or p | 0.9 |
| $CuGaS_2$ | 5.33 | 5.32 | 0.01 | p | 2.1 |
| $CuGaSe_2$ | 5.60 | 5.58 | 0.1 to 100 | p | 1.5 |
| $CuIn(S_{0.5}Se_{1.5})$ | 5.70 | — | 0.01 to very high | p | 1.0 |
| $CuIn(SSe)$ | 5.66 | — | 0.3 | p | 1.2 |
| $CuIn(S_{1.5}Se_{0.5})$ | 5.61 | — | 4 to very high | n and p | 1.3 |
| $Cu(Ga_{0.5}In_{0.5})S_2$ | 5.50 | — | 0.02 | p | 1.4 |
| $Cu(Ga_{0.5}In_{0.5})Se_2$ | 5.73 | — | 20 | p | 1.1 |
| $Cu(Ga_{0.75}In_{0.25})Se_2$ | 5.70 | — | 0.1 | p | 1.35 |
| $Cu(Ga_{0.25}In_{0.75})Se_2$ | 5.76 | — | Very high | — | 1.0 |
| $Cu(Ga_{0.5}In_{0.5})(SSe)$ | 5.68 | — | 20 | p | 1.2 |
| $Cu(Ga_{0.25}In_{0.75})(S_{1.5}Se_{0.5})$ | 5.59 | — | 0.1 | p | — |
| $Cu(Ga_{0.25}In_{0.75})(S_{0.5}Se_{1.5})$ | 5.71 | — | 20 | p | 1.0 |
| $Cu(Ga_{0.75}In_{0.25})(S_{0.5}Se_{1.5})$ | 5.70 | — | 0.04 | p | 1.1 |
| $Cu(Ga_{0.75}In_{0.25})(S_{1.5}Se_{0.5})$ | 5.56 | — | 0.1 | p | — |

Copyrolysis has been successfully utilized by a number of workers (*4, 16, 18, 25, 27, 28*) to dope sulfide and selenide films of Cd, Zn, and Pb. The versatility of the technique in yielding ternary, quaternary, and quinary compounds has been demonstrated by Pamplin and Feigelson (*71, 72*). These authors have deposited a large variety of stannite compounds, adamantine materials (e.g., I III IV $\Box$ VI$_4$; I III VI$_2$–III$_2$ $\Box$ VI$_3$), and chalcopyrites, which are listed in Table III.

It is important to note that since it is not possible to have exactly the same thermodynamic parameters for different pyrolytic reactions, the composition of the film is not simply related to the composition of the spray solution. Indeed, the correlation can only be established empirically for each system. In some systems, such as mixed chlorides of Cd and Zn, for example, single-phase Cd$_x$Zn$_{1-x}$S films having the same composition as that of the solution have been prepared (*27*) over the whole composition range by choosing an appropriate substrate temperature for each composition. Deviations from these conditions result in Zn- or Cd-rich films. If one of the elements has a low vapor pressure or forms a volatile gaseous product, its incorporation into the film is very limited. This is the case when SnO$_x$ films are doped with fluorine. Typically, a F:Sn atomic ratio of 0.38 in solution yields films containing less than a few percent of F (*73*). On the other hand, in Sb-doped TO and Sn-doped IO films, the ratios Sb:Sn and Sn:In are the same as in the solution. The nature of the substrate has, however, been shown to affect these ratios. As shown in Fig. 7, the atomic ratio Sn:In in ITO films deposited at 500°C on Si is the same (*50*) as that in the spray solution. But, this ratio is significantly different for films deposited on borosilicate glass (*50*), presumably because of contributions of impurities such as B, Na, Al, K, Si, and the oxides from the glass surface.

Copyrolysis has been utilized very effectively to prepare doped and multicomponent oxide films (see Table IV) of Sn, In, Cd, Zn, V, Ti, and Co. Tin oxide films have been doped with cationic impurities of Sb (*26*), In (*74*),

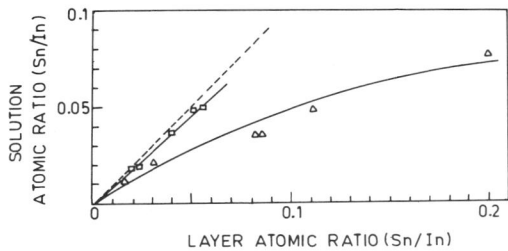

FIG. 7. Dependence of the Sn:In atomic ratio in ITO films on the spray solution composition (*50*): ($\triangle$) substrate pyrex; ($\Box$) substrate silicon.

TABLE IV

Physical Parameters of Various Undoped, Doped, and Multicomponent Oxide Films

| Oxide | Dopant | Concn. (mol %) | Sheet resistivity ($\Omega/\square$) | Carrier concn. ($cm^{-3}$) | Transmission (%) | Plasma edge $\lambda_p$ ($\mu m$) | Reflectance (%) in IR ($\lambda = 5\ \mu m$) | Reference |
|---|---|---|---|---|---|---|---|---|
| $SnO_x$ | — | | 1,000 | $8 \times 10^{18}$ | 80 (at 2.5 $\mu m$) | — | — | (66) |
| | Sb | 2.5 | 100 | $2 \times 10^{20}$ | 5 (at 2.5 $\mu m$) | — | — | (66) |
| | Sb | 10 | 600 | $1.5 \times 10^{19}$ | 16 (at 2.5 $\mu m$) | — | — | (66) |
| | Sb | 40 | 25,000 | — | 55 (at 2.5 $\mu m$) | — | — | (66) |
| | P | 6 | 300 | $2 \times 10^{20}$ | 75 (at 2.5 $\mu m$) | — | — | (66) |
| | P | 40 | 10,000 | $1 \times 10^{17}$ | 90 (at 2.5 $\mu m$) | — | — | (66) |
| | Tl | 1 | 3,000 | — | 85 (at 2.5 $\mu m$) | — | — | (66) |
| | In | 2 | $10^7$ | — | 85 (at 2.5 $\mu m$) | — | — | (66) |

| | | | | | | | |
|---|---|---|---|---|---|---|---|
| Sb | 1.4 | 60 | $2 \times 10^{20}$ | 87–96 (0.35–1.1 μm) | 2.7 | 66 | (30) |
| Sb | 3 | 40 | $5.7 \times 10^{20}$ | 80–90 (0.35–0.7 μm) | 2.1 | 80 | (30) |
| Sb | 10 | 275 | $7.9 \times 10^{20}$ | 25–50 (0.35–1.0 μm) | 1.44 | 55 | (30) |
| F | 10 | 19 | $2.5 \times 10^{20}$ | 70–87 (0.35–1.4 μm) | 2.4 | 55 | (73) |
| F | 30 | 13 | $3.7 \times 10^{20}$ | 76–86 (0.35–1.4 μm) | 1.9 | 66 | (73) |
| F | 65 | 9 | $4.8 \times 10^{20}$ | 72–84 (0.35–1.1 μm) | 1.7 | 90 | (73) |
| Sb + F | 1.4 + 0.7 | — | $3 \times 10^{20}$ | 78–86 (0.4–1.3 μm) | 1.9 | 70 | (49) |
| Sb + F | 1.4 + 10 | — | $3.9 \times 10^{20}$ | 70–86 (0.4–1.2 μm) | — | 85 | (49) |
| F + Sb | 65 + 0.7 | — | $4.2 \times 10^{20}$ | 54–76 (0.4–1.2 μm) | 1.8 | 85 | (49) |
| F + Sb | 65 + 1.4 | — | $4.8 \times 10^{20}$ | 58–77 (0.4–1.2 μm) | — | 88 | (49) |
| $In_xO_y$ | Sn | 4 at. % | 19 | $7.4 \times 10^{20}$ | 85–95 (0.4–1.5 μm) | — | 80 | (50) |
| $ZnO_x$:In | In | 2 at. % | 20 | $\leq 10^{21}$ | 80–85 (0.5–1.1 μm) | — | — | (54a) |

TABLE V
Some Commonly Used Spraying Solution Compositions

| Oxide | Salt for Matrix | Dopant | Carrier solution | Additive | References |
|---|---|---|---|---|---|
| $SnO_x$:Sb | $SnCl_4 \cdot 5H_2O$ | $SbCl_3$, 0.135 g (1.4 mol %) dissolved in conc. HCl | Isopropyl alcohol, 150 cm$^3$ | — | (26, 30) |
| $SnO_x$:F | $SnCl_4 \cdot 5H_2O$, 5 cm$^3$ of 2.85 M | $NH_4F$, 1.0 g (65.4 mol %) | Isopropyl alcohol, 5 cm$^3$ | — | (49, 73) |
| $In_xO_y$:Sn | $InCl_3$ 8.2 g | $SnCl_4 \cdot 5H_2O$ (2 at. %) 0.25 g | $H_2O$ and $C_2H_5OH$, 42 g each | 7.5 g HCl | (50) |
| $CdSnO_3$ | $CdCl_2 \cdot 5H_2O$, 100 g of 1.67 M and $SnCl_4 \cdot 5H_2O$, 10 g of 1 M (aq. solution) | — | $H_2O$ (as existing in $CdCl_2$ and $SnCl_4$ solutions) | $InCl_3$, 3 g of 1 M, 10 g HCl (conc.) | (93) |
| $ZnO_x$ | $ZnCl_2$, 100 cm$^3$ of 0.1 M | — | $H_2O$ (as existing in $ZnCl_2$ aq. solution) | 1.2 cm$^3$ of $H_2O_2$ | (54) |
| $ZnO_x$:In | $Zn(CH_3COO)_2$ 25 cm$^3$ of 0.5 M | $InCl_3$ 0.4 cm$^3$ of 0.5 M | Isopropyl alcohol 75 cm$^3$ | Acetic acid (few drops) | (54a) |

Cd (75), Bi (76), Mo (78), B (77), P (78, 79), Te (66), and W (80), anionic impurities of F (73) and Cl, and mixed F–Sb (81). Similarly, indium oxide films have been doped with Sn (50, 82), Ti (82), Sb (82), F (83, 84), and Cl, and mixed impurities of Sn and F (85). Zinc oxide films have been doped with indium (54a).

Some commonly used spraying solution compositions to yield high optical transmission and low ohm/□ for TO and ITO films are listed in Table V. Note that the dopants used (group V for tin oxide and group III for indium oxide) increase the conductivity of these so-called controlled valence semiconductors by regulating the oxygen vacancy concentration. For example, $Sb^{3+}$ ions are substituted for the fraction $\delta$ of $Sn^{4+}$ sites in $SnO_2$, by mixing $\delta/2$ moles of $Sb_2O_3$ and $(1 - \delta)$ moles of $SnO_2$ to yield

$$Sn^{4+}_{(1-\delta)}Sb^{3+}_{\delta}O^{2-}_{(2-\delta/2)}$$

Addition of $\delta/4$ moles of oxygen in a reducing atmosphere will give a controlled valence semiconductor according to the reaction

$$Sn^{4+}_{(1-\delta)}Sb^{3+}_{\delta}O^{2-}_{(2-\delta/2)} + \delta/4\ O_2 \rightarrow Sn^{4+}_{(1-\delta)}Sb^{4+}_{\delta}O^{2-}_{2} \rightarrow Sn^{4+}_{(1-\delta)}e^{-}_{\delta}Sb^{5+}_{\delta}O^{2-}_{2}$$

Thus, $\delta$ moles of electrons are donated to the conduction band. If complete oxidation of $Sb^{3+}$ to $Sb^{5+}$ takes place, then

$$Sn^{4+}_{(1-\delta)}Sb^{3+}_{\delta}O^{2-}_{(2-\delta/2)} + \delta/2\ O_2 \rightarrow Sn^{4+}_{(1-\delta)}Sb^{5+}_{\delta}O^{2-}_{(2-\delta/2)}$$

which is an insulator. Therefore, reduction of oxygen is essential to get high conductivity in these films. Consequently, alcohol, a small quantity of concentrated HCl, and 1.2% $H_2O_2$ (for ZnO films) are known (54) to decrease the sheet resistivity of the films.

Copyrolysis can also be used to obtain a heterogeneous mixture of oxide and sulfide films, by choosing spray solutions that yield sulfides and hydroxides/oxides. For example, cosprayed solutions of $CdCl_2$, $AlCl_3$, and thiourea yield CdS films with segregated $Al_2O_3$ formed (86–88) at the grain boundaries. This technique can obviously be extended to yield a microscopic mixture of a number of semiconducting chalcogenides and metal oxides. Thus, ZnO-based ceramic semiconductors for varistor applications can be easily prepared.

## V. Structural Properties

Sprayed films are invariably polycrystalline, with a small grain size. The microstructure must depend on the spray parameters, which determine the surface mobility of sprayed droplets, the formation of disk-like clusters,

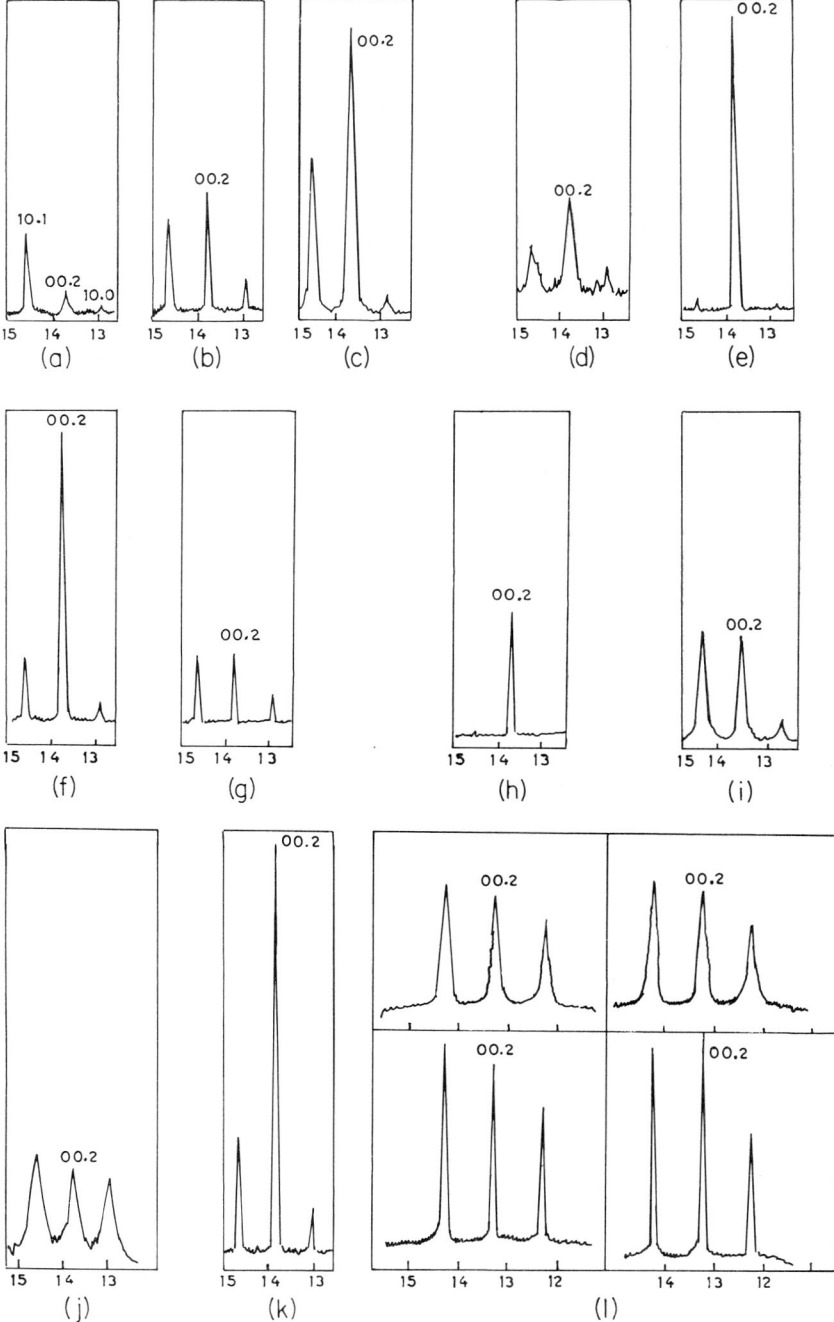

and the subsequent sintering and recrystallization processes involving overlapping disks. Significant recrystallization is not possible due to the involvement of various pyrolytic reaction products at relatively low pyrolytic decomposition temperatures. The situation is very similar to that in the codeposition of vapors of multicomponent materials (89). One expects a microstructure ranging from amorphous to micropolycrystalline, depending on the droplet mobilities and chemical reactivities of various constituents. Typical grain sizes of sulfide and selenide films range from 0.2 to 0.5 $\mu$m (27, 28), and those of oxide films range from 0.1 to 0.2 $\mu$m (30).

As a result of the growth process and consequent microstructure, sprayed films are generally strongly adherent, mechanically hard, pinhole free, and stable with time and temperature (up to the spray temperature). Clearly, if the deposition conditions are not optimum, one can get powdery, nonadherent, and low-density films. Postdeposition annealing of films generally affects the oxygen-dominated electrical properties significantly but not the microstructure (29). At annealing temperatures above the spray temperature, or in some reactive environments, recrystallization increases the grain size and may produce some preferential orientation effects (29).

Although the parameter that dominates the microstructure of sprayed films is substrate temperature during spray pyrolysis, other factors, such as type of salts, ratio of cations to anions, and dopants, also affect the grain size and any orientation effects. As an example, Fig. 8 shows the effect of Cd:S ratio, substrate temperature, film thickness, multilayers, and postdeposition annealing on the $c$-axis orientation of CdS films. If deposited

FIG. 8. Effects of Cd:S ratio, substrate temperature, film thickness, impurities, multilayer deposition, and postdeposition annealing on the crystal ($c$-axis) orientation of spray-deposited CdS films (14, 22, 88):

| Fig. part | Cd:S | $T_s$ (°C) | $t$ ($\mu$m) | Fig. part | Cd:S | $T_s$ (°C) | $t$ ($\mu$m) |
|---|---|---|---|---|---|---|---|
| a | 1/2 | 340 |  | g |  | 380 | 1.8 |
| b | 2 | 340 |  | h | 1 | 400 | 6 |
| c | 1 | 340 |  | i[a] | 1 | 400 | 6 |
| d | 1 | 300 |  | j[a] |  | 380 | 4.3($t_1$) |
| e | 1 | 420 |  | k[b] |  | 380 | 9($t_{total}$) |
| f |  | 380 | 4.3 | l[c] | 1/2 |  |  |

[a] Dopant, Al, 5%.
[b] $t_1$ + pure CdS.
[c] Cd:S = 1/2: (1) pure, unannealed; (2) Zn doped, unannealed; (3) pure annealed; (4) Zn doped, annealed.

from acetate solution, CdS films have a very small grain size (4) (amorphous-like). On the other hand, chloride solutions yield (4, 27) larger grains, as well as a $c$-axis orientation perpendicular to the substrate. Impurities such as In and Ag facilitate (69, 70) recrystallization, whereas insoluble impurities such as $Al_2O_3$ inhibit it (86–88) and thus reduce the grain size of CdS films drastically and also destroy the preferred $c$-axis orientation. The segregated $Al_2O_3$ at the grain boundaries produces a characteristic sepentine structure on the surface of CdS films. The surface topographic features of pure CdS and $Al_2O_3$ and of composite $CdS:Al_2O_3$ films are shown in Fig. 9.

The crystal structure of the sprayed films is generally the same as that of the corresponding bulk material. This may not be the case, however, when the material exhibits polymorphism, or when different components are formed during pyrolysis. Thus, the existence of sphalerite and wurtzite phases (14) in CdS films deposited at low ($\lesssim 400°C$) and high ($>400°C$) temperatures, respectively, is understandable in much the same way as the presence of metastable structures in vapor-deposited films. With increasing interest in these studies, one should expect to see a variety of new metastable structures frozen in the sprayed films. For example, presence of SnO in $SnO_x$ films has been reported by Manifacier et al. (50). Also, an abundance of $Sn^+$ and $O^{2-}$ and small amounts of $Sn^{2+}$ and $Cl^-$ ions have been reported in $SnO_x$ films by various workers (66, 90, 91), using photoelectron spectroscopy measurements.

Doped TO and ITO films exhibit (30, 50) rutile (tetragonal) and cubic (bcc) structures, respectively, which correspond to the bulk $SnO_2$ and $In_2O_3$ structures. Clearly, the dopants are incorporated substitutionally. Dopants do, however, affect the recrystallization process. Antimony in TO, for example, increases the grain size, but Tl and P decrease it (66). With increasing grain size, preferred orientation effects are observed. Addition of Sn to IO yields some preferred [111] orientation (50). A strong $c$-axis orientation having an angular distribution sharply centered around $90°$ to the substrate plane has been observed in hcp ZnO films (54).

As expected, the sprayed films have in general a rough surface topography. The roughness depends on spray conditions and particularly on the substrate temperature. At low deposition temperatures, the film surface has submicron- to micron-size particles scattered on a smooth background. Under optimum conditions of deposition, the rms roughness of CdS and TO films deposited in our laboratory is $\sim 1$ $\mu$m in a total thickness of 10 $\mu$m and $\sim 0.1$ $\mu$m in a thickness of 0.5 $\mu$m, respectively. The roughness is readily seen in its effect on the optical quality of the films. Addition of impurities, such as Sb in TO, decreases (30) the roughness up to 3 mol% of Sb. With increasing concentration, the films become coarser again. Figure 10 shows the surface topography of some of the films.

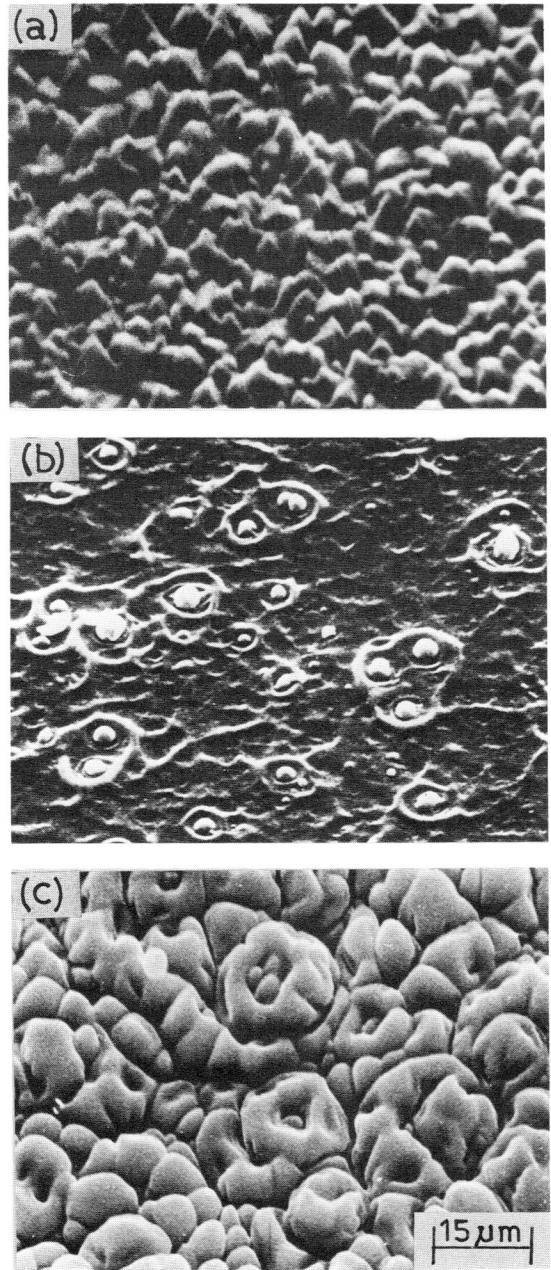

FIG. 9. SEM micrographs of (a) pure CdS, (b) pure $Al_2O_3$, and (c) composite CdS:$Al_2O_3$ films (88).

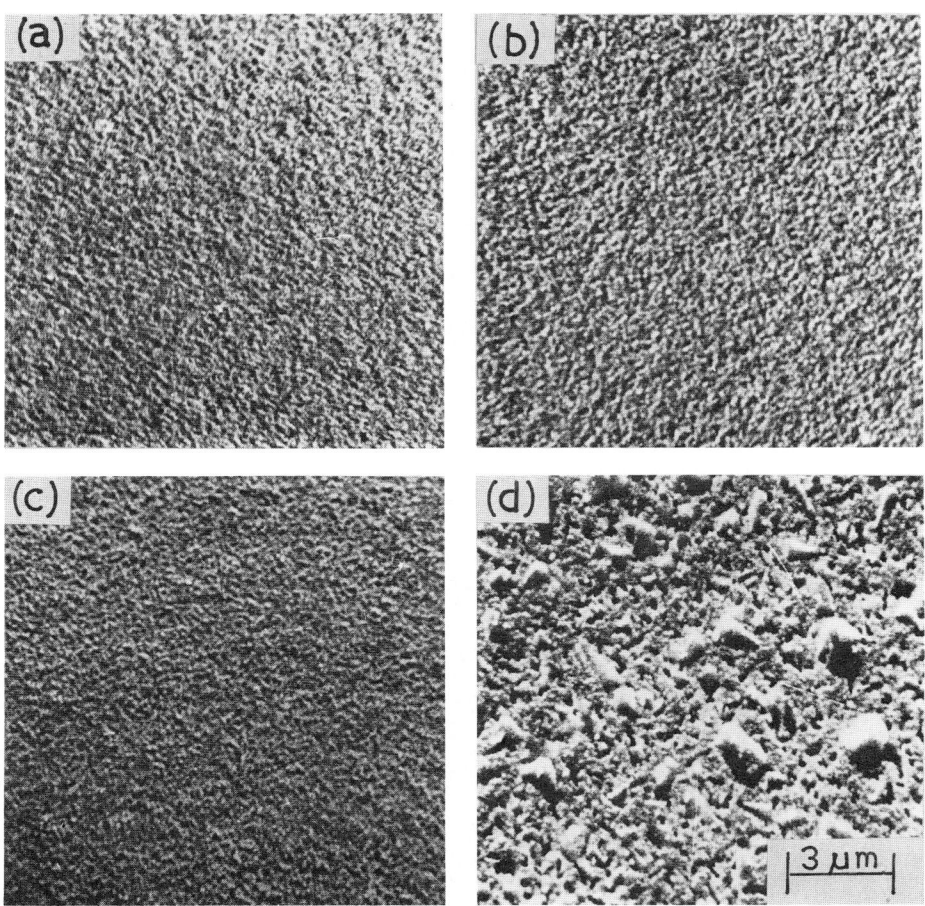

Fig. 10. Surface topography of TO films for various Sb concentrations: (a) 0.7 mol %, (b) 1.4 mol %, (c) 3.0 mol %, and (d) 10.0 mol % (*30*).

## VI. Electrical and Optical Properties

Like multiple-component films prepared by other techniques, the properties of sprayed films are sensitively dependent on deposition conditions. Even if the deposition conditions are given, one can at best have a very qualitative and empirical idea of the expected physical properties. The properties of each system must therefore be studied.

Since electrical and optical properties are of major interest in large-area sprayed films, we describe them for the most extensively studied films of CdS, TO, and ITO.

As expected, the band gap and the fundamental optical absorption edge of sprayed chalcogenide films are not affected (27) by the microstructure. The diffuse scattering, and hence the transmittance, depends on film thickness as well as substrate temperature, as illustrated for CdS films in Figs. 11 and 12. Whereas diffuse scattering increases with film thickness, it is reduced with increasing deposition temperature (due to increased grain size and orientation effects). At very high temperatures (above ~500°C), the films become rough and translucent, presumably because of major changes in the growth kinetics of the films.

The transmittance of single-phase, ternary, and quarternary solid-solution films is generally poor (72) due to considerable diffuse scattering in the fine-grained microstructure. The absorption gap or optical band gap depends on the composition, as shown in Figs. 13 and 14 for variable-composition $Cd_{1-x}Pb_xS$ and $Cd_{1-x}Zn_xS$ films, respectively.

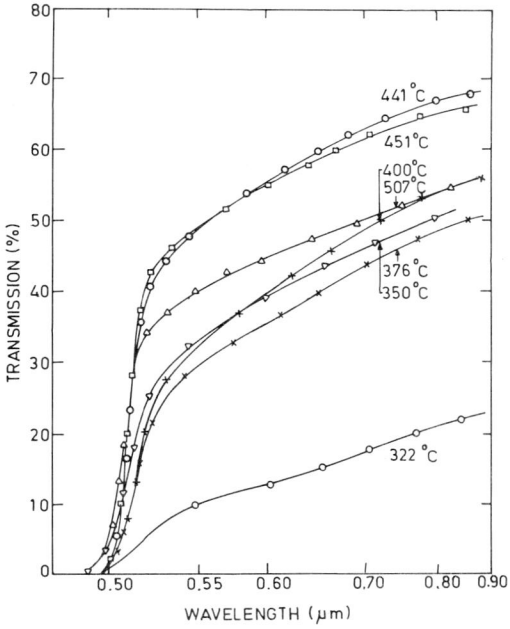

FIG. 11. Optical transmission of CdS films as a function of substrate temperature during spray (14).

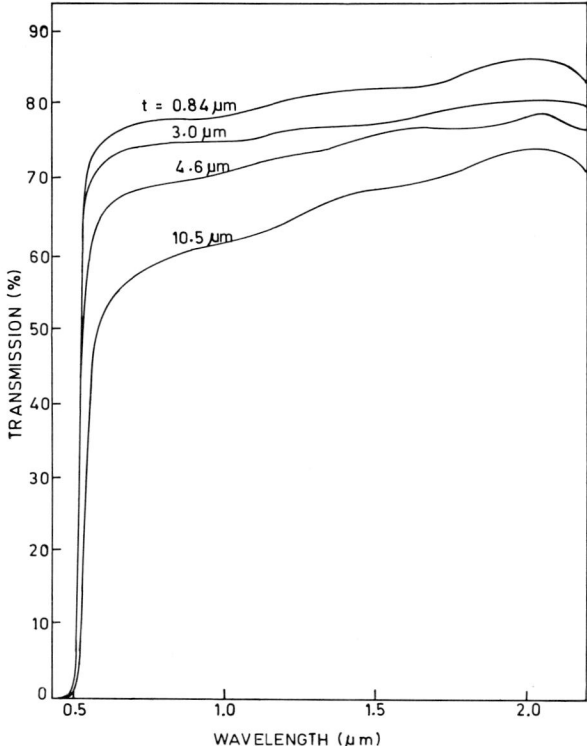

FIG. 12. Transmission spectra of spray-deposited CdS films as a function of thickness (22); $T_s = 320°C$.

Spray-deposited sulfide and selenide films of Cd and Zn are invariably n type due to the sulfur/selenium vacancies. The resistivity ($\rho$) of the films depends on the deposition conditions and can be varied by as much as a factor of $10^8$ (23, 27). Detailed electron transport measurements on CdS films have been reported by several workers (9, 11, 13, 14, 27, 92). It should be noted that the oscillatory dependence of dark and light values of electrical conductivity, carrier concentration ($n$), and mobility ($\mu$) on the deposition temperature as reported by Ma and Bube (14) is of doubtful nature.

Postdeposition annealing in air increases the resistivity of CdS films to $\sim 10^7$ $\Omega$ cm and makes them highly photoconducting. A photoconductivity gain of $\sim 10^6$–$10^7$ with response time of $\sim 1$ msec under $\sim 50$ mW cm$^{-2}$ illuminated from a tungsten lamp has been reported by Gogna et al. (23). The high resistivity and photoconductivity are due to the chemisorbed oxygen (27, 92) at the grain boundaries, which reduces both $n$ and $\mu$. On

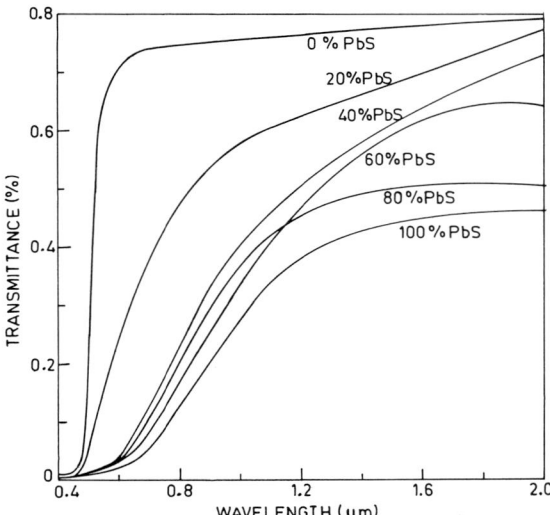

FIG. 13. Transmission edge of variable-composition $Cd_{1-x}Pb_xS$ films in the Pb concentration range $0 \leq x \leq 1$.

vacuum annealing, the resistivity is decreased to $1–10\ \Omega$ cm and the photoconductivity is quenched, indicating the reversibility of chemisorption and desorption in oxygen processes (92). The dependence of $\rho$, $\mu$, and $n$ of as-deposited films on annealing temperature is shown in Fig. 15. The reduction in $n$ is due to the increase in the density of oxygen states, whereas the

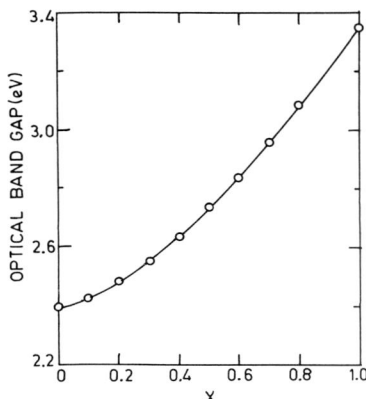

FIG. 14. Band-gap variation of $Cd_{1-x}Zn_xS$ ($0 \leq x \leq 1$) films with Zn composition ($x$) (24).

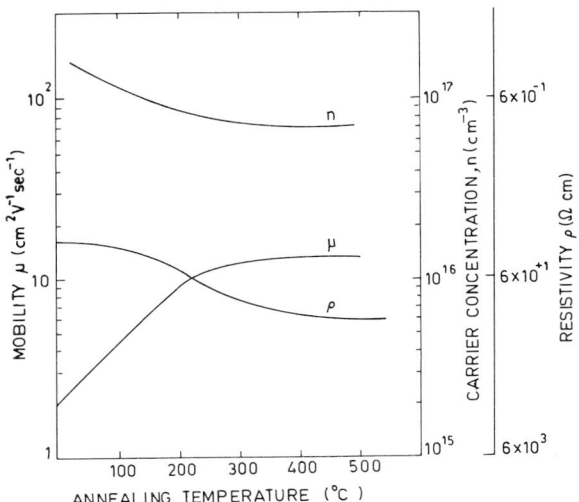

Fig. 15. The dependence of resistivity, mobility, and carrier concentration of CdS films on annealing temperature (22).

increase in $\mu$ is due to the grain growth and reduction of defects, which reduce scattering (29). The chemisorption kinetics are also affected by the rate of cooling of an as-deposited film. The consequent changes in the electron transport properties of CdS films have been observed by Ma and Bube (14). A slow cooling rate has been found to be equivalent to an air-bake, resulting in additional chemisorption of oxygen.

It should be pointed out that transport measurements in high-resistivity films are very difficult. Photo-Hall effect and photothermoelectric power measurements by a few investigators (12, 27) have shown that on illumination $n$ or $\mu$, or both, change. Which parameter changes more depends on the relative roles of the microstructure (grain size) and postdeposition treatment (chemisorbed oxygen) in the electrical conduction processes. It is, therefore, clear that the photoconductivity mechanism is rather complicated, although definitely related to chemisorbed oxygen and microstructure.

On addition of Zn to CdS films (94), the transport parameters change, as shown in Fig. 16. The photoconductive gain decreases rapidly with increasing zinc concentration. The observation has been attributed (24, 27) to the decrease in the number of acceptor-like oxygen states with increasing Zn concentration, for which no concrete evidence is available.

Sprayed oxide films of tin, indium, cadmium stannate, and zinc are transparent and conducting (n type, degenerate). Both transparency and sheet resistance are critically dependent on deposition conditions; in par-

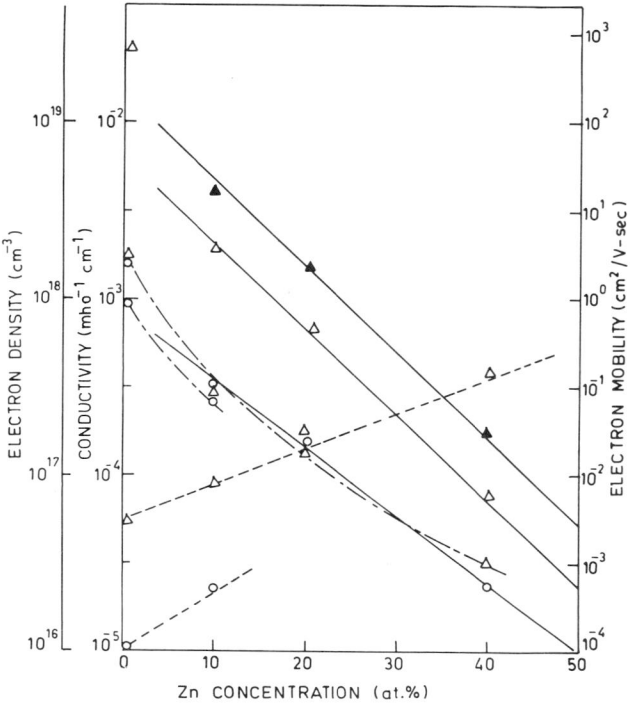

FIG. 16. Electron transport parameters $\sigma$ (—), $n$ (---), and $\mu$ (-·-) of $Cd_{1-x}Zn_xS$ films under various conditions (27, 94): (○) dark; (△) light; (▲) after heat treatment, 250°C, 3 hr forming gas.

ticular, substrate temperature, spray solutions, dopants, and postdeposition treatments. The effect of the deposition temperatures on $\Omega/\square$, $n$, $\mu$, and $T$ (transmissivity) of TO films is shown in Fig. 17. Under optimum conditions, 100–200 $\Omega/\square$ sheet resistance and 80–85% transmission in the visible region are obtained. The addition of Sb and F impurities decreases the resistivity sharply, producing a minimum at 3 mol % Sb concentration (26) as shown in Fig. 18. The addition of In, an acceptor impurity, to TO increases the resistivity but improves the transmittance in the visible (26). Doping of IO films with Sn (ITO) yields results (50) similar to those with $SnO_x:Sb$ and $SnO_x:F$ (73), and $CdSnO_3$ and $Cd_2SnO_4$ (93) films show 1–2 $\Omega/\square$ sheet resistance and $\sim 92\%$ peak transmission in the visible region. Results on indium doped ZO films show a sheet resistance of 10–20 $\Omega/\square$ and peak transmission of 85% in the visible region. The onset of transmission in these films occurs at 0.38 $\mu$m and a sharp decrease in $T$ is observed beyond 1.1 $\mu$m.

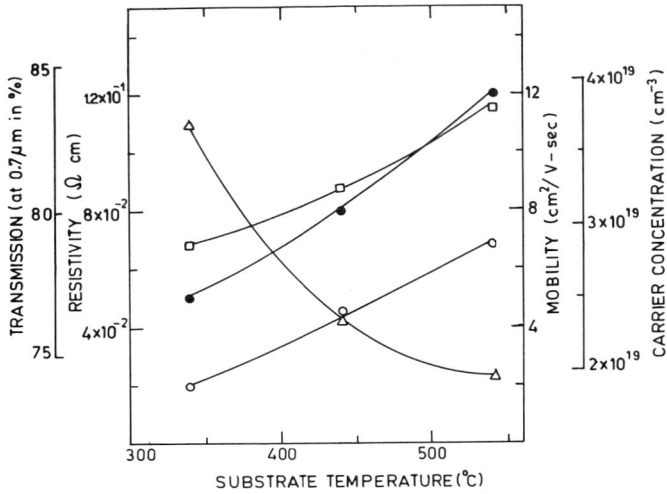

Fig. 17. The effect of the deposition temperature on resistivity (△), electron density (□), mobility (○), and transmission (●) at 0.7 μm of TO films (30, 49).

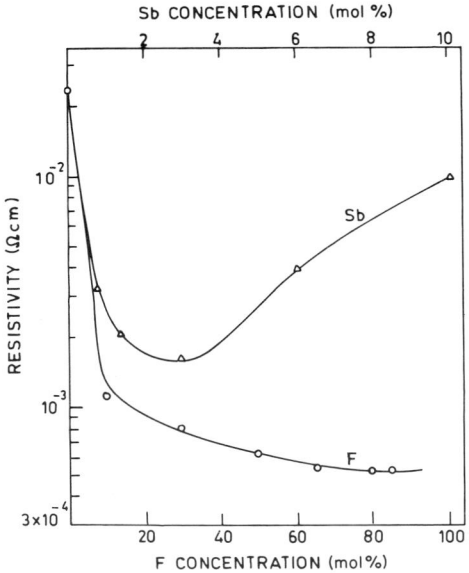

Fig. 18. Resistivity dependence on impurity (Sb, F) concentration in TO films (30, 73).

Because of large carrier concentration, the transparent oxide films exhibit a plasma resonance edge in the near IR, followed by a high IR reflectivity, as seen in Fig. 19. Since the position of the edge and the transmittance in the visible and near-IR regions are determined by the carrier concentration and the effective mass of the carriers, a strong dependence of plasma wavelength on the impurity concentration is observed (*30, 49*).

Although the electrical and optical properties of oxygen-deficient oxide films have been studied by numerous workers, only a few systematic studies of both the optical and transport properties on the same films have been undertaken. Studies (*29, 30, 49*) on TO films show that these are degenerate n-type semi-conductors with complex energy-band diagrams. The addition of donor impurities drastically shifts the Fermi level in TO films, and there is a corre-spondingly large change in the effective mass of the carriers and a Burstein–Moss shift.

Postdeposition annealing of undoped TO and IO films and ITO films for 1 hr in air at 400°C produces an irreversible decrease ($\sim 20\%$) in sheet resistance (*92*). However, $SnO_x$:F (*73, 92*) films show no such behavior. On the other hand, Shanthi *et al.* (*29*) have demonstrated that the resistivity of $SnO_x$ and $SnO_x$:Sb films increases on annealing in oxygen, and that this

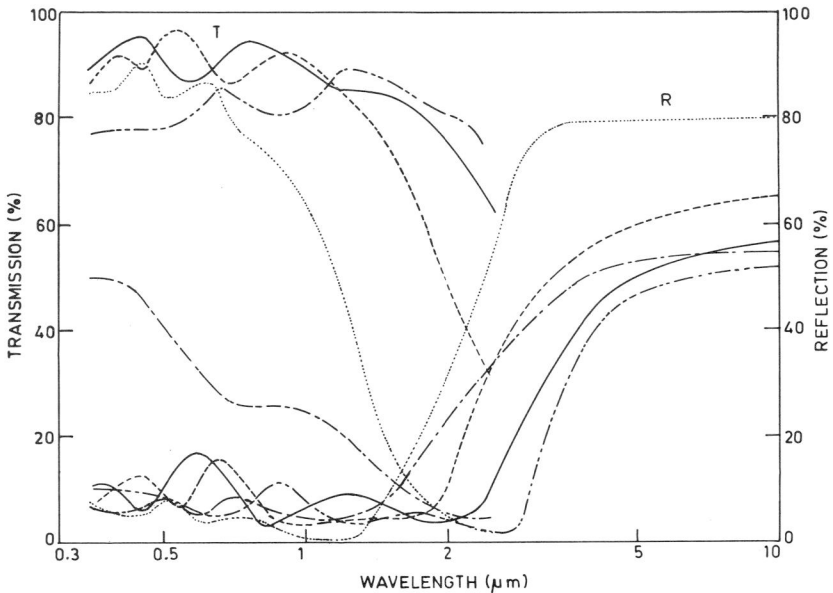

FIG. 19. Reflection and transmission spectra of antimony-doped TO films as a function of Sb concentration (*30*): (- - - -) 0, (——) 0.7, (- - -) 1.4, (· · ·) 3.0, (– – –) 10.0 mol %.

TABLE VI

SOME USEFUL PROPERTIES OF COMMONLY USED TRANSPARENT CONDUCTING OXIDES

| Oxide | Thickness (Å) | Sheet resistance ($\Omega/\square$) | Resistivity ($\Omega$ cm) | Mobility ($cm^2 V^{-1} sec^{-1}$) | Carrier concentration ($cm^{-3}$) | Transmission (%) ($\lambda$ range) | Optical gap (eV) | Refractive index | Ref. |
|---|---|---|---|---|---|---|---|---|---|
| $SnO_x$ | 3500 | 700 | $2.4 \times 10^{-2}$ | 6.4 | $3.6 \times 10^{19}$ | 78–89 (0.35–2.0 $\mu$m) | 4.0 | 2.0 | (30) |
| $SnO_x$:Sb (1.4 mol %) | 5000 | 20 | $10^{-3}$ | 15 | $4 \times 10^{20}$ | 75–87 (0.4–1.0 $\mu$m) | 4.3 | 1.96 | (30) |
| $SnO_x$:F (65 mol %) | 6000 | 5 | $6 \times 10^{-4}$ | 23 | $6 \times 10^{20}$ | 82–92 (0.4–1.2 $\mu$m) | 4.6 | 1.84 | (73) |
| $In_xO_y$ | 5000 | 250 | — | 30 | $4 \times 10^{19}$ | — | 3.5 | — | (50) |
| $In_xO_y$:Sn (2 at. %) | 5000 | 10 | — | 16 | $7.4 \times 10^{20}$ | 85–90 (0.4–1.5 $\mu$m) | 3.85 | 1.9–1.97 | (50) |
| $CdSnO_3$ | — | 14–20 | $10^{-3}$ | — | — | 75–80 (0.5–0.9 $\mu$m) | — | — | (93) |
| $Cd_2SnO_4$ | — | 120 | $10^{-3}$ | — | — | 80–90 (0.4–1.6 $\mu$m) | — | — | (93) |
| $ZnO_x$ ($H_2$ treated) | — | — | $10^{-3}$ | 10 | — | 88–96 (0.4–0.8 $\mu$m) | 3.3 | — | (54) |
| $ZnO_x$:In (2 at. %) | 6000 | 20 | $\lesssim 10^{-3}$ | ~10 | $\gtrsim 10^{21}$ | 80–85 (0.5–1.1 $\mu$m) | — | — | (54a) |

increase can be reversed by subsequent annealing in vacuum, hydrogen, nitrogen, and argon. The annealing effects are obviously determined by the oxygen chemisorption and desorption processes, which presumably have different temperature dependences under different environmental conditions. This conclusion is further supported by the observed decrease (54), by three orders of magnitude, in the resistivity of ZnO films, on annealing in hydrogen at 350°C for 5 min. These changes are largely in carrier mobility, which suggests the important role of surface traps due to chemisorbed oxygen in grain boundaries.

The chlorine incorporated in as-deposited oxide films obtained by spray pyrolysis of corresponding chlorides acts as a donor impurity and thus helps improve the conductivity. But after heat treatment, for example in ZnO films (54), the chlorine's effects as a donor are negligible compared to the increased electron density due to desorbed oxygen, and only its harmful effects, in terms of forming features that limit the transmission by acting as scattering centers, remain. This argument finds support from the observed (54) decrease in the as-deposited conductivity of ZnO films in which the concentration of electrically active Cl is reduced by addition of $H_2O_2$ to the spraying solution. A corresponding increase in transmission of such films has also been observed.

Some of the useful properties of doped and undoped oxide films are listed in Table VI. It should be pointed out that of all the oxide films, TO films are the least attacked and etched by any acid and are thus best suited for many applications.

## VII. Solution Growth Process

As already pointed out in the introduction to this review, the pyrolytic reaction resulting in chalcogenide films can also be effected with the help of $OH^-$ ions in the same chemical solutions. First developed during World War II to deposit PbS films, this technique, called the solution growth technique, has been the subject of pioneering work by G. A. Kitaev et al. (34–39) at the Ural Polytechnic Institute, USSR. D. E. Bode et al. (31–33) of the Santa Barbara Research Center optimized conditions for the deposition of PbS and PbSe films for IR detection. The understanding of the basic processes of growth of a thin film by this technique and the extension of the technique to the preparation of variable composition multicomponent films have resulted from extensive work in the authors' laboratory (40–43).

## 1. Solution Growth Chemistry

The technique is basically a modification of the well known process of chemical precipitation of insoluble sulfides obtained by bubbling $H_2S$ through an aqueous solution of a metal salt containing metal ions (95). The $H_2S$ hydrolyses in water to generate $S^{2-}$ ions according to the reactions

$$H_2S + OH^- \rightleftharpoons HS^- + H_2O, \quad HS^- + OH^- \rightleftharpoons S^{2-} + H_2O \tag{1}$$

The metal sulfide is formed by the reaction of the two ions when the requirements of the solubility product principle (defined below) are fulfilled. In a saturated solution of a slightly soluble compound, the product of the molar concentrations of its ions (each concentration term being raised to a power equal to the number of ions of that kind as given by the formula for the compound) is a constant at a given temperature. According to the solubility product principle, a definite numerical relationship exists between the concentrations of the ions in a saturated solution of an electrolyte that is in contact with its solid phase. Equilibrium does not exist if this relationship is not satisfied. Consequently, either the dissolved compound will precipitate or the solid will dissolve, until the ion concentrations satisfy the equation for the solubility product constant. Precipitation occurs when the ionic product (IP) exceeds the solubility product (SP), so that the solution contains more ions than are required for saturation. When IP is less than SP, precipitation cannot take place.

In order to form a thin film by a controlled ion-by-ion reaction, it is necessary to eliminate spontaneous precipitation. This can be achieved by having a fairly stable complex of the metal ions, which provides a controlled number of the free ions according to an equilibrium reaction of the type

$$M(A)^{2+} \rightleftharpoons M^{2+} + A \tag{2}$$

The concentration of the free metal ions at a particular temperature is given by

$$[M^{2+}][A]/[M(A)^{2+}] = K_i \tag{3}$$

where $K_i$ is known as the instability constant of the complex ion. The smaller this constant, the greater is the stability of the complex and hence the smaller is the concentration of metal ions in the solution. The concentration of the metal ions is controlled by the concentration and temperature of an appropriate complexing agent. Table VII lists various ions and their complexing agents.

Localized spontaneous precipitation of a sulfide can occur if a high concentration of $S^{2-}$ ions exists locally, so that the solubility product is exceeded. One may overcome this problem by generating chalcogen ions slowly and

TABLE VII

IONS AND THEIR COMMON COMPLEXING AGENTS

| Ion | Complexing agent |
|---|---|
| $Ag^+$ | $CN^-$, $NH_3$, $Cl^-$ |
| $Cd^{2+}$ | $CN^-$, $NH_3$, $Cl^-$, $C_6H_5O_7^{3-}$, $C_4H_4O_6^{2-}$, EDTA |
| $Co^{2+}$ | $NH_3$, $CN^-$, $SCN^-$, $C_6H_5O_7^{3-}$, $C_4H_4O_6^{2-}$ |
| $Cu^{2+}$ | $NH_3$, $Cl^-$, $CN^-$, EDTA |
| $Hg^{2+}$ | $NH_3$, $Cl^-$, $CN^-$, EDTA |
| $Mn^{2+}$ | $C_2O_4^{2-}$, $C_6H_5O_7^{3-}$, $C_4H_6O_6^{2-}$, $CN^-$, EDTA |
| $Ni^{2+}$ | $CN^-$, $SCN^-$, EDTA, $NH_3$ |
| $Pb^{2+}$ | EDTA, $C_6H_5O_7^{3-}$, $C_4H_6O_6^{2-}$, $OH^-$ |
| $Sn^{2+}$ | $C_6H_5O_7^{3-}$, $C_4H_6O_6^{2-}$, $C_2O_4^{2-}$, $OH^-$ |
| $Zn^{2+}$ | $CN^-$, $NH_3$, EDTA, $C_4H_6O_6^{2-}$, $C_6H_5O_7^{3-}$ |

uniformly throughout the volume of the solution. This takes place, for example, by having thiourea $(NH_2)_2CS$ in an alkaline aqueous solution, in accordance with the reaction

$$(NH_2)_2CS + OH^- \rightleftharpoons CH_2N_2 + H_2O + HS^-$$
$$HS^- + OH^- \rightleftharpoons H_2O + S^{2-} \quad (4)$$

In place of thiourea, its derivatives, such as allylthiourea and $N,N$-dimethylthiourea, can also be used.

By replacing thiourea or its derivatives by selenourea or its derivatives, $Se^{2-}$ ions can be generated to yield selenide films (35, 96). Following the work of Kitaev et al. (38), Kainthla et al. (42) have generated $Se^{2-}$ ions by using inorganic sodium selenosulfate compound in an alkaline solution in accordance with the reaction

$$Na_2SeSO_3 + 2OH^- \rightleftharpoons Na_2SO_4 + H_2O + Se^{2-} \quad (5)$$

Sodium selenosulfate can be easily synthesized by dissolving Se in $Na_2SO_3$ solution. Since selenourea and its derivatives are not easy to synthesize, and since their aqueous solutions have to be stabilized by using antioxidants such as $Na_2SO_3$, the technique developed by Kitaev et al. (38) and Kainthla et al. (42) is very useful contribution to the field.

Tellurium-bearing compounds, being very unstable, cannot be synthesized. However, it may be possible to obtain $Te^{2-}$ ions in a solution by using inorganic compounds such as sodium dithionite $(Na_2S_4O_6)$ to dissolve Te.

When the IP of the metal and the chalcogen ions exceeds the SP of the corresponding chalcogenide, metal chalcogenide is formed by an ion-by-ion combination process on a surface dipped in the solution to provide nuclea-

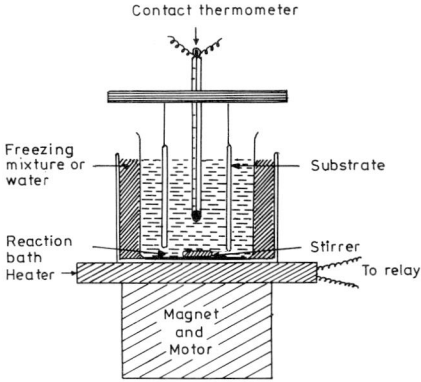

FIG. 20. Experimental arrangement for the deposition of chalcogenide films by solution growth technique (*43*).

tion centers. The film deposition is performed in a simple set-up of the type shown in Fig. 20. The substrates are immersed vertically in the reaction mixture, which is stirred continuously with a magnetic stirrer.

The composition and temperature dependence of the chemical reactions (2)–(5) for obtaining chalcogenide films can be worked out thermodynamically. Such an analysis has been carried out for the following systems:

(1) $Cd(NH_3)_4^{2+}-(NH_2)_2CS-OH^-$ (*34, 43*)
(2) $Cd(en)_3^{2+}-(NH_2)_2CS-OH^-$ (*97*)
(3) $Cd(C_6H_5O_7)^- -(NH_2)_2CSe-OH^-$ (*98*)
(4) $Cd(NH_3)_4^{2+}-SeSO_3^{2-}-OH^-$ (*38, 42*)

The deposition conditions for CdS and CdSe films in the systems 1 and 4 are determined by a graphical solution of the following equations:

$$Cd^{2+} + 2OH^- \rightarrow Cd(OH)_2 \qquad (6)$$

with $[Cd^{2+}][OH^-]^2 = 2.2 \times 10^{-14}$ and

$$Cd(NH_3)_4^{2+} \rightleftharpoons Cd^{2+} + 4NH_3 \qquad (7)$$

with $[Cd^{2+}][NH_3]^4/[Cd(NH_3)_4^{2+}] = 7.56 \times 10^{-8}$. If we plot pH against $p[Cd^{2+}]$, Eqs. (6) and (7) (at constant $pc_{salt}$), should both yield straight lines, called the hydroxide line and the complex line, respectively, as shown in Fig. 21. The shaded regions $B_I$ and $B_{II}$ that lie above the hydroxide line correspond to the presence of $Cd(OH)_2$ in the solution, which, according to Kitaev et al. (*34*), must be present in the solution for the deposition of CdS films. In these regions, thin ($\sim 600$ Å on a glass substrate), hard, physically

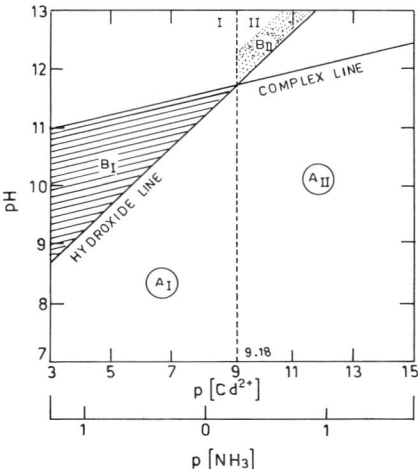

FIG. 21. Plot of Eqs. (6) and (7) (43).

coherent, and specularly reflecting films having a wurtzite structure are obtained by an ion-by-ion reaction process. However, Kaur et al. (43) have established that the CdS films can also be obtained even when the solution contains no hydroxide. In regions $A_I$ and $A_{II}$, where no $Cd(OH)_2$ can exist in the solution, powdery films with a sphalerite structure are obtained, presumably as a result of adsorption of colloidal particles of CdS. For conditions corresponding to the points on the complex line in region II, the film formation takes place only at temperatures above 45°C, since the IP at lower temperatures does not exceed the SP of the cadmium chalcogenides. Reaction starts as the concentration of the $Cd^{2+}$ ions increases with temperature, so that the IP exceeds the SP. The films obtained by using the conditions corresponding to the complex line are hard, physically coherent, and specularly reflecting, with mixed sphalerite and wurtzite structures (43).

In the case of CdSe films obtained by using $Na_2SeSO_3$ at room temperature, films are obtained only in the region $B_{II}$ (38). These films have mixed sphalerite and wurtzite structures. As in the case of CdS, CdSe films can be obtained with a pure sphalerite structure under conditions on the complex line at or above 45°C (42).

The solution growth process as described is expected to have the following characteristic features:

(1) Under given conditions, a film should reach a terminal thickness determined by the availability of the total number of metal and chalcogen ions, such that IP < SP.

(2) Both the rate of deposition and the film thickness should depend on the chemical nature of the solutions, complexing agent and substrate, concentration, and temperature.

(3) Although the crystallographic structure and microstructure of the films would depend on the energetics of the deposition process, the stoichiometry of the films would be relatively insensitive to the deposition conditions.

(4) If the IP of any of the insoluble compounds of impurities present in the solutions does not exceed the SP under given deposition conditions, one does not expect the solution impurities to be incorporated into the films.

(5) By incorporating cations that also form insoluble chalcogenides in the solution, or by carrying out simultaneous solution growth processes of two or more chalcogenides, doping as well as formation of multiple compounds/alloys should be possible. Similarly, it should be possible to sequentially deposit multilayer films of different chalcogenides. Some of these aspects are discussed further in the following sections.

## 2. Film Growth and Deposition Parameters

The kinetics of growth of a thin film in this process are determined by the ion-by-ion deposition of the chalcogenide on nucleating sites on the immersed surfaces. Initially, the film growth rate is negligible due to an "incubation" period required for the formation of critical nuclei from a homogeneous system onto a clean surface. Once the nuclei are formed, by a homogeneous nucleation process, the rate rises rapidly until the rate of deposition equals the rate of dissolution, that is IP < SP. Consequently, the film growth stops and one obtains a "terminal" thickness, as shown in Fig. 22. If a substrate is sensitized with a film of $Sn(OH)_2$ by dipping it in $SnCl_2$ and then washing it in distilled water, or a thin film of a chalcogenide is predeposited on the substrate, no incubation period for nucleation is observed, since nucleation centers already exist on the substrate (see Fig. 22). Also, when the substrates are suspended in the container before forming the complex in the solution, film thickness grows in a way similar to that of the $Sn(OH)_2$ sensitized surface, thereby showing that the nuclei for the formation of the film are provided by the solution itself (Fig. 23).

The rate of deposition and terminal thickness depend on the number of nucleation centers, supersaturation of the solution (defined as the ratio of IP to SP), and stirring. The larger the number of nucleation centers, the more rapid is the deposition and the larger is the terminal thickness. By using larger surface area it is possible to collect more chalcogenide on the surface in the form of a film. Dipping the coated surfaces again in a fresh solution causes further deposition to take place. Thick and multilayer films can thus be deposited by sequential dippings.

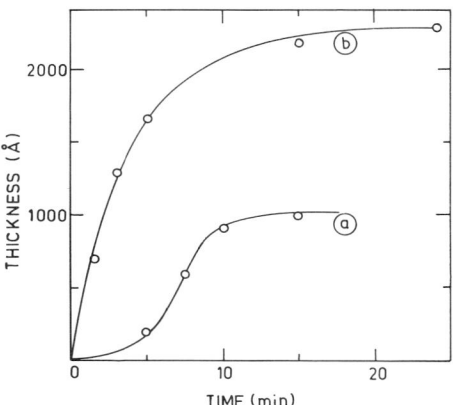

FIG. 22. Typical growth kinetics for solution-grown films deposited on (a) unsensitized substrate and (b) sensitized substrate (99).

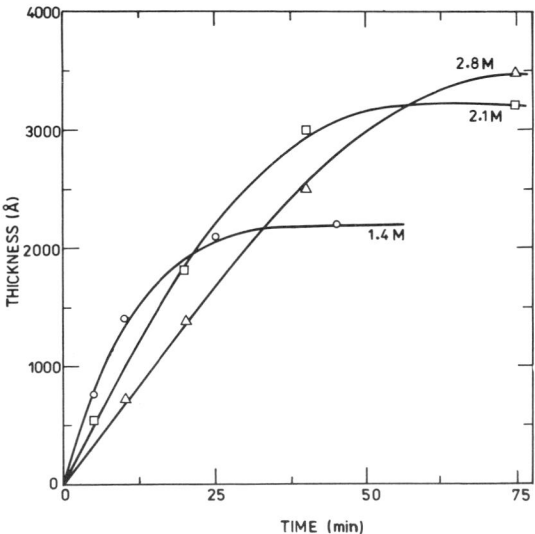

FIG. 23. CdSe film thickness as a function of time for a reaction mixture at 77°C and of composition $4.8 \times 10^{-2}$ $M$ Cd(CH$_3$COO)$_2$, $5.3 \times 10^{-2}$ $M$ Na$_2$SeSO$_3$, and various NH$_3$ concentrations (42).

The growth kinetics are expected to depend on the concentration of ions, their velocities, and nucleation and growth processes on the immersed surfaces. The effect of various deposition conditions on these parameters is discussed next.

*a. Nature of the Salt.* The growth kinetics depend on the salts/compounds used for metal and chalcogenide ions. Although no systematic study has been made in this area, it is expected that the rate of deposition will decrease and terminal thickness will increase if metal sulfate is used to deposit metal selenide films by using sodium selenosulfate. Similar results are expected if $CdCl_2$ is used to deposit CdS and CdSe films. In the first case, the $SO_4^{2-}$ ions obtained from the metal sulfate reduce the concentration of $Se^{2-}$ ions, while the $Cl^-$ ions formed by dissolution of $CdCl_2$ reduce the concentration of $Cd^{2+}$ ions by forming the complex $CdCl_4^{2-}$. In general, the rates and terminal thickness are higher for sulfide films than for selenide films under similar deposition conditions.

An increase in the chalcogen ion concentration initially increases the deposition rate and terminal thickness. At high concentrations precipitation becomes more significant, leading to decreased film thickness on the substrate. This effect is illustrated in Fig. 24 for CdS films.

*b. Complexing Agent.* The concentration of $M^{2+}$ ions decreases with increasing concentration of the complexing ions. Thus, the rates of reaction and hence precipitation are reduced, leading to a larger terminal thickness of the film. Such behavior has been observed in the case of CdSe, CdS, PbSe, and ZnS films. The effect of $NH_3$ concentration on the rate of deposition and terminal thickness of CdSe films is shown in Fig. 25.

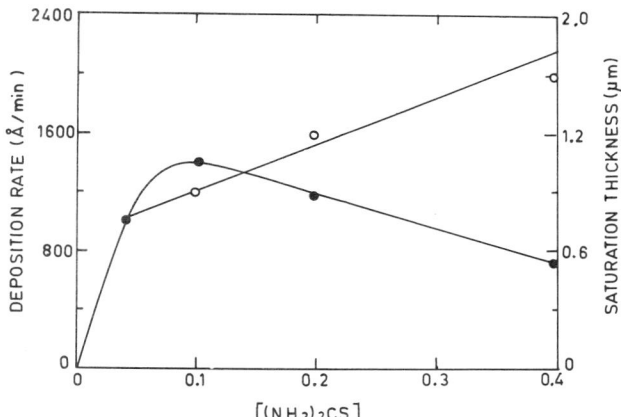

FIG. 24. Rate of deposition (○) and terminal thickness (●) of CdS films as a function of $(NH_2)_2CS$ concentration (*103*).

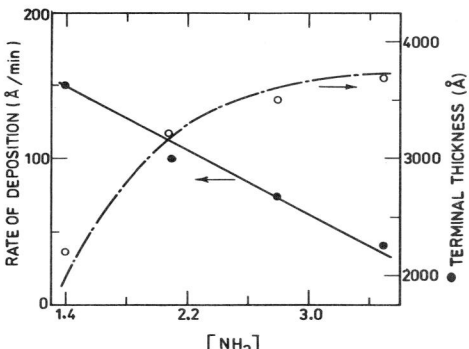

FIG. 25. Rate of deposition (●) and terminal thickness (○) of CdSe films as a function of $NH_3$ concentration for a reaction mixture of composition $4.8 \times 10^{-2}$ $M$ $Cd(CH_3COO)_2$ and $5.3 \times 10^{-2}$ $M$ $Na_2SeSO_3$ at 77°C (42).

c. *pH Value.* If $OH^-$ ions take part in the complex formation (as in $Pb(OH)C_6H_5O_7^{2-}$), the addition of $OH^-$ (i.e., increase in pH) makes the complex more stable, thereby reducing the concentration of free $M^{2+}$ ions. Thus the deposition rate decreases and the terminal thickness increases with increasing pH. The dependence of the rate of deposition and terminal thickness on pH for PbSe films is shown in Fig. 26.

When the $OH^-$ ions do not take part in the complex formation (as in the case of $Cd(NH_3)_4^{2+}$), the addition of $OH^-$ precipitates the corresponding hydroxide. As more of $Cd(OH)_2$ is formed in the bulk of solution, most of

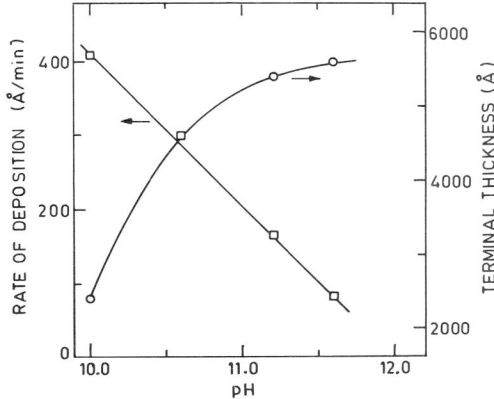

FIG. 26. Rate of deposition (□) and terminal thickness (○) of PbSe films as a function of pH for a reaction mixture of composition $7 \times 10^{-2}$ $M$ $Pb(CH_3COO)_2$, $1.2 \times 10^{-1}$ $M$ $Na_3C_6H_5O_7$, and $5 \times 10^{-2}$ $M$ $Na_2SeSO_3$ (42). Bath temperature = 60°C.

the CdSe is precipitated on addition of $Na_2SeSO_3$, resulting in lower terminal thicknesses at high pH. Similar results have been obtained in the case of CdS films, where the film thickness decreases as the distance of the selected point increases from the hydroxide line (42).

A very significant result obtained by Kitaev et al. on CdSe films is that the Cd:Se ratio is not dependent on pH. This result is supported by the following data (38):

| pH | 12.85 | 12.98 | 13.10 | 13.41 |
|---|---|---|---|---|
| Cd:Se | 1.02 | 1.0 | 1.07 | 1.05 |

*d. Substrate.* When the lattice and lattice parameters of the deposited material match well with those of the substrate, the free-energy change of nucleation is smaller, which facilitates nucleation. Consequently, higher deposition rate and terminal thickness are observed for such substrates. This effect is illustrated in Fig. 27 for PbSe films deposited on glass, copper,

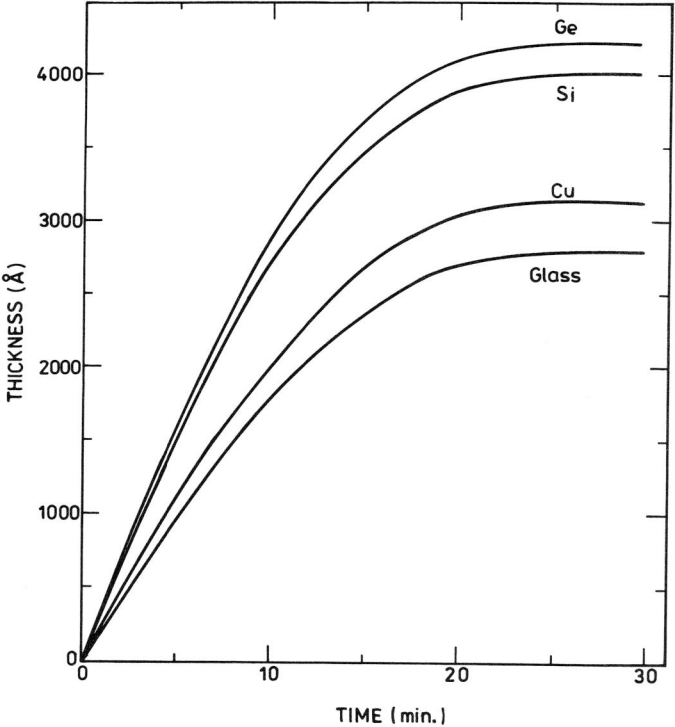

FIG. 27. PbSe film thickness as a function of time for a reaction mixture at 28°C and pH 10.4, of composition $7 \times 10^{-2}$ $M$ $Pb(CH_3COO)_2$, $1.2 \times 10^{-1}$ $M$ $Na_3C_6H_5O_7$, and $5 \times 10^{-2}$ $M$ $Ne_2SeSO_3$ on various substrates (42).

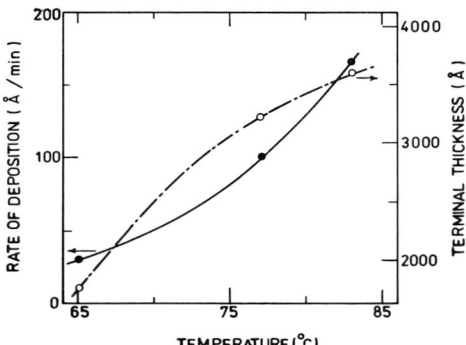

Fig. 28. Rate of deposition (●) and terminal thickness (○) of CdSe films as a function of temperature of deposition for a reaction mixture of composition $4.8 \times 10^{-2}$ $M$ Cd(CH$_3$COO)$_2$, $5.3 \times 10^{-2}$ $M$ Na$_2$SeSO$_3$, and $2.1$ $M$ NH$_3$ (42).

and polished single-crystal Si and Ge. For example, under similar conditions, higher rate and thickness have been observed on Ge than on Si because of better matching of the lattice parameters of PbSe with those of Ge. One can similarly understand (99) higher rate and thickness on single-crystal Ge than on polycrystalline and amorphous Ge surfaces.

*e. Bath/Substrate Temperature.* With increasing solution temperature, the dissociation of the complex and the chalcogen-bearing compound increases. The increased concentration of metal and chalcogen ions, coupled with higher kinetic energy of the ions, results in increased interaction and hence rate of deposition of the metal chalcogenide film. On the other hand, the terminal thickness may increase or decrease with increasing bath temperature, depending on the degree of supersaturation. The terminal thickness first increases with increasing supersaturation (due to increased concentration of ions) and then decreases at sufficiently high supersaturations, at which precipitation dominates. The supersaturation may, however, be controlled not only by the bath temperature, but also by the complexing agent concentration. This effect of temperature on the rate of deposition and terminal thickness of CdSe films is shown in Fig. 28.

## VIII. Impurity and Dopant Effects

The deposition of the film by an ion-by-ion reaction process on the substrate surface ensures the stoichiometric composition of the binary chalcogenides. As already pointed out, the metal:chalcogen ratio in the film is not significantly dependent on various deposition conditions.

The impurities in the starting chemicals can be incorporated into the films only if the impurities can form insoluble chalcogenides under the same conditions of deposition and their corresponding IP is greater than SP. In general, these conditions are satisfied by a few dopants. Thus the degree of purity of the starting chemicals is not so important a factor in determining the purity of the resulting film if the impurity concentration is low and the corresponding IP < SP.

Sodium and potassium salts, such as hydroxides, are commonly used to adjust the pH of the bath for the deposition of CdS, PbS, CdSe, and PbSe films. Also, $Na_2SO_3$ is used as antioxidant in selenourea baths. Since Na, K, and Li do not form insoluble chalcogenides or other compounds, these impurities are not expected to be incorporated into the films. The reported doping of CdS with Li by Shikalgar et al. (100) is therefore questionable.

## IX. Multicomponent Films

By appropriate choice of reacting one or more complexed metal ions with $S^{2-}$ and/or $Se^{2-}$ ions in a solution, it should be possible to form multicomponent chalcogenide films over a wide composition range. We have studied this process in detail and have successfully prepared a variety of variable-composition ternary alloy films, which are listed in Table VIII.

The alloy films in the non-isoelectronic systems are prepared by reacting sodium selenosulfate or thiourea with the mixture of different complexed ions. If two noninterfering, independent complexing agents are used for complexing the two cations, then the ions dissociate in an aqueous solution to give free metal ions according to the reactions

$$MA_n^{2+} \rightleftharpoons M^{2+} + nA, \qquad M'B_m^{2+} \rightleftharpoons M^{2+} + mB$$

In this case, by controlling the initial salt concentration, complexing salt concentration, and temperature of the bath, the composition ($x$) of the films

TABLE VIII
ALLOYS FORMED BY SOLUTION-GROWTH TECHNIQUE

| Alloy | Composition | Ref. | Alloy | Composition | Ref. |
|---|---|---|---|---|---|
| $Pb_{1-x}Hg_xS$ | $0 \leq x \leq 0.33$ | (104) | $Cd_{1-x}Zn_xS$ | $0 \leq x \leq 1.0$ | (103) |
| $Pb_{1-x}Hg_xSe$ | $0 \leq x \leq 0.35$ | (102) | $Cd_{1-x}Pb_xSe$ | $0 \leq x \leq 1.0$ | (105) |
| $Cd_{1-x}Hg_xS$ | $0 \leq x \leq 0.20$ | (103) | $CdSe_{1-x}S_x$ | $0 \leq x \leq 1.0$ | (106) |
| $Cd_{1-x}Hg_xSe$ | $0 \leq x \leq 0.15$ | (102) | $PbSe_{1-x}S_x$ | $0 \leq x \leq 1.0$ | (106a) |

can be varied. The composition can also be varied by changing the concentration of the complexing agents without altering the ratio of the salt concentrations. This has been utilized in the case of $Cd_{1-x}Pb_xSe$ films, in which the composition has been varied by adding different amounts of $NH_4Cl$ solution to the reaction mixture (105). The variable composition in $Cd_{1-x}Hg_xSe$, $Cd_{1-x}Zn_xS$, $Cd_{1-x}Hg_xS$, and $Pb_{1-x}Hg_xS$ films has been obtained by changing relative amounts of the salts in the solution, using the same complexing agent for the two cations (102–104).

In the preparation of $MSe_{1-x}S_x$ films, the fraction of $S^{2-}$ ions in the solution is expected to be more than the fraction of thiourea in the solution, since thiourea has a higher dissociation constant. As the SP's of sulfides and selenides do not differ much, the films are expected to be sulfur rich, compared in each case to the fraction of thiourea in the solution. Thus, if the films are given the formula $CdSe_{1-y}S_y$ and $x$ represents the fraction of thiourea in the solution, then $y > x$ (106).

As in the atom-by-atom deposition process, the solubility conditions of multicomponents in an ion-by-ion reaction process are expected to be relaxed. Similarly, one would expect stabilization of high-temperature and/or high-pressure polymorphs of the chalcogenide under certain deposition conditions. These expectations have indeed been realized experimentally. For example, $Pb_{1-x}Hg_xS$ ($0 \leq x \leq 0.33$) and $Cd_{1-x}Pb_xSe$ ($0 \leq x \leq 1$) alloy films have been obtained even though solid solubilities of the two chalcogenides in each case are not known to exist in bulk form (104, 105). Furthermore, either the room-temperature-stable $\alpha$ phase (2.0 eV band gap, trigonal) or the high-temperature $\beta$ phase (cubic, 0.1 eV band gap, stable above 280°C) of HgS can be stabilized in yielding uniphase fcc structure of $\alpha'$- or $\beta'$-$Pb_{1-x}Hg_xS$ (104).

## X. Oxide Films

The solution growth technique has been extended to the growth of oxide films in our laboratory. Basically, this is possible because some cations in an alkaline medium are precipitated during hydrolysis as hydrous oxides instead of as hydroxides. If a substrate is immersed in the solution, a film of the hydrous oxide is obtained by an ion-by-ion deposition process. On subsequent heating in a controlled oxygen atmosphere, dissociation takes place, resulting in a pure oxide film. For example, $Mn_2O_3$ films have been obtained by immersing glass substrates in an aqueous solution containing $Mn^{2+}$ ions to which $NH_3$ is added. A film of $Mn(OH)_2$ is formed and oxidizes rapidly in air to form $Mn_2O_3 \cdot nH_2O$. In the presence of $NH_3$ only, the reaction is very fast and uncontrolled, leading to the formation of precipitates. But the addition of $NH_4Cl$ before $NH_3$ controls the rate and one gets good optical quality films. On drying $MnO_3 \cdot nH_2O$ at 100°C, one first gets $\gamma$-MnO(OH), which on further dehydration at 250°C in vacuum gives $\gamma$-$Mn_2O_3$. Prolonged heating in vacuum at 500°C yields $\gamma$-$MnO_2$, $\gamma$-$Mn_2O_3$, $Mn_3O_4$, and MnO films (103).

Similarly, $SnCl_4$ complexed with $NH_4F$ yields free $Sn^{4+}$ ions at a controlled rate determined by the solution temperature. In the presence of an alkaline medium and oxygen these metal ions undergo hydrolysis and give $SnO_2$ films (103a). The growth behavior of solution-grown oxide films is similar to those of sulfide and selenide films. Under given growth conditions of metal-ion concentration, nature and concentration of complexing agent, and pH and temperature of solution, the films acquire a terminal thickness. A maximum terminal thickness of 4500 Å has been obtained (103a).

With proper choice of metal salts and complexing agent, the same technique can be used for the preparation of other oxides such as Cr, Fe, Mo, Nb, Ta, Ti, V, W, and Zn. Some of the commonly used complexing agents are $NH_4Cl$, EDTA, and $C_6H_5O_7^{-3}$.

## XI. Structure

Transmission electron microscopy studies of solution grown films have established that film formation proceeds via nucleation and growth processes in a way similar to vapor-deposited film (42). The sequence of micrographs of PbSe films at different stages of growth shown in Fig. 29 illustrates this point.

Generally speaking, the films are micropolycrystalline, with typical grain sizes ranging from 300 to 1000 Å for PbSe films. The grain size depends on the composition and temperature of the bath and the nature of the substrate. The grain size is larger at lower deposition rates (i.e., at lower supersaturations), higher bath temperatures, and for lattice-matched substrates.

A detailed study of the dependence of the microstructure of selenide films on the deposition parameters has been reported by Kainthla et al. (102). A finer microstructure is obtained (102, 107) when two or more chalcogenides are codeposited, as is the case in vapor-deposited films. The microstructure in multicomponent materials is dominated by the sizes of and interaction between the various ions.

Under suitable deposition conditions, the adion mobility is large enough to yield well oriented epitaxial films on single-crystal substrates. Epitaxial growth of $Pb_{1-x}Hg_xS$ films on Ge substrates below 20°C bath temperature (108) and PbS films on Si substrates (108a) has been reported.

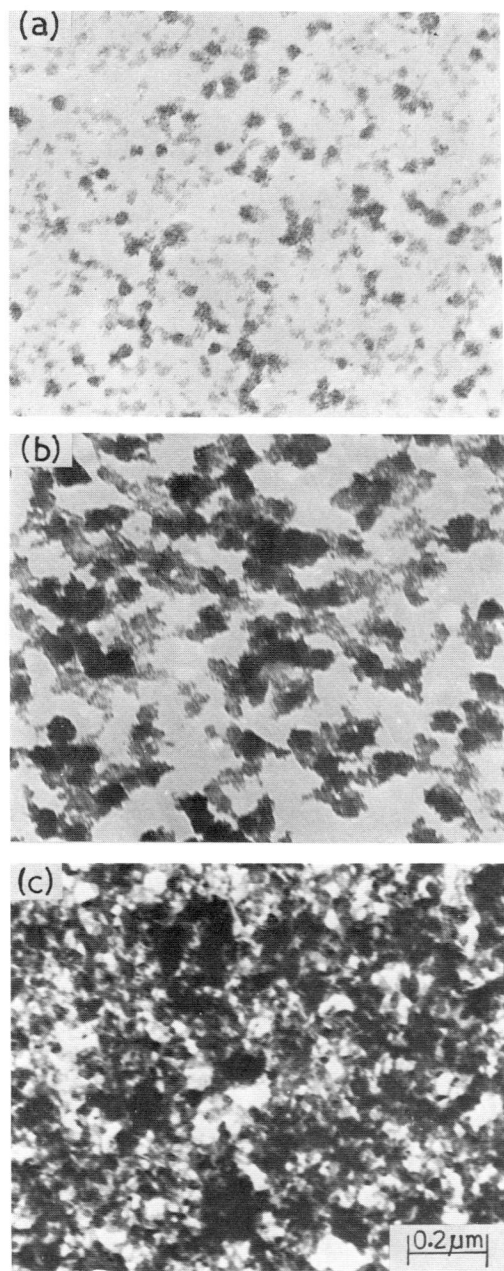

FIG. 29. Electron micrographs showing growth of PbSe films after (a) 3 min, (b) 6 min, and (c) 12 min of immersion time (42).

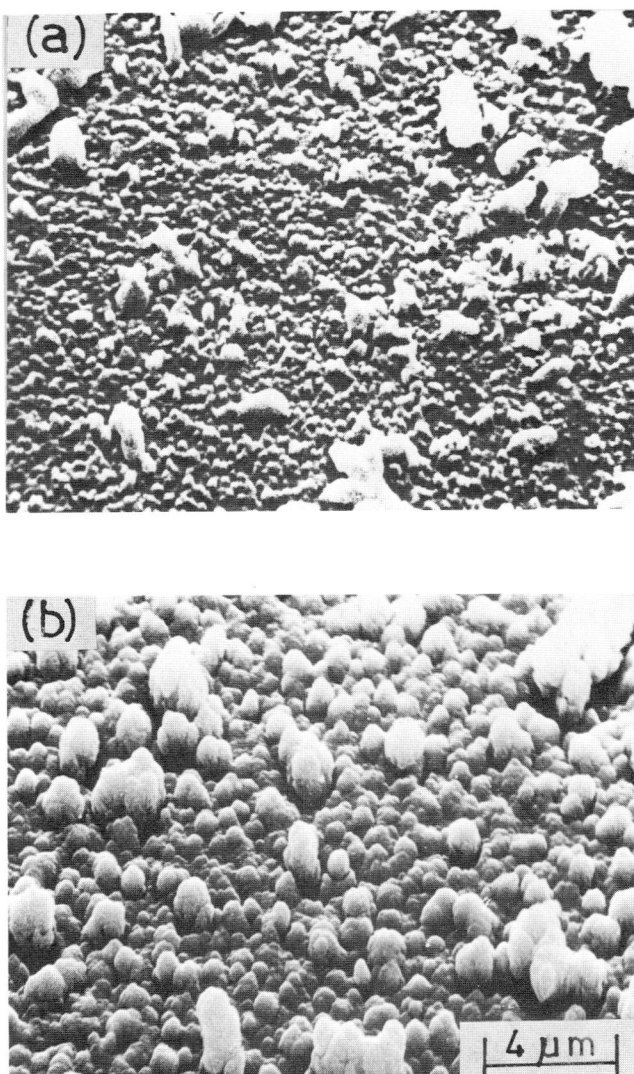

Fig. 30. SEM micrographs of CdSe films (a) as deposited and (b) after recrystallization by heating at 350°C for 30 min in a CdS powder containing CuCl (*102*).

Most chalcogenides are known to have polymorphic structures. As already pointed out (Section VII,2), films of these chalcogenides can be obtained in different structures, depending on the deposition conditions. Generally, the more stable phase is obtained when the ion concentration in the solution and the bath temperature are low and the solution is constantly agitated. The precipitates and precipitate-like (powdery) films tend to have the high-temperature/pressure polymorph. Additive impurities may help to stabilize one or the other phase. For example, the presence of $FeCl_3$ in the deposition bath during the deposition of $Pb_{1-x}Hg_xS$ films stabilizes the high-temperature $\beta$ phase of HgS to yield uniphase fcc $\beta'$-$Pb_{1-x}Hg_xS$ films (40, 104).

The complexing agent has also been observed to affect the CdS film structure. CdS films obtained from the $Cd(NH_3)_4^{2+}$ complex have sphalerite, wurtzite, or mixed structure, depending on the deposition conditions, while those obtained from $Cd(CN)_4^{2-}$ and $Cd(en)_3^{2+}$ complexes always exhibit wurtzite structure, with $c$ axis perpendicular to the substrate (43, 97, 103). The formation of a unique crystal structure from cyanide and ethylene–diamine baths is not understood clearly and requires further investigation.

As mentioned earlier, the grain size of solution-grown films is generally limited to $\sim 0.1$ μm. Postdeposition heat treatment leads to grain growth. To obtain an appreciable rate of grain growth, the films have to be annealed at temperatures greater than 400°C for a sufficiently long time. Higher impurity-induced growth rates at lower temperatures ($\sim 300°C$) have been achieved by using the embedding technique (102, 109), in which films are embedded in powder of the corresponding chalcogenide containing a calculated amount of a suitable metal halide to provide the dopant metal. Figure 30 shows scanning electron microscope (SEM) micrographs of (a) as-deposited CdSe films and (b) the same films with considerable grain growth after heat treatment at 350°C for 30 min on embedding in CdS powder containing CuCl. The CdSe and CdS films have also been recrystallized by depositing a thin layer of copper or silver on the film and heating at temperatures greater than 350°C (102). The embedding technique is very effective in obtaining controlled doping and recrystallization. However, the kinetics of the reactions involved have not been studied in detail.

## XII. Transport Properties

We confine our brief description to the electrical and optical properties, which are of interest for large-area device applications. As in sprayed films, both these properties are sensitively dependent on the microstructure and changes therein on postdeposition annealing. Since the microstructures of the

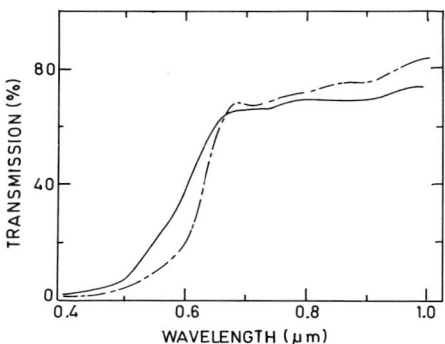

FIG. 31. Optical transmission of CdSe films, as deposited (—) and after recrystallization (---) (99).

sprayed and solution-grown films are very similar, the similarity of their electrical and optical properties is not surprising.

## 1. Optical Properties

The optical absorption edge of solution-grown chalcogenide films is the same as that of the corresponding bulk material (102, 103, 107). However, the sharpness of the edge is considerably reduced, due largely to the diffuse scattering of light as a result of the fine-grained microstructure of the films.

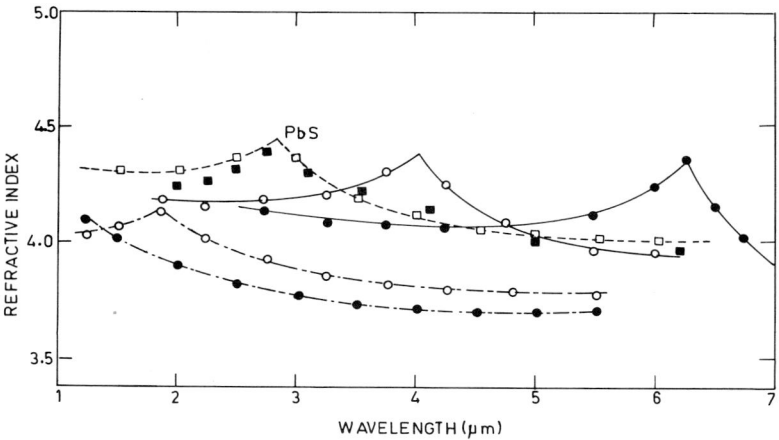

FIG. 32. Wavelength dependence of the refractive index of $\alpha'$-$Pb_{1-x}Hg_xS$ (---) and $\beta'$-$Pb_{1-x}Hg_xS$ (—) films (110, 111). (○) $x = 0.14$; (●) $x = 0.33$; (□) PbS solution-grown; (■) PbS epitaxial.

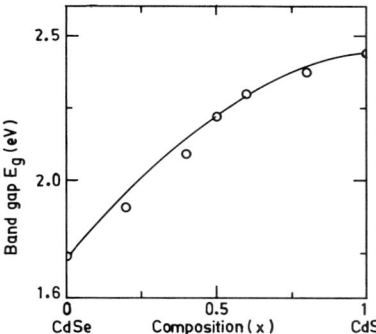

FIG. 33. Optical band-gap ($E_g$) variation of $CdSe_{1-x}S_x$ films with composition ($x$) (*106*).

On postdeposition annealing, the sharpness of the edge improves significantly, as shown for CdSe films in Fig. 31.

The optical constants and their dispersion behavior are similar to those of the corresponding evaporated films and bulk materials. Figure 32 shows the wavelength dependence of the refractive index of solution-grown $\alpha'$- and $\beta'$-phase $Pb_{1-x}Hg_xS$ films and epitaxial PbS films (*110, 111*). The agreement of the data on PbS films with those for epitaxial films and the occurrence of the dispersion peak at optical band-gap energies for the multicomponent films should be noted.

The optical band gap of the multicomponent alloy films varies continuously with composition between the values for its constituent compounds. This variation is shown for $CdSe_{1-x}S_x$ films in Fig. 33 (*102, 106*). The dispersion behavior of the optical constants of these alloy films is similar to that of $Pb_{1-x}Hg_xS$ films (see Fig. 32). These results demonstrate clearly that variable-composition films allow variable and predictable optical properties to be obtained.

## 2. Electrical Properties

Films of CdS, CdSe, ZnS, and ZnSe are invariably n type, whereas those of PbS and PbSe are always p type. The n- and p-type conductivities are due to the vacancies and excess of the chalcogen, respectively. On postdeposition annealing at temperatures above ~435 K, the PbS and PbSe films change over to n-type conduction, which is maintained on cooling the film to room temperature (Fig. 34) (*102, 111*). The films can be reversed back to p-type conductivity on annealing in $O_2$ or S/Se vapors. Figure 35 shows the effect of annealing PbSe films in Se vapors and in vacuum. Annealing and photoconductivity studies on these films suggest that the change in conductivity is predominantly due to the desorption of oxygen (*102, 112*).

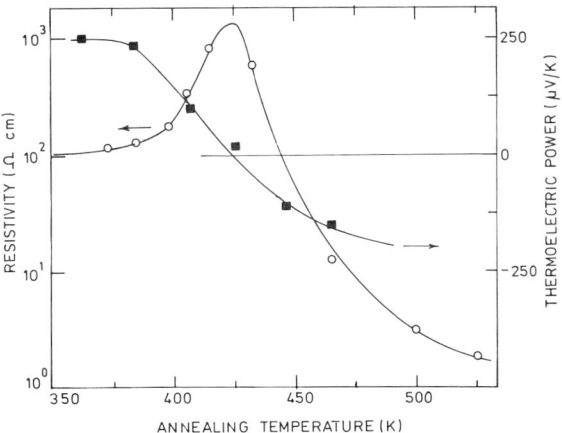

FIG. 34. Dark resistivity (○) and thermoelectric power (■) of PbSe films (measured at room temperature) as a function of annealing temperature. Annealing time, 5 min (*112*).

The type of conductivity of alloy films depends on the constituent compounds and the composition of the films. For example, $CdSe_{1-x}S_x$ ($0 \leq x \leq 1$), $Cd_{1-x}Hg_xS$ ($0 \leq x \leq 0.2$) and $Cd_{1-x}Zn_xS$ ($0 \leq x \leq 1$) are always n type, while $Pb_{1-x}Hg_xS$ ($0 \leq x \leq 0.33$) and $Pb_{1-x}Hg_xS$ ($0 \leq x \leq 0.35$) films are p type. On the other hand, $Cd_{1-x}Pb_xSe$ ($0 \leq x \leq 1$) are n type for $x < 0.2$ and p type for $x \gtrsim 0.2$. The oxygen chemisorption effects are important in alloy films also. Annealing of the alloy films in vacuum changes them from p to n type; they can be converted back to p type by subsequent annealing in oxygen or chalcogen vapors (*102*).

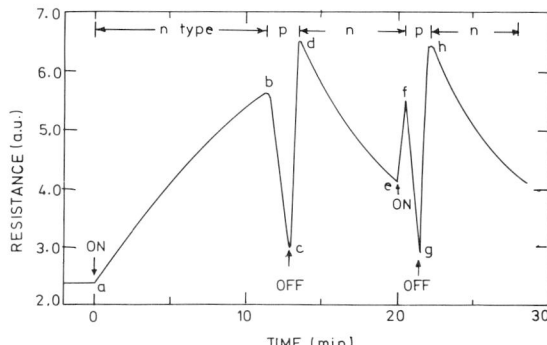

FIG. 35. Effect of annealing PbSe films at 475 K in Se vapors (a–b–c, e–f–g) and in vacuum (c–d–e, g–h–i) (*112*).

TABLE IX

DEPENDENCE OF RESISTIVITY OF PbSe FILMS ON THE TEMPERATURE
OF DEPOSITION AND pH OF THE BATH (*102*)

| Temperature of deposition 28°C | | Temperature of deposition 60°C | |
|---|---|---|---|
| pH | $\rho$ ($\Omega$ cm) | pH | $\rho$ ($\Omega$ cm) |
| 9.1 | $1.25 \times 10^4$ | 9.8 | $10^3$ |
| 10.0 | $10^4$ | 10.3 | $2.5 \times 10^2$ |
| 10.7 | $6.6 \times 10^3$ | 10.7 | $1.6 \times 10^2$ |
| — | — | 11.8 | $10^2$ |

The resistivity of the films depends on deposition conditions. It decreases with increasing pH and temperature of the bath. A systematic variation with these parameters has been established only in PbSe films, as shown by the data in Table IX (*112*). Whereas as-deposited lead chalcogenide films have a resistivity of $\sim 10^2$–$10^4$ $\Omega$ cm, cadmium and zinc chalcogenides have values of $\sim 10^7$–$10^9$ $\Omega$ cm. On annealing in vacuum ($\sim 10^{-5}$ Torr), resistivities of cadmium and lead chalcogenides decrease to $\sim 1$–$10$ $\Omega$ cm. This reduction (Fig. 36) is attributed to the desorption of oxygen from the films. The original resistivity values can be recovered by subsequent heating in air/

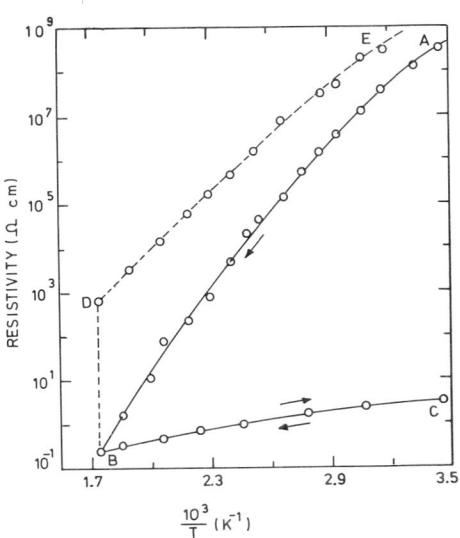

FIG. 36. Dark resistivity of CdSe films as a function of temperature cycling (ABCBDE) (*102*).

oxygen. The process of adsorption and desorption is completely reversible (102, 103). A similar oxygen adsorption–desorption-controlled variation of resistivity is obtained (103a) in pure and indium-doped $SnO_2$ films. Pure $SnO_2$ films have an as-deposited resistivity of 200 Ω-cm which decreases to 40 Ω-cm on vacuum annealing at 450°C. On the other hand, indium-doped $SnO_2$ films show a variation from $1.2 \times 10^{-2}$ to $6.0 \times 10^{-3}$ Ω-cm on vacuum annealing at 350°C. In both cases the as-deposited value of resistivity is restored by annealing in air at 500 and 400°C, respectively.

It is extremely difficult to measure the Hall coefficient in the high-resistivity, low-mobility films. Such a measurement is possible either on lower-resistivity vacuum-annealed films, or under illumination of sensitized films of high photoconductive gain. The carrier concentration can also be estimated from photo-thermopower data. The resistivity, carrier concentration, and mobility of some of the chalcogenide films are listed in Table X. It is noteworthy that all vacuum-annealed films have a mobility range of $\sim 10^{-1}$–$10^1$ cm$^2$ V$^{-1}$ sec$^{-1}$ and carrier concentration range is $10^{16}$–$10^{18}$ cm$^{-3}$. The temperature dependence studies of these parameters clearly show that the mobility is thermally activated. Electrical conduction in these films is dominated by grain-boundary scattering, chemisorption of oxygen, and by the electronic effects of chemisorbed oxygen on grain boundaries. The conduction mechanism in these films is discussed elsewhere in more detail (102).

TABLE X

Electron Transport Parameters of Various Solution-Grown Films[a]

| Film | $\rho$ (Ω cm) | n or p (cm$^{-3}$) | $\mu$ (cm$^2$ V$^{-1}$ sec$^{-1}$) |
|---|---|---|---|
| $\alpha'$-Pb$_{1-x}$Hg$_x$S ($0 \leq x \leq 0.33$) | $10^1$–$10^3$ (ad) | $10^{15}$–$10^{17}$ | $10^{-1}$ |
| $\beta'$-Pb$_{1-x}$Hg$_x$S ($0 \leq x \leq 0.33$) | $10^1$–$10^2$ (ad) | $10^{16}$–$10^{17}$ | 1 |
| Cd$_{1-x}$Pb$_x$Se | | | |
| ($0 < x < 0.2$) | $10^8$–$10^9$ (ad) | — | — |
| ($0.2 < x \leq 1$) | $10^2$–$10^5$ (ad) | $10^{16}$–$10^{17}$ | $10^{-1}$–10 |
| CdSe | $10^8$–$10^9$ (ad) | — | — |
|  | 1–10 (an) | $10^{17}$–$10^{18}$ | 1–10 |
| CdS | $10^8$–$10^9$ (ad) | — | — |
|  | 40–50 (an) | — | — |
| ZnS | $10^8$–$10^9$ (ad) | — | — |
|  | $10^7$–$10^8$ (an) | — | — |
| SnO$_2$ | $2 \times 10^2$ (ad) | — | — |
| SnO$_2$:In | $10^{-2}$ (ad) | — | — |
|  | $6 \times 10^{-3}$ (an) | $10^{20}$ | 3 |

[a] ad = as deposited; an = annealed.

## 3. Photoconductivity

As-deposited films are weakly photoconducting, having photoconducting gain of 10 or less. Lead chalcogenide films can be sensitized by heating in oxygen/air only (*102*) On the other hand, cadmium chalcogenide films can be activated by heating in air, or by doping with such acceptor impurities as Cu, Ag, and Li (*102, 113*). Doping can be achieved during film deposition or by postdeposition heat treatment involving embedding of the films in the corresponding chalcogenide powder containing a calculated amount of the required impurities. By doping, a photoconducting gain of up to $\sim 10^8$ has been obtained in CdSe and CdS films (*114*). The response time increases with photoconducting gain and ranges from 0.2 to 5 msec in CdSe and CdS films (*102, 106*).

The spectral response of photoconductivity depends on the photoconductivity processes. Being essentially a band-to-band excitation process in the Cd and Pb chalcogenide films, the spectral response exhibits a peak at energies nearly equal to the optical band gap of the films. This point is well illustrated by the spectral response of $CdSe_{1-x}S_x$ films shown in Fig. 37. Determination of the hole ionization energy of the sensitizing centers shows that the ionization energies are 0.58 and 1.2 eV for CdSe and CdS films, respectively. The values are comparable to those obtained in bulk (*102, 106*).

## XIII. Some Large-Area Applications

The chemical solution techniques have made it possible to develop a number of large-area thin-film devices. Some of these devices are described

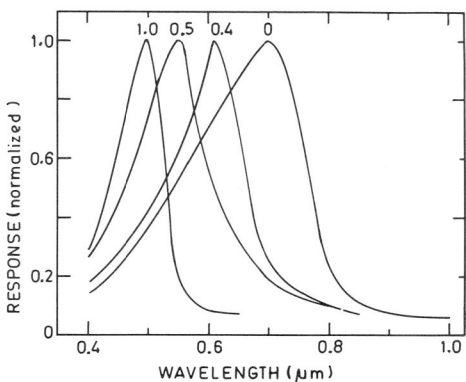

FIG. 37. Spectral response of photoconductivity in $CdSe_{1-x}S_x$ films (*106*).

briefly below, with the primary objective of highlighting the role of the films, rather than discussing the characteristics of the devices.

### 1. Thin-Film Solar Cells

Spray-deposited films of CdS, TO, ITO, and ZnO have been utilized in the fabrication of $\sim 6$–$8\%$ CdS/Cu$_2$S (*28, 115, 116*), $\sim 12.3\%$ TO/Si (*117*), $\sim 11.0\%$ ITO/Si (*118*), and $4.6\%$ n-ZnO/p-CdTe (*119*), and $4\%$ CdS/CdTe (*48a*) heterojunction solar cells. The efficiencies under AM1 conditions are for $\sim 3.0$ cm$^2$ area for Si solar cells. High-efficiency CdS/Cu$_2$S cells have been obtained with $\sim 1$ cm$^2$ area in our laboratory as well as by Photon Power, Inc. As the area is increased, the cell efficiency decreases. Photon Power has obtained panel efficiency of $\sim 2.5\%$ in their production facility for continuous processing of $50 \times 60$ cm panel, although the individual cells of $\sim 50$ cm$^2$ area ($50 \times 1$ cm) have efficiency of $\sim 4$–$5\%$. We have established that the decrease of efficiency is due to high series resistance and considerable spatial variation of the physical and chemical properties of CdS films and hence of the cell. As the spray technology is better understood these difficulties are being overcome, and one expects to see marked improvements in the uniformity of performance of the large-area cells.

A schematic diagram of the CdS/Cu$_2$S solar cell fabrication process (*120*) is shown in Fig. 38. Typical cross-sectional views of the CdS/Cu$_2$S and

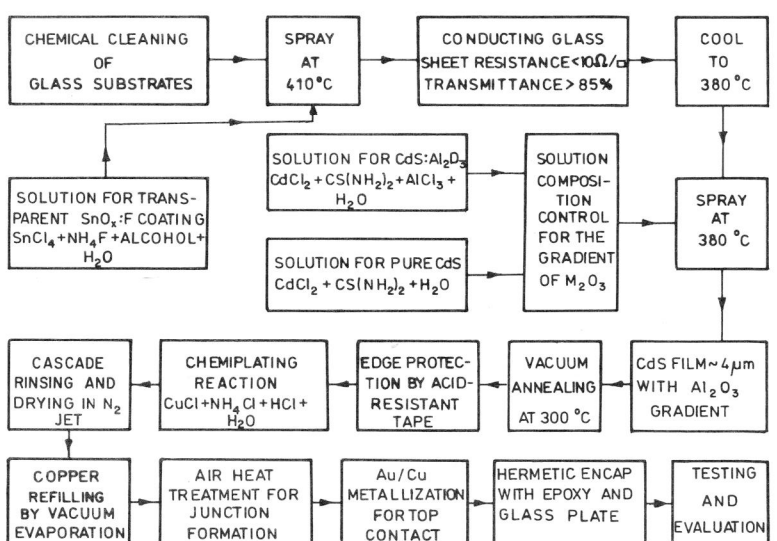

Fig. 38. Block diagram of the fabrication process for spray pyrolytically deposited CdS-based thin-film solar cells (*48*).

TO/Si cells are shown in Fig. 39. The sprayed CdS solar cells have the following structural features derived from the spray process.

(1) The columnar structure, a characteristic of evaporated CdS solar cells, is inhibited partly by the fine-grained structure, and largely by the incorporation of dopants (such as Zn) or $Al_2O_3$ in grain boundaries (obtained by copyrolysis of $AlCl_3$). The resulting serpentine structure of CdS film on the surface is compact and void free and prevents the penetration of $Cu_2S$ along the grain boundaries and voids, which normally results in the shorting of cells.

(2) Double-layer and gradient microstructure CdS films have been exploited to improve the topography of the junction and its performance.

(3) CdS alloy films (e.g., $Cd_{1-x}Zn_xS$) have been utilized (28) to enhance the open-circuit voltage from 0.45 to $\sim 0.66$ V. The problem of lower current density due to the high resistivity of these films continues to pose a challenge for further research. The high ($>90\%$) solar transmittance and low sheet resistance ($\sim 2$–$5\ \Omega/\square$) obtainable on large areas in thin films of such inexpensive materials as $SnO_x:F$ and $Cd_2SnO_4$ have inspired hopes for the economic viability of both the "modified" $CdS/Cu_2S$ and MIS-type Si cells. Because of the simplicity of forming a junction and the high chemical and mechanical stability of $SnO_x:F$ films, single-crystal, polycrystalline, and amorphous silicon solar cells of the MIS type offer a very attractive alternative to the conventional diffused p–n junction technology. Many problems relating to the spatial uniformity of the ideal properties of the oxide coatings and the formation of a controlled high-resistance oxide thickness of 15–50 Å undoubtedly remain to be solved before the technology is universally accepted.

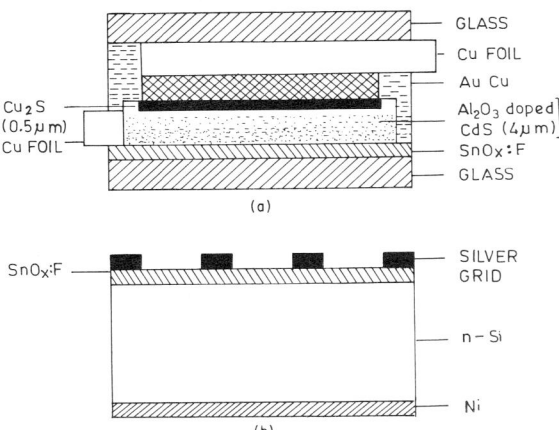

FIG. 39. Cross-sectional view of (a) $CdS/Cu_2S$ and (b) $SnO_x/Si$ solar cells.

## 2. Photoelectrochemical Solar Cells

A major breakthrough in economic solar energy conversion would be possible if stable semiconductor/electrolyte junction solar cells could be obtained. The electrochemical and photodissolution processes determine the stability of these cells. Both spray pyrolytic and nonpyrolytic solution growth processes make it possible to obtain very strongly adherent and coherent chalcogenide films of a composition that is optimal from both optical-gap and chemical-stability points of view. Preliminary work in this direction has already demonstrated (*121*) the usefulness of CdSe-based films for obtaining large-area ($\sim 2$ cm$^2$) solar cells of $\sim 3\%$ efficiency. These cells consist of doped CdSe films deposited on SnO$_x$:F-coated glass as one electrode, and graphite as the other electrode in an electrolyte of Na$_2$S$_x$. Figure 40 shows the cell geometry as well as the *I–V* characteristic of such a cell. It is hoped that higher efficiencies and chemical stability will be obtained by incorporating an appropriate gradient of sulfur in CdSe films. An efficiency of $\sim 5\%$ for CdSe cells has been obtained (*121a*) on $\sim 3$ cm$^2$ active area. Larger area ($\sim 20$ cm$^2$) cells show efficiencies of $\sim 3\%$. The main drawback of these large-area cells is their poor fill factor.

## 3. Selective Coatings

The oxide coatings mentioned in the preceding sections are selective in that they transmit the visible part of the solar spectrum and reflect the IR part. A more useful large-area selective coating for photothermal conversion of solar energy is one that absorbs the visible part (0.3–2.0 $\mu$m) and reflects the IR part (beyond 2.0 $\mu$m), so that the coating heated by absorption of visible light does not radiate away the thermal energy in the IR (because of its low emissivity). Such coatings of oxides, carbides, nitrides, sulfides, and phosphides are produced by chemical solution growth and surface chemical conversion techniques. Spray pyrolysis is useful for obtaining such films as PbS (*4*), CoO (*55*), MoO (*56*), VC (*122*), and Cu$_2$S (*123*). The nonpyrolytic solution growth technique is ideally suited for obtaining large-area, adherent, and stable multilayer interference stacks of multicomponent chalcogenide films as selective coatings. A multilayer stack of metal–PbS–CdS–PbS $\cdots$ has been studied in our laboratory (*124*, *125*). Figure 41 shows the stack and its optical and thermal properties. With absorptance of over 90% and emissivity of 0.12 at 100°C, a very attractive performance is obtained. One can tailor-make a variety of interference stacks of different materials to obtain as high an absorptance as 95% and as low an emissivity as 0.05. Besides being simple, inexpensive, and energy-intensive, this technique allows the deposition of films on a variety of substrates, including plastic foils.

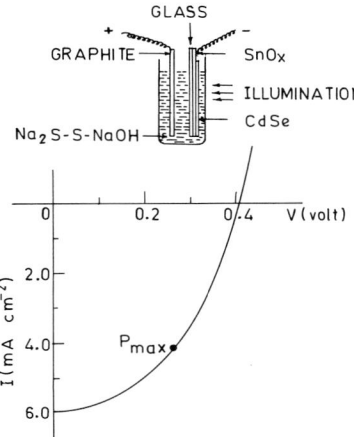

FIG. 40. Schematic diagram and $I-V$ characteristics of a thin-film photoelectrochemical cell (121): $V_{oc} = 410$ mA; $I_{sc} = 6$ mA/cm$^2$, FF = 0.45, $\eta = 2.2\%$.

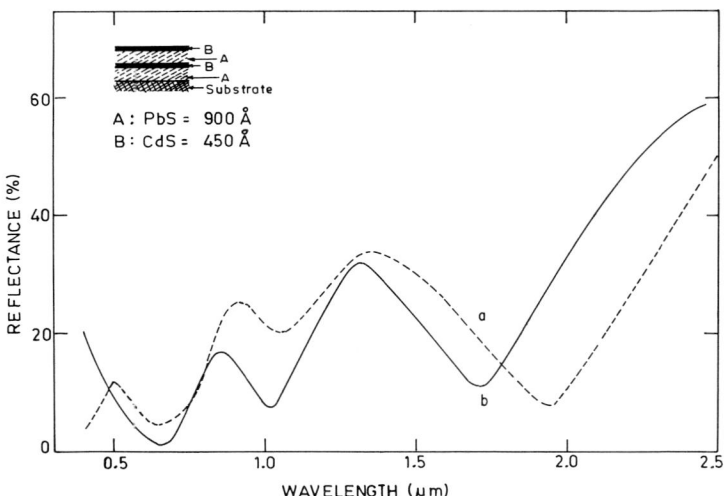

FIG. 41. Optical performance of a four-layer PbS–CdS solution-grown multilayer stack deposited on stainless steel. Theoretically calculated and experimentally evaluated values of $\alpha_s$ and $\epsilon_{100}$ are indicated (124): (a) $\alpha_s = 0.90$ (calculated), (b) $\alpha_s = 0.91$ (measured), $\epsilon = 0.12$.

## 4. Transparent Conducting Coatings

Besides solar cells, spray-deposited transparent conducting coatings are already of vital importance for a number of other applications, such as deicing and defogging of aircraft and other windows, antistatic instrument panels, electroluminescent displays and lamps, liquid crystals and electrochromic displays, optoelectronic functions in a host of devices, vidicons, etc. The desirable properties of coatings vary considerably from application to application and are not so stringent as those for solar cells. Consequently, the present-day technology for obtaining over 90% transmission in the region 0.35–1.2 $\mu$m in doped oxide films of tin, indium, and Cd–Sn suffices for these applications.

## 5. Heat Mirrors

The transparent conducting oxide films also exhibit high reflectivity (over 80%) in the IR. The wavelength at which the high reflectivity sets in is determined by the plasma resonance frequency, $\omega_p$. ($\omega_p^2 = 4\pi n e^2/\epsilon_\infty m^*$, where $n$ is the carrier concentration, $m^*$ the effective mass, $e$ the electronic charge, and $\epsilon_\infty$ the dielectric constant). By appropriate doping, both $n$ and $m^*$ can be changed over a large range so that desired $\lambda_p (= 2\pi c/\omega_p)$ can be obtained. For example, by doping $SnO_2$ with Sb and F, $\lambda_p$ can be varied from 2.8 $\mu$m (0 mol % Sb) to 1.3 $\mu$m (3 mol % Sb) (30). Similar results can be obtained with $Cr_2O_3$, $Fe_2O_3$, and CoO coatings.

The high IR reflectivity is useful for elimination of heat flow through windows in buildings so that heat input from outside in summer and heat output from within during winter through the windows is minimized. The thermal energy conservation property of these coatings is utilized in air-conditioned residential buildings, greenhouses, and flat-plate and tubular solar collectors.

## 6. Photon Detectors (Visible and IR)

As already pointed out in Sections IV and VIII, both chemical solution growth techniques are ideally suited for the preparation of larger-area coatings of multicomponent and variable-composition chalcogenide films. By doping, as well as by postdeposition treatment, the resistivity and the photoconductivity and its response time can be varied over a large range. Thus by choosing materials of appropriate band gap and optical absorption processes, a variety of photon detectors with peak sensitivity ranging from the visible to the far IR is possible. For example, sensitized films of CdS, CdSe, and $CdS_xSe_{1-x}$ with photoconductive grains of $\sim 10^6$, $10^5$, and $10^8$, respectively, with corresponding response times of 5, 0.2, and 0.1 msec in 50 mW cm$^{-2}$ solar radiation are easily obtained (*102, 103, 106, 114*).

Solution-grown and sensitized films of PbS and PbSe have been used as IR detectors for a very long time. With peak sensitivities at 3.0 and 5.0 μm for PbS and PbSe, respectively, corresponding detectivity ($D^*$) values as high as $10^{11}$ and $10^{10}$ cm Hz$^{1/2}$ W$^{-1}$ at 300 K are obtained. The $D^*$ value at 77 K is higher by a factor of up to 10. The use of variable-gap ternary alloys of the type PbHgS, CdSSe, PbSnS, PbSnSe, PbCdSe, and PbSSe for photon detection has been studied by Sharma *et al.* and Kainthla *et al.* in our laboratory (*41, 102, 107*). By an appropriate choice of deposition conditions, either the large band gap ($\sim 2.0$ eV) trigonal phase or the small band gap ($\sim 0.1$ eV) cubic phase of HgS can be stabilized in $Pb_{1-x}Hg_xS$ alloy films. Thus, films with band gap varying from 0.2 to 1.2 eV have been prepared (*108*). By sensitizing, photoconductive detectors with peak response from 1.0 to 6.0 μm have been obtained. The band gap and thermal response of variable-composition $Pb_{1-x}Hg_xS$ detectors are shown in Fig. 42 (*107*). A value of

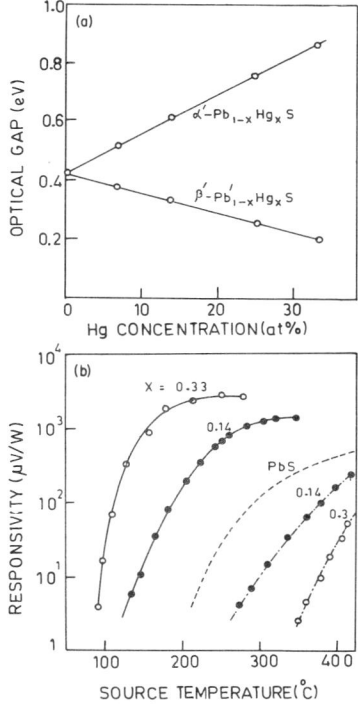

FIG. 42. (a) Composition dependence of the band gap of $Pb_{1-x}Hg_xS$ films and (b) thermal response of $\alpha'$-$Pb_{1-x}Hg_xS$ (—·—) and $\beta'$-$Pb_{1-x}Hg_xS$ (——) photoconductive detectors (*107, 111*).

$D^* \sim 1 \times 10^9$ cm $Hz^{1/2}$ $W^{-1}$ at 300 K has been obtained for these detectors, and a typical response time is $\sim 100$ μsec. By depositing the chalcogenide films on single-crystal Ge or Si slices, heterojunctions are very conveniently formed. These heterojunctions are well suited for photovoltaic detection of IR with response time in the range $10^{-6}$–$10^{-9}$ sec (107). Very little work has been done in this area so far. Considerable improvement in the performance of these photovoltaic devices is expected, once these heterojunctions are well understood.

Table XI lists some of the useful properties of the thin-film photon detectors.

## XIV. Concluding Remarks

It has been shown in this review that the two very old and empirical techniques of growth of inorganic films from chemical solutions have now been established on scientific foundations. The films obtained by these techniques have already found interesting and useful large-area applications. Some significant advantages of the two techniques over other physical and chemical vapor deposition techniques are: (1) simplicity and inexpensiveness, (2) easy control of deposition rates and film thicknesses, (3) inexpensive experimental setups under atmospheric conditions, (4) uniform coating of large areas, (5) efficient utilization of materials to form thin films, (6) con-

TABLE XI

PHOTOCONDUCTING CHARACTERISTICS OF SOLUTION-GROWN CHALCOGENIDE FILMS

| Film | Gain or responsivity | Peak response frequency (μm) | Response time (msec) | Reference |
|---|---|---|---|---|
| CdS | $10^6$ | 0.5 | 5 | (102, 103) |
| CdSe | $2 \times 10^3$ | 0.7 | 0.2 | (103, 106) |
| CdSe:Cu | $10^7$ | 0.7 | 0.1 | (103, 114) |
| CdSe:Ag | $5 \times 10^3$ | 0.7 | 0.2 | (103) |
| CdSe:Li | $10^6$ | 0.7 | 0.2 | (103) |
| PbS | 0.7 V $W^{-1}$ | 3.0 | 0.24 | (33, 107) |
| PbSe | 0.5 V $W^{-1}$ | 5.0 | 0.20 | (33, 103) |
| $CdSe_{1-x}S_x$ | $2 \times 10^3$–$10^6$ | 0.5–0.7 | 0.2–5 | (103, 106) |
| $PbSe_{1-x}S_x$ | 0.5–0.7 V $W^{-1}$ | 3.0–5.0 | 0.2 | (99) |
| $\alpha'\text{-}Pb_{1-x}Hg_xS$ | 0.7–60 V $W^{-1}$ | 1.2–3.0 | 0.2 | (107) |
| $\beta'\text{-}Pb_{1-x}Hg_xS$ | 0.7–80 V $W^{-1}$ | 3.0–6.0 | 0.1–0.2 | (107) |
| $Cd_{1-x}Pb_xSe$ | 0.5–10 V $W^{-1}$ | 0.7–5.0 | 0.2–0.5 | (103) |

venient doping with a predetermined profile, (7) codeposition and/or sequential deposition of different compounds/alloys to yield a gradient composition or a multilayer structure, (8) multicomponent uniphase or heterogeneous alloys/compounds, and (9) metastable and unusual crystal structures.

A number of aspects of the two techniques are not known or are poorly understood. Some of these aspects that deserve investigation are:

(1) To obtain uniform size and uniformly distributed droplets of micron or submicron size, the spray and atomization processes need to be studied more thoroughly. Alternative atomization techniques, such as ultrasonic cavitation (46, 126) and airless spray (127), have shown promising preliminary results.

(2) The chemistry of reaction kinetics and pyrolytic processes in micron-size droplets and particles is at best poorly understood at present. The "microchemistry" can be easily studied in a spray pyrolytic process.

(3) The thermodynamics of copyrolysis and sequential pyrolysis and the associated intermediate reactions are not well known.

(4) The use of the spray pyrolytic technique to obtain nonstoichiometric and doped oxide films and its extension to stoichiometric electronic oxides is a technologically useful development. The possibility of forming carbide and nitride films has also been demonstrated. An understanding of the chemistry of the processes involved and conditions for obtaining desired films is urgently needed.

(5) Our experiments show that the growth rate is influenced significantly by the topographical and chemical nature of the substrate and the presence of such metallic impurity ions in spray solution which modify the topography of the films. This area needs to be investigated. Also, the substrate–pyrolytic products interactions are not known to any great extent. Questions such as substrate etching, lack of adhesion on some substrates (e.g. metallic ones), and impurity diffusion from the substrate remain unanswered.

(6) The small grain size and the consequent high density of structural defects limit the usefulness of the films for electronic applications. The grain-growth and orientation effects induced by suitable dopants and by post-deposition annealing have received little attention so far.

(7) It is not possible at present to obtain very thick films by the solution growth technique. This can be achieved if precipitation can be minimized by using appropriate chemical inhibitors, or if the solution is continuously replenished with the requisite ions. The effect on the terminal thickness of nucleation centers and impurities on the substrate surface should be examined.

(8) In principle, the solution growth technique can be extended to form telluride and arsenide films if corresponding stable complexes can be synthesized.

(9) The kinetics of nucleation and growth in codeposition and sequential deposition processes in the solution growth technique need to be established in more detail.

(10) The applicability of the solution growth technique for forming oxide films has been demonstrated. A lot more work is required to establish the technique for more systems.

(11) The effects of an ac electric field and nucleation centers on the substrate on the ion-by-ion condensation process and microstructure of the films need further study.

(12) The dependence of the formation of metastable structures and relaxation of solubility limits on deposition conditions and catalytic agents requires detailed investigation.

(13) Although the solution growth technique has so far been used with cations of group II and IV elements, it should, in principle, be possible to extend it to elements in other groups such as Cu Ag, Al, Ga, In, As, Sb, Bi, Mo, and W. Preparation of chalcogenides of Sb and Bi has already been reported (*128*).

In conclusion, the versatility and usefulness of the spray pyrolytic and solution growth techniques for obtaining large-area coating of a wide variety of multicomponent oxides and chalcogenides have been established. It is hoped that this review will inspire further understanding and development of the techniques and their exploitation for large-area electronic, electro-optical, and solar energy conversion applications.

## References

1. M. Foex, *Bull. Soc. Chim. Fr.* **11,** 6 (1944).
2. R. R. Chamberlin, WPAFB Contract No. AF331657-7919 (1962).
3. J. E. Hill and R. R. Chamberlin, U.S. Patent 3,148,084 (1964).
4. R. R. Chamberlin and J. S. Skarman, *J. Electrochem. Soc.* **113,** 86 (1966).
5. R. J. Cashman, *J. Opt. Soc. Am.* **36,** 356 (1946).
6. J. L. Vossen, "Physics of Thin Films," (G. Hass, M. H. Francombe, and R. W. Hoffman, eds.), Vol. 9, p. 1. Academic Press, New York, 1979.
7. G. Haacke, *Annu. Rev. Mater. Sci.* **7,** 73 (1977).
8. T. S. Moss, *Proc. I.R.E.* **43,** 1869 (1955).
9. J. S. Skarman, *Solid State Electron.* **8,** 17 (1965).
10. R. R. Chamberlin, *Ceram. Bull.* **15,** 698 (1966).
11. R. R. Chamberlin and J. S. Skarman, *Solid State Electron.* **9,** 819 (1966).
12. C. S. Wu, R. S. Feigelson, and R. H. Bube, *J. Appl. Phys.* **43,** 756 (1972).
13. C. S. Wu and R. H. Bube, *J. Appl. Phys.* **45,** 648 (1974).
14. Y. Y. Ma and R. H. Bube, *J. Electrochem. Soc.* **124,** 1430 (1977).
15. R. H. Bube, F. Buch, A. L. Fahrenbruch, Y. Y. Ma, and K. W. Mitchell, *IEEE Trans. Electron Dev.* **ED-24,** 487 (1977).
16. R. S. Feigelson, A. N. Diaye, S. Yin, and R. H. Bube, *J. Appl. Phys.* **48,** 3162 (1977).

17. F. Buch, A. L. Fahrenbruch, and R. H. Bube, *J. Appl. Phys.* **48,** 1596 (1978).
18. S. Y. Yin, A. L. Fahrenbruch, and R. H. Bube, *J. Appl. Phys.* **49,** 1294 (1978).
19. J. Bougnot, M. Perotin, J. Marucchi, M. Sirkis, and M. Savelli, *Proc. IEEE Photovolt. Special. Conf. 12th* **519** (1976).
20. M. Savelli, *Proc. Workshop II–VI Solar Cells, Montpellier* p. I-1 (1979).
21. J. Bougnot, M. Savelli, J. Marucchi, M. Perotin, M. Marjin, O. Maris, C. Grill, and R. Pommier, *Proc. Workshop II–VI Solar Cells, Montpellier* p. II-1 (1979).
22. Oudeacoumar, Ph.D. Thesis, Université des Sciences et Techniques du Languedoc, Montpellier 1979.
23. P. K. Gogna, L. K. Malhotra, and K. L. Chopra, *Res. Ind.* **22,** 74 (1977).
24. A. Banerjee, Prem Nath, V. D. Vankar, and K. L. Chopra, *Phys. Status Solidi A* **46,** 723 (1978).
25. A. Banerjee, Prem Nath, S. R. Das, and K. L. Chopra, *Proc. Int. Solar Energy Congr. 7th New Delhi* (1978), p. 698.
26. E. Shanthi, D. K. Pandya, and K. L. Chopra, *Proc. Int. Solar Energy Congr., 7th, New Delhi* (1978).
27. A. Banerjee, Ph.D. Thesis, IIT Delhi, New Delhi (1978).
28. A. Banerjee, S. R. Das, A. P. Thakoor, H. S. Randhawa, and K. L. Chopra, *Solid State Electron.* **22,** 495 (1979).
29. E. Shanthi, A. Banerjee, V. Dutta, and K. L. Chopra, *Thin Solid Films* **71,** 237 (1980).
30. E. Shanthi, V. Dutta, A. Banerjee, and K. L. Chopra, *J. Appl. Phys.* **51,** 6243 (1980).
31. D. E. Bode, *Proc. Natl. Electron. Conf.* **19,** 630 (1963).
32. D. E. Bode, T. H. Johnson, and B. N. Maclean, *Appl. Opt.* **4,** 327 (1965).
33. D. E. Bode, in "Physics of Thin Films" (G. Hass and R. E. Thun, eds.), Vol. 3, p. 275. Academic Press, New York, 1966.
34. G. A. Kitaev, A. A. Uritskaya, and S. G. Mokrushin, *Russ. J. Phys. Chem.* **39,** 1101 (1965).
35. A. B. Lundin and G. A. Kitaev, *Inorg. Mater.* **1,** 2107 (1965).
36. G. A. Kitaev, S. G. Mokrushin, and A. A. Uritskaya, *Colloid J.* **27,** 38 (1965).
37. G. M. Fofanov and G. A. Kitaev, *Russ. J. Inorg. Chem.* **14,** 322 (1969).
38. G. A. Kitaev and T. S. Terekhova, *Russ. J. Inorg. Chem.* **15,** 25 (1970).
39. G. A. Kitaev and T. P. Sokolova, *Russ. J. Inorg. Chem.* **15,** 167 (1970).
40. N. C. Sharma, D. K. Pandya, H. K. Sehgal, and K. L. Chopra, *Mater. Res. Bull.* **11,** 1109 (1976).
41. N. C. Sharma, R. C. Kainthla, D. K. Pandya, and K. L. Chopra, *Thin Solid Films* **60,** 55 (1979).
42. R. C. Kainthla, D. K. Pandya, and K. L. Chopra, *J. Electrochem. Soc.* **127,** 277 (1980).
43. I. Kaur, D. K. Pandya, and K. L. Chopra, *J. Electrochem. Soc.* **127,** 943 (1980).
44. Spraying Systems Co., Wheaton, Illinois, 60187.
45. C. M. Lampkin, *Prog. Crystal Growth Charact.* **1,** 405 (1979).
46. F. Dutault, Ph.D. thesis, L' Universite de Haute Alsace et L' Universite Louis Pasteur de Strasbourg, 1979.
47. F. Dutault and J. Lahaye, *Proc. Photovolt. Solar Energy Conf., Berlin* (*1979*) p. 898.
48. K. Yokoto, Y. Yoshimatsu, S. Katayam, and T. Kariya, *Japan J. Appl. Phys.* **18,** 1957 (1979).
48a. H. B. Serreze, S. Lis, M. R. Squillante, R. Turcolte, M. Talbot, and G. Entine, *Proc. 15th IEEE Photovoltaic Specialists Conf. Kissimmee, Florida, Aug. 1981*, p. 1068 (1981).
49. B. R. Mehta, A. P. Thakoor, D. K. Pandya, and K. L. Chopra (unpublished).
49a. E. Shanthi, Ph.D. Thesis, Indian Institute of Technology, New Delhi (1980), unpublished.
50. J. C. Manifacier, L. Szepessy, J. F. Bresse, M. Perotin, and R. Stuck; *Mater. Res. Bull.* **14,** 109 (1979).
51. B. R. Mehta, A. P. Thakoor, D. K. Pandya, and K. L. Chopra (unpublished).
52. O. V. Varob'eva and E. S. Bessonova, *Steklo Kepan.* **21,** 9 (1964).

53. K. Chidambaram, L. K. Malhotra, and K. L. Chopra, *Thin Solid Films* **87**, 365 (1981).
54. J. Aranovich, A. Ortiz, and R. H. Bube, *J. Vac. Sci. Tech.* **16**, 994 (1979).
54a. S. Major, A. Banerjee, and K. L. Chopra (unpublished).
55. I. I. Borisova and O. K. Botvinkin, *Steklo Keram.* **22**, 15 (1965).
56. G. E. Carve, *Solar Energy Mater.* **1**, 357 (1979).
57. H. J. Hovel, *J. Electrochem. Soc.* **125**, 983 (1978).
58. T. R. Viverito, E. W. Rilu, and L. H. Slack, *Am. Ceram. Soc. Bull.* **54**, 217 (1955).
59. J. Kane, H. P. Schweizer, and W. Kein, *J. Electrochem. Soc.* **122**, 144 (1978).
60. V. F. Korzo and L. A. Ryabova, *Sov. Phys. Solid State* **9**, 745 (1967).
61. V. F. Korzo and V. N. Chernayev, *Phys. Status Solidi A* **20**, 695 (1973).
62. J. Kane, H. P. Schweizer, and W. Kern, *J. Electrochem. Soc.* **29**, 155 (1975).
63. E. Horvath and A. J. Perry, *Thin Solid Films* **65**, 309 (1980).
64. V. N. Semenov, and Yu. E. Babenko, *Inorganic Mater.* **14**, 193 (1978).
65. R. C. Kainthla, A. Banerjee, D. K. Pandya, and K. L. Chopra, (unpublished).
66. A. Rohtagi, T. R. Viverito, and L. H. Slack, *Int. Q. Sci. Rev. J.* (1975) p. 139.
67. C. A. Vincent, *J. Electrochem. Soc.* **119**, 515 (1972).
68. Marcel van der Leij, Ph.D. thesis, Delft University, 1979.
69. W. J. Deshotels, F. Augustine, and A. Carlson, *2nd Q. Rep.* Contract NAS 7-203, Clevite Corp. (1963).
70. W. J. Deshotels, F. Augustine, A. Carlson, J. Koening, and M. P. Makowski, *3rd Q. Rep.* Contract NAS-7-203, Clevite Corp. (1963).
71. B. R. Pamplin, *Prog. Cryst. Growth Charact.* **1**, 395 (1979).
72. B. R. Pamplin and R. S. Feigelson, *Mater. Res. Bull.* **14**, 1 (1979); *Thin Solid Films* **60**, 141 (1979).
73. E. Shanthi, A. Banerjee, and K. L. Chopra, *Thin Solid Films* **88**, 93 (1982).
74. R. E. Aitchison, *Aust. J. Appl. Sci.* **5**, 10 (1954).
75. A. Fischer, *Z. Naturforsch.* **9** A, 508 (1954).
76. H. Ladwig, *Silikattechnik* **15**, 182 (1964).
77. I. Golovcenco, Gh. I. Rusu, V. Stefan, and M. Rusu, *Iasi. Sect. Ib. Fiz.* **11**, 77 (1965).
78. P. W. Haayman, P. C. Vander Linden, D. Veeneman, and G. H. Janssen, U.S. Patent 2,772,190 (1956).
79. Philips Electricals Industries Ltd., British Patent 73,2566 (1955).
80. J. W. McAuley, U.S. Patent 2,692,836 (1954).
81. Union des Verreries Macaniques Belges, British Patent 892708.
82. R. Groth, *Phys. Status Solidi* **14**, 69 (1966).
83. B. P. Kryzhanoyskii and M. A. O Katoy, *Zhur. Prik. Khim* **39**, 2832 (1966).
84. W. O. Lytle and A. E. Wagner, U.S. Patent 2,740,731 (1956).
85. M. S. Tarnopol, U.S. Patent 2,694,649 (1954).
86. R. S. Berg, R. D. Nasby, and C. Lampkin, *J. Vac. Sci. Tech.* **15**, 359 (1978).
87. A. P. Thakoor, B. R. Mehta, D. K. Pandya, and K. L. Chopra, *Thin Solid Films* **83**, 231 (1981).
88. B. Raj, D. K. Pandya, and K. L. Chopra, *Proc. Nat. Solar Energy Conv. 1981, Bangalore, India*, p. 7.001. Allied Publishers, New Delhi (1982).
89. For example, K. L. Chopra, in "Thin Film Phenomena." McGraw-Hill, New York, 1969.
90. D. L. Zelmer, Ph.D. thesis, University of Illinois, 1969.
91. D. Elliott, D. L. Zellmer, and H. A. Laitiner, *J. Electrochem. Soc.* **117**, 1343 (1970).
92. F. B. Micheletti and P. Mark, *Appl. Phys. Lett.* **10**, 136 (1967).
93. G. Haacke, H. Ando, and W. E. Mealmaker, *J. Electrochem. Soc.* **124**, 1923 (1977).
94. H. L. Kwok and Y. C. Chau, *Thin Solid Films* **66**, 303 (1980).
95. For example, F. J. Welcher and R. B. Hahn, in "Semi-Micro Qualitative Analysis." Reinhold (Van Nostrand) Princeton, New Jersey, 1963.

96. D. O. Skovlin and R. A. Zingaro, *J. Electrochem. Soc.* **111**, 42 (1964).
97. G. A. Kitaev, V. Ya Shcherbakova, V. I. Dvoinin, and N. N. Belyaeva, *Zh. Prikl. Khim.* **51**, 18 (1978).
98. T. M. Racheva, I. D. Dragieva, D. H. Djoglev, and P. P. Dimitrova, *Thin Solid Films* **17**, 85 (1973).
99. R. C. Kainthla, D. K. Pandya, and K. L. Chopra (unpublished).
100. A. G. Shikalgar and S. H. Powar, *Solid State Commun.* **32**, 361 (1979).
101. N. R. Pavaskar, C. A. Memzes, and A. P. B. Sinha, *J. Electrochem. Soc.* **124**, 743 (1977).
102. R. C. Kainthla, Ph.D. thesis, Indian Institute of Technology, Delhi, 1980.
103. I. Kaur, Ph.D. thesis, Indian Institute of Technology, Delhi, 1980.
103a. R. D. Pachori, I. J. Kaur, and K. L. Chopra (unpublished).
104. N. C. Sharma, D. K. Pandya, H. K. Sehgal, and K. L. Chopra, *Thin Solid Films* **42**, 383 (1977).
105. R. C. Kainthla, D. K. Pandya, and K. L. Chopra, (to be published).
106. R. C. Kainthla, D. K. Pandya, and K. L. Chopra, *J. Electrochem. Soc.* **129**, 99 (1982).
106a. Y. S. Sarma, H. N. Acharya, and N. K. Misra, *Thin Solid Films* **90**, L43 (1982).
107. N. C. Sharma, Ph.D. thesis, Indian Institute of Technology, Delhi, 1978.
108. N. C. Sharma, D. K. Pandya, H. K. Sehgal, and K. L. Chopra, *Thin Solid Films* **59**, 157 (1979).
108a. H. Rahnamai, H. J. Gray, and J. N. Zemel, *Thin Solid Films* **69**, 347 (1981).
109. A. Vecht, "Physics of Thin Films" (G. Hass and R. E. Thun, eds.), Vol. 3, p. 165. Academic Press, New York, 1966.
110. J. N. Zemel, J. D. Jensen, and R. B. Schoder, *Phys. Rev. A* **140**, 330 (1965).
111. N. C. Sharma, D. K. Pandya, H. K. Sehgal, and K. L. Chopra, *Thin Solid Films* **62**, 97 (1979).
112. R. C. Kainthla, V. Dutta, A. K. Mukerjee, D. K. Pandya, and K. L. Chopra, (to be published).
113. R. C. Kainthla, D. K. Pandya, and K. L. Chopra, (to be published).
113a. R. H. Bube, "Photoconductivity of Solids," Wiley, New York (1960).
114. S. V. Svechnikov and E. B. Kagonovich, *Thin Solid Films* **66**, 41 (1980).
115. V. P. Singh, Photon Power Inc., El Paso, USA (personal communication).
116. M. Savelli, *Proc. IEEE Photovolt. Special. Conf., 14th (1980)*.
117. T. Feng, A. K. Ghosh, and C. Fishman, *Appl. Phys. Lett.* **35**, 266 (1979).
118. S. Ashok, P. P. Sharma, and S. J. Fonash, *IEEE Trans. Electron Devices* **ED-27**, 725 (1980).
119. J. Aronovich-Magran, *Prog. Cryst. Growth Charact.* **1**, 4 (1979).
120. K. L. Chopra and D. K. Pandya, CdS-Based Thin Film Solar Cells. A Department of Science and Technology Project Report, Dec. (unpublished) (1980).
121. M. S. Yadav, M.Tech. Thesis, Department of Physics, Indian Institute of Technology, Delhi (unpublished) (1981).
121a. R. C. Kainthla and D. Hanneman (unpublished).
122. S. N. Kumar, L. K. Malhotra, and K. L. Chopra, (unpublished).
123. J. Vedel, P. Cowache, and M. Dachraoui, (to be published).
124. G. B. Reddy, V. Dutta, D. K. Pandya, and K. L. Chopra, *Proc. Nat. Solar Energy Conversion, Annamalainagar, India, 1980*, p. 260. Allied Publishers, New Delhi (1980).
125. G. B. Reddy, V. Dutta, D. K. Pandya, and K. L. Chopra, *Solar Energy Materials* **5**, 187 (1981).
126. G. Blandenet, M. Court, and Y. Lagarde, *Thin Solid Films* **77**, 81 (1981).
127. J. Vedel, B. Thiebaut, and M. Levart, (to be published).
128. P. Pramanik and R. N. Bhattacharya, *J. Electrochem. Soc.* **127**, 1857 and 2087 (1980); *Solar Energy Materials* **6**, 317 (1982); and *J. Solid State Chem.* (1982), in press.

# Plasma-Enhanced Chemical Vapor Deposition of Thin Films

S. M. OJHA

*Standard Telecommunications Laboratories Limited*
*Harlow, Essex, England*

| | |
|---|---|
| I. Introduction | 237 |
| II. Deposition Techniques and Systems | 238 |
| III. Preparation and Properties of Films | 246 |
|     1. Elements | 246 |
|     2. Oxides | 258 |
|     3. Nitrides | 265 |
|     4. Carbides | 272 |
| IV. Plasma Oxidation | 275 |
|     1. GaAs Oxide | 277 |
|     2. Silicon Oxide | 282 |
|     3. rf Oxidation | 282 |
| V. Plasma Carburizing | 283 |
| VI. Glow-Discharge Nitriding | 286 |
| VII. Conclusions | 289 |
|     References | 289 |

## I. Introduction

In recent years there has been growing interest in using low-pressure glow discharges for promoting chemical reactions that are otherwise difficult or indeed impossible to achieve by conventional techniques. This has resulted in the development of novel processes and materials. Plasma etching of materials for fabricating submicron semiconductor devices is a well-established example. Plasma oxidation, which has been known for a long time, has now become important for producing device-quality oxide layers on GaAs, for making metal–oxide–semiconductor field-effect transistors (MOSFET), and for fabricating Josephson junction devices. Oxygen or hydrogen glow discharges are used to clean, decarburize, and passivate metallic surfaces in ultrahigh-vacuum (UHV) systems such as Auger chambers and fusion tokamaks. Glow discharges are also used commercially for cleaning electrical contacts on printed-circuit boards. Deposition of solid films by glow discharge decomposition of gases has also been known for some time. The nonequilibrium nature of the plasma (high electron temperature and low gas temperature) leads to dissociation of a gas and deposition of a film at much lower temperatures than are possible with pyrolytic chemical vapor deposition (CVD). This is one of the main reasons for the sustained interest in this technique. Thus a wide range of elemental and compound materials can be deposited on temperature-sensitive substrates. At the same time, the materials deposited by plasma-enhanced CVD (hereinafter referred to as plasma CVD) have been found to have physical, structural, and chemical properties different from similar materials deposited by other techniques. Novel materials have thus been deposited. Plasma-deposited films are conformal in nature and produce a lower pinhole density compared to "line-of-sight" deposition techniques such as evaporation and sputtering. The recent interest in plasma-induced deposition has stemmed from the possibility of using amorphous hydrogenated silicon as a cheap solar cell material (*1*) and the commercial exploitation of plasma–silicon nitride for passivation of silicon devices. Insulating and amorphous C(H) films are being commercially used for protective and antireflection coatings on infrared lenses. Besides producing films of commercial value, plasma CVD has given rise to new materials of novel physical properties, for example, amorphous phosphorous, insulating carbon, phosphorous nitride, and silicon carbide. Polycrystalline silicon films are reported to have been grown by chemical transport in a low-pressure plasma at temperatures as low as 80°C (*2*). In this review various recent developments that have taken place in the field of plasma-induced CVD are briefly discussed. Emphasis is given to new materials that have not been discussed elsewhere.

## II. Deposition Techniques and Systems

In the majority of work reported on plasma CVD the discharge is excited by an rf field, though dc and microwave fields have also been used. In the low-pressure range ($10^{-3}$–10 Torr) typically used for plasma deposition the average electron energy is in the range 2–12 eV and is sufficient for ionizing and dissociating most types of gas molecules. The degree of ionization, extent of dissociation, and excitation of the gas atoms or molecules varies from gas to gas. Although electrons are the ionizing source, collisions involving excited species also assist the ionization process and radical formation.

For rigorous discussion of the ionization processes and plasma chemistry the reader is referred to various articles in the literature (*3–8*). The electron energy distribution and its density not only influence the homogeneous reactions in a discharge, but also affect the heterogeneous reactions occurring on the substrate surface (*9–11*). The substrates immersed in a plasma acquire a negative floating potential with respect to the plasma, the magnitude of which depends on the electron energy and the ion mass. The condensing species on the substrates would be subjected to ion bombardment, which can influence the properties of the growing film. In addition to the discharge parameters, experimental conditions such as pressure, flow rate, power, and substrate temperature also affect the heterogeneous reactions (*12*).

Plasma-CVD techniques can be divided broadly into two types, depending on the way the electrical energy is coupled to the discharge (Fig. 1). The potential distributions inside the tunnel and the parallel-plate reactor are

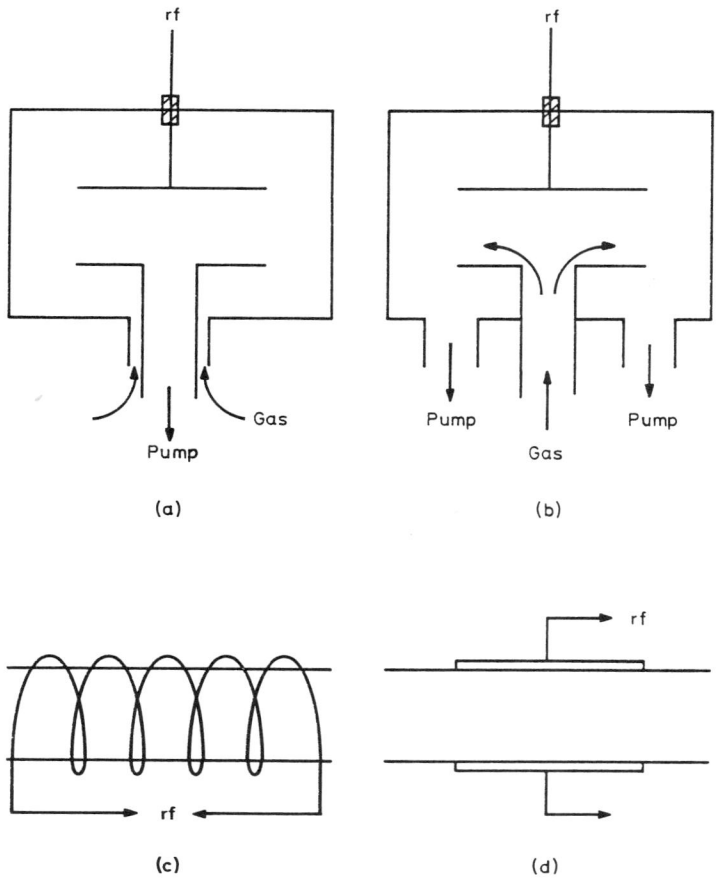

FIG. 1. Basic plasma-CVD reactors: (a, b), parallel-plate, radial flow, or flat-bed reactors; (c, d), tunnel reactors.

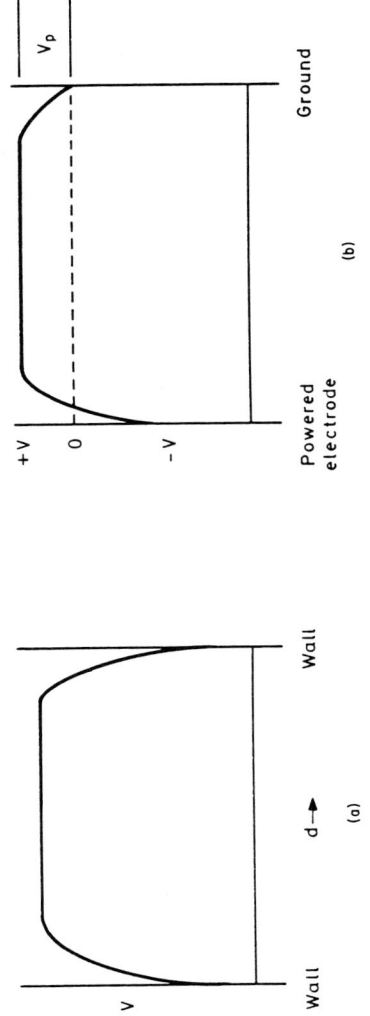

FIG. 2. Distribution of potential in space in (a) tunnel and (b) flat-bed reactors. [After Vossen (*13*). Reprinted by permission of the publisher, The Electrochemical Society, Inc.]

TABLE I

Plasma Deposition Reactor Potentials[a,b]

| rf[c] voltage (P–P V) | rf current (A) | $V_p^d$ (V) | $V_f^d$ (V) | Substrate and Al$_2$O$_3$-coated wall[e] (V) |
|---|---|---|---|---|
| 880 | 0.53 | +190 | + 80 | −110 |
| 1200 | 1.2 | +200 | +100 | −100 |
| 1300 | 3.5 | +210 | +110 | −100 |

[a] After Vossen (13). Reprinted by permission of the publisher, The Electrochemical Society, Inc.
[b] Pressure: 0.2 Torr. Gas flows (cm$^3$ min$^{-1}$): SiH$_4$, 74.2; N$_2$, 469; NH$_3$, 185.
[c] Frequency: 50 kHz; single-probe measurements.
[d] $V_p$ = plasma potential with respect to ground; $V_f$ = floating potential with respect to ground.
[e] Average value with respect to $V_p$.

different (13), as seen in Fig. 2. The plasma potential and floating substrate and wall potentials in a parallel-plate deposition system (Table I) and an "inductively" coupled tunnel reactor (Table II) can vary substantially, depending on the discharge conditions (13). The high wall potentials can lead to sputtering of the walls, thus contaminating a growing film (229). The uniformity of film thickness and properties in parallel-plate reactors is reported to be better than in tunnel reactors (12). Tunnel reactors, on the

TABLE II

Tunnel Reactor Potentials[a]

| Gas | Pressure (Torr) | Rf[b] power (W) | Peak coil voltage (V) | Wall[c] potential (V) | Substrate potential (V) |
|---|---|---|---|---|---|
| CF$_4$ | 3 | 50 | 500 | − 40 | − 8 |
|  |  | 100 | 560 | −100 | − 30 |
|  |  | 200 | 930 | −200 | −110 |
| O$_2$ | 7 | 50 | 450 | − 40 | − 5 |
|  |  | 100 | 600 | −100 | − 26 |
|  |  | 200 | 950 | −320 | −160 |

[a] After Vossen (13). Reprinted by permission of the publisher, The Electrochemical Society, Inc.
[b] Frequency: 50 kHz; double-probe measurements.
[c] With respect to the plasma potential.

other hand, can take a higher throughput of wafers than parallel-plate reactors. Rosler and Engle (*14*) have developed a plasma deposition system that combines the advantages of parallel-plate and tubular reactors (Fig. 3). The system is claimed to deposit uniform films over a wide range of deposition conditions. Vertical positioning of the wafers also reduces contamination of the films from particulates. It is worth noting that the emphasis of plasma-CVD systems has so far been directed toward coating two-dimensional substrates (Si wafers). Not much has been reported about the coating systems and deposition conditions that can deposit uniform films over three-dimensional substrates. Tubular reactors with external electrodes for exciting the discharge are likely to have obvious advantages.

Although it is possible to produce films at near ambient temperatures by plasma CVD, in many cases the films can be under high stress and the film growth rate very low. By raising the electrical power applied to the discharge it is possible to achieve a higher degree of gas dissociation and thus greater deposition rates; this can also, however, lead to excessive substrate heating. Different approaches to solving this problem have been taken (*15–18*). Dissociation of nitrogen in small cells prior to the films' entering the deposi-

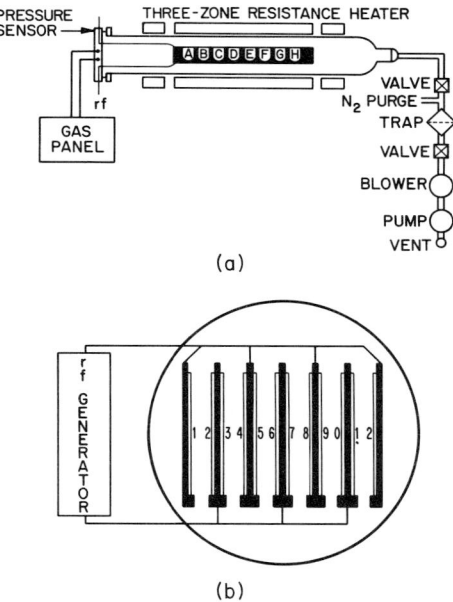

FIG. 3. Plasma deposition system of Roster and Engle (*14*). (a) Side view cross section of system; (b) front view cross section of reactor internals. (Courtesy Solid State Technology.)

tion zone is claimed to have resulted in silicon nitride films of low stresses at deposition rates of about 200–300 Å min$^{-1}$ at substrate temperatures only a few degrees above room temperature (Fig. 4).

Another novel method for restricting the substrate to room temperature during deposition uses a high-energy pulsed plasma rather than a continuous discharge (16). The discharge between two coaxial electrodes is struck by discharging the capacitors (Fig. 5). The high temperatures produced during the pulsed discharge evaporate and ablate the electrode material, which reacts with the gas in the chamber to form compound films. Materials having high melting points, such as $Al_2O_3$, AlN, $Ta_2O_5$ and Borazone, are reported to be deposited by pulsing gases such as oxygen, nitrogen, or mixtures of nitrogen and hydrogen during the discharge (17). Alternatively, gases such as methane or a mixture of diborane and nitrogen are pulsed during the discharge, and dissociated products condense on substrates. One of the stated disadvantages of this method is that the materials deposited are not pure and contain other phases.

Deposition of films by plasma-chemical transport is another interesting technique for producing films at low substrate temperatures (18). In this technique (Fig. 6), a material is transported from a solid phase by plasma-dissociated gas, such as hydrogen, and decomposed by thermal and plasma activation in another part of the reactor. Various materials, including phosphorus, silicon, germanium, TiN, AlN, and phosphorus nitride, have been grown by this method. Polycrystalline silicon is claimed to have been deposited at temperatures as low as 80°C.

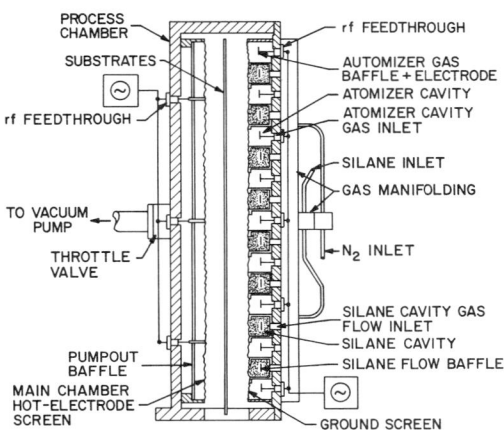

FIG. 4. Silicon nitride deposition system with separate nitrogen dissociation cells (15).

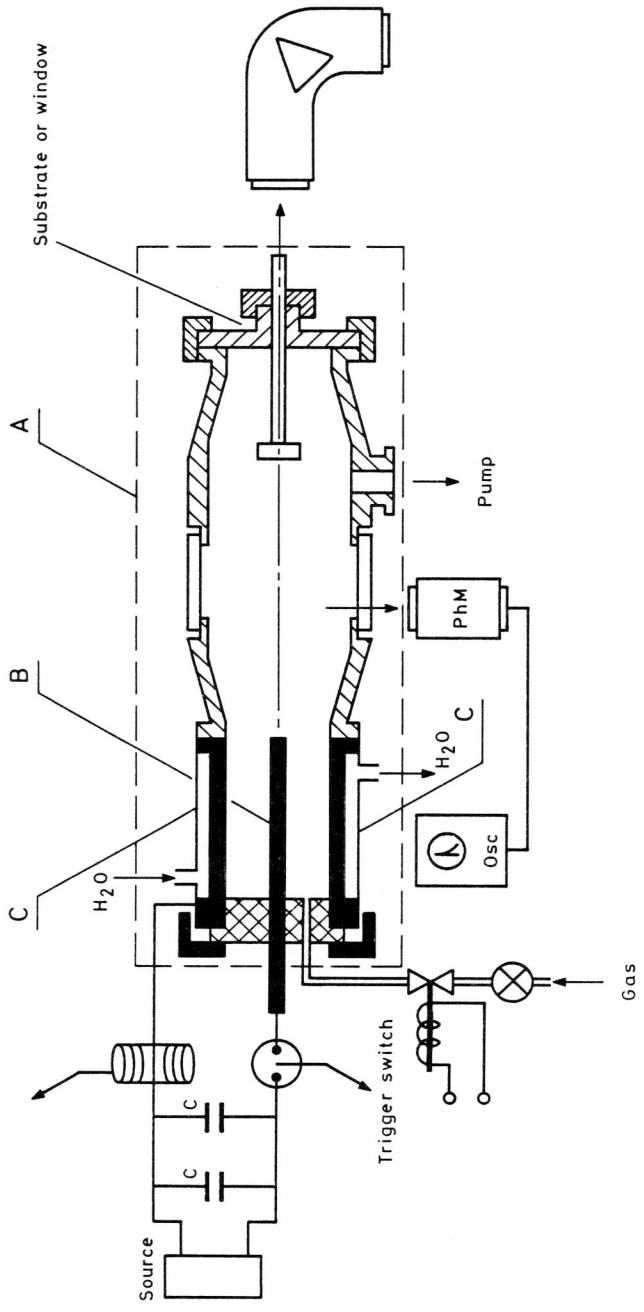

FIG. 5. Pulsed-plasma deposition system. [After Sokolowski (16).]

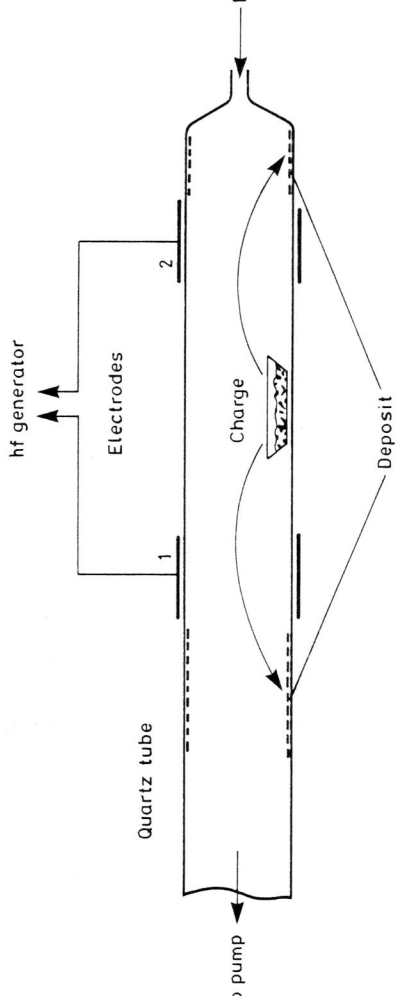

FIG. 6. Experimental arrangement for chemical transport of red phosphorous with hydrogen. [After Veprek (*18*).]

## III. Preparation and Properties of Films

### 1. ELEMENTS

*a. Carbon.* Carbon coatings have been deposited by the following techniques:

(1) Thermal (arc, electron, or laser beam) evaporation (*19, 20*)
(2) Sputtering (*21, 22*)
(3) Pyrolytic decomposition of organic compounds (*23, 24*)
(4) CVD on diamond seed crystals (*25–29*)
(5) Ion beam deposition (*31–35*)
(6) Ion plating (*30*)
(7) Plasma CVD

Each of these techniques is known to produce carbon films of different physical properties. However, on the basis of electrical conductivity and mechanical hardness the carbon coatings can be broadly divided into two groups:

(i) Conducting ($\leq 10^2$ Ω cm) and soft
(ii) Insulating and mechanically hard

Carbon films prepared by vacuum evaporation, sputtering, or pyrolytic decomposition are known to fall into group (i), and the films prepared by ion beam deposition, ion plating, and plasma CVD fall into group (ii). Insulating carbon films have also been epitaxially grown on diamond seed crystals by thermal decomposition of methane. For example, Angus and co-workers (*27–29*) claim deposition of epitaxial diamond films on diamond seed crystals (0–5 μm size) by decomposing methane and hydrogen mixture over the crystals heated to 1253–1338 K. The ion beam and plasma-CVD techniques, however, can deposit insulating carbon films at much lower temperatures and as such have recently been widely investigated. In the following these techniques are briefly discussed and properties of the films produced by the same are compared.

*i. Ion-Beam Deposition.* The deposition of films by extracting ions of a material from an ion source and subsequent condensation of these ions on substrates has been reported in the past for various materials, such as Si, Ge, and C. One of the main features of this technique is the manner in which the growth conditions can be controlled. For example, the nucleation and growth conditions of a film grown on a substrate maintained at a constant temperature can be altered merely by changing the biasing voltage on the substrates, which in turn alters the kinetic energy of the condensing ions.

The increased mobility of the atoms on substrates maintained at near-ambient or lower temperatures can produce films of novel physical properties.

Aisenberg and Chabot (32) made the first detailed study of carbon film deposition by C-ion beam. The authors used a carbon-sputter ion source (Fig. 7) in which argon gas, typically at a partial pressure of $\sim 2 \times 10^{-3}$–$50 \times 10^{-3}$ Torr was used for producing carbon ions. The deposition chamber, separated from the ion source by a constrictor ($\sim 1.5$ mm diameter) was maintained at much lower pressure ($10^{-6}$ Torr). An axial magnetic field was used for enhancing the discharge in the ion source and deposition chamber. Thus the substrates were observed to be surrounded by a weak plasma. The carbon and argon ions were extracted from the source by applying either a dc or a rf bias voltage to the substrates, which was also necessary for neutralizing the impinging positive ions. Prior to deposition the substrates were sputter-etched by applying $-400$ V to the substrates; the voltage was increased to $-40$ V during the deposition of films. The film deposition rate was reported to be around 50 Å min$^{-1}$ and the rise in substrate temperature was claimed to be no more than 10°C above the ambient.

The carbon films grown in this manner were optically transparent in the visible region, with a refractive index of 2. Electrical resistivity of the films deposited on single-crystal silicon (100) and stainless-steel substrates was typically $10^{11}$ Ω cm. The dielectric constant determined from capacitance measurements was $\sim 8$–$14$. The dielectric constant of diamond is reported to be 16.5 (56). The density of states determined from $C$–$V$ measurements was calculated to be about $5 \times 10^{11}$ cm$^{-2}$. The $C$–$V$ characteristics of the films showed less than 0.5 V shift when exposed to a $^{60}$C source ($10^6$ rad). Thin-film transistors of coplanar structure were fabricated and a trans-

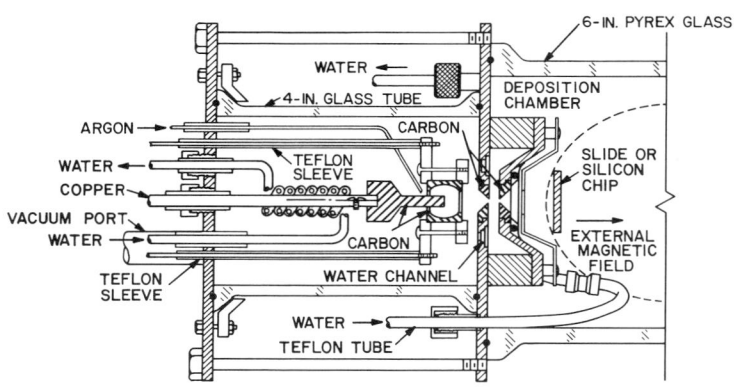

FIG. 7. Ion beam source for depositing carbon films. [After Aisenberg and Chabot (32).]

conductance value of 75 μmhos was reported. The films, several microns thick, had an amorphous surface layer and crystalline bulk.

A more detailed study of the structure of similarly deposited films was done by Spencer et al. (33). X-ray diffraction studies on carbon films about 2 μm thick with the exposure time extending up to 24 hr showed the presence of very weak lines corresponding to diamond. The absence of any lines from graphite suggested tetragonal bonding of carbon atoms. The average grain size in the films, estimated from line broadening, was ~50–100 Å. Electron diffraction analysis of the approximately 100-Å-thick carbon films showed that they were polycrystalline, with a cubic diamond structure. Bright-field imaging of the films revealed 1–5 μm single crystals. Deposition of epitaxial layers on single-crystal diamond substrates by carbon ion beam has also been reported by Freeman et al. (36). They claim to have grown crystalline films up to 10 μm thick. The carbon ion energy was typically 900 eV, which is much higher than those mentioned earlier. The substrates were heated to about 700°C. The films contained carbonaceous inclusions whose density depended on the substrate cleanliness.

*ii. Ion Plating.* The effect of argon ion bombardment on the mechanical and structural properties of electron-beam-evaporated carbon films during their growth has been studied by Teer and Salma (30). The substrate (copper or mild steel) bias voltage during the deposition was varied from $-500$ to $-5000$ V. The ion-plated films had better adhesion and lower internal stresses than evaporated films. Carbon films deposited by either technique were graphitic in nature, but ion-plated films had a better crystallinity. Fujimari and Nagai have also observed changes in the physical properties of laser-evaporated carbon films due to argon ion bombardment during the deposition. The deposition rate was approximately 10Å $\text{min}^{-1}$ at a partial pressure of $10^{-5}$ Torr of argon, and the argon ion current density was typically 20 μA $\text{cm}^{-2}$. The ion-bombarded films were found to have higher transmittance at wavelengths of 0.5–1.4 μm, better adhesion, and a higher electrical resistivity, typically $5 \times 10^{-1}$ Ω cm, compared to unbombarded films. The changes in the physical properties were attributed to the breaking up of larger carbon clusters of graphitic structure into single atoms or smaller clusters by the impinging argon ions, thus producing a random structure with a mixture of graphitic and diamond bonds. By carefully balancing the carbon growth rate and energy of the bombarding ions, Weissmantel and co-workers (34, 35) were able to grow films that were much closer to the properties of the diamond films. Carbon was sputtered by an argon ion beam and the growing film was simultaneously bombarded by another argon ion beam ($<1$ keV, 120 μA $\text{cm}^{-2}$). Films deposited on silicon and quartz substrates were insulators and were transparent in the visible region. Films deposited on metallic substrates were black and conducting. Electron microscopy studies showed the films to contain crystallites about 100 Å in diameter. The crystallite size and distribution were, however, found to vary from sample to sample. The films were under high compressive stress.

*iii. Plasma CVD.* Unlike the previously mentioned techniques, deposition of carbon films by the glow discharge decomposition of hydrocarbon gases is a much simpler and more versatile technique for depositing insulating carbon films.

Both dc and rf discharges (internally or externally excited) have been used for depositing films. However, for depositing films with mechanical and optical properties approaching those of diamonds, specific deposition conditions are required, as discussed below.

Polymerization of a hydrocarbon when introduced into a low-pressure glow discharge is a well-documented process *(38–40)*. The collisions between the energetic electrons in the plasma and the monomer lead to the production of active species, such as free radicals, excited molecules, and ions. The ratio of the ionic and neutral fragments in a dissociated gas depends on the electron energy. For example, for methane it has been observed that below 50 eV most of the dissociated fragments are uncharged ground state or long-lived excited-state molecules *(41)*.

Normally these species form polymers, either in the gas phase or on the surfaces inside the reactor. The condensing species on the surfaces exposed to glow discharge undergo very weak ion bombardment, which is typical in the case of an electrically floating substrate. However, if the substrates placed in such a discharge are deliberately biased negatively with respect to ground, then the ionized species will impinge energetically on the growing film. The growth process and properties of the films will then depend on the energy of the ions. It is worth noting here that during the initial stages of the film growth, deposition of carbonaceous species is preceded by sputter-etching of the substrates by the impinging energetic particles *(42–45)*. However, since carbon has a very low sputtering yield, the subsequent film growth predominates over etching. Dc *(46)* and rf sources *(47)* (Figs. 8 and 9) have been used for electrically biasing the substrates at a negative potential to ground during the growth of carbonaceous films. The films grown under different ion energies have been divided, depending on their optical and mechanical properties, into three categories *(51)*.

(1) Films deposited at high monomer pressure (e.g., $>0.5$ Torr), low substrate bias (e.g., $\leq -100$ V to ground), and power density (e.g., $<1$ W cm$^{-2}$). The films thus grown are mechanically soft, yellow in transmission, and show optical absorption peaks in the infrared (IR) region of the spectrum that are characteristic of degraded polymers (Fig. 10). Growth and proper-

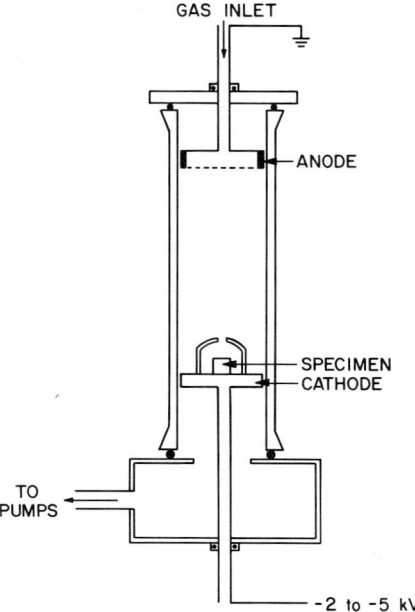

FIG. 8. dc System for depositing carbon films. [After Whitwell and Williamson (46).]

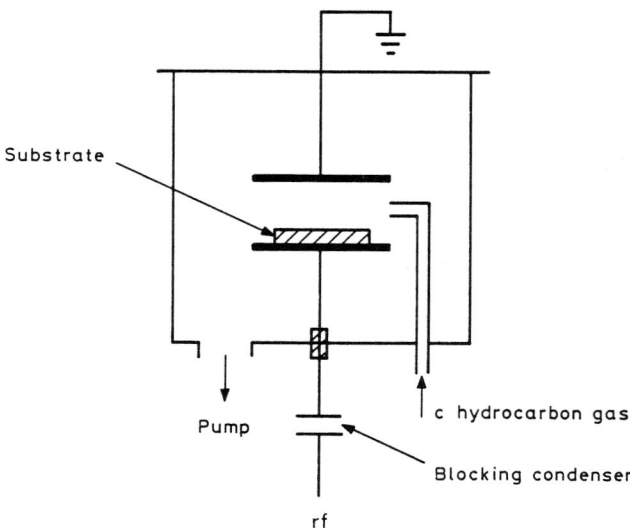

FIG. 9. rf System for depositing carbon films.

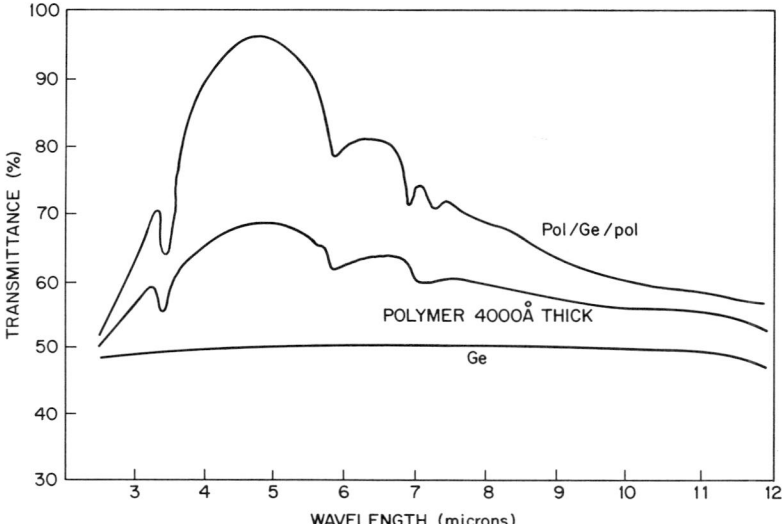

FIG. 10. The transmittance of germanium blank before and after coating one and two surfaces with polymer films obtained at low target power and high hydrocarbon pressure (30 W, 0.6 Torr). [After Holland and Ojha (51).]

ties of polymerized hydrocarbons have been reported by various workers in the past (38–40).

(2) Films deposited at relatively lower pressures (e.g., <0.5 Torr), higher substrate potentials (e.g., >500 V), and power density (e.g., ~4 W cm$^{-2}$). The films thus grown show dramatic change in their properties compared to condition (1). They are extremely hard (hardness >2000 VHN), electrically insulating (e.g., ~$10^{12}$ Ω cm), with good adhesion to the substrates, and have far fewer and weaker absorptions in the IR due to bonded hydrogen (Fig. 11).

(3) Films grown at low pressures (e.g., <0.5 Torr) and at high power inputs (≥6 W cm$^{-2}$) and substrate potentials. Joule heating of the growing film due to excessive ion current and ion energy results in films that are mechanically soft, absorbing in the IR (Fig. 12), and have a much higher electrical conductivity compared to films in category (2).

The properties of the carbonaceous films deposited under condition (2) have been the subject of considerable investigation, mainly due to some similarities between their physical properties and those of diamond (42–52, 62, 63). Electron and X-ray diffraction analysis of the films indicate that the films are amorphous in structure (49), unlike the ion-beam deposited

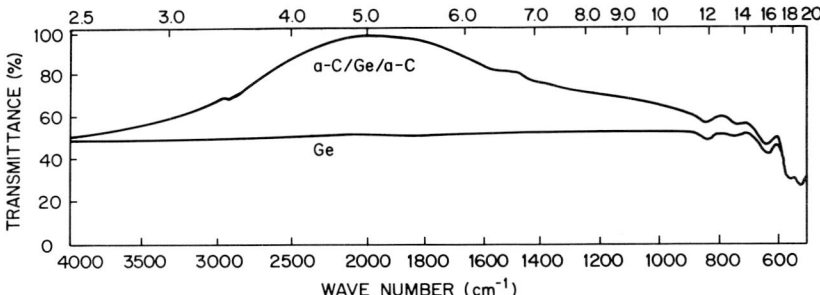

FIG. 11. The transmittance of germanium blank coated on both sides with $\lambda/4$ a-C films for $\lambda = 5$ μm. [After Holland and Ojha (51).]

films (33). In addition to the IR spectroscopic studies mentioned earlier, Rutherford backscattering analysis of the films by 2-MeV $^4$He$^+$ and 0.9-MeV protons show that as much as 25 at. % hydrogen can be present in the films (52). The films deposited under condition (2) would be referred to as a-C(H) films. It is worth noting that unlike the a-Si(H) films, a-C(H) coatings are grown under relatively high-energy ion impact.

The a-C(H) films tend to absorb in the visible spectrum, the absorption decreasing towards the higher wavelengths. The films appear brownish in transmission. In reflection the color of the films changes, depending on their thickness (54). In the IR region from 2.5 to 20 μm the films are reported to be virtually absorption free (Fig. 13). The refractive index of the films, deduced from the transmission curves, is 2.0, and is sensitive to the prepara-

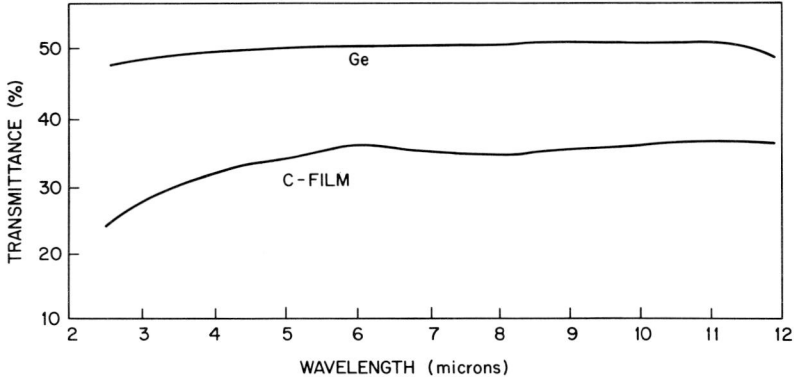

FIG. 12. The transmittance of a germanium blank with one side coated with a carbon film (800 Å thick) that was partially graphitized at high power (420 W, 0.15 Torr). [After Holland and Ojha (51).]

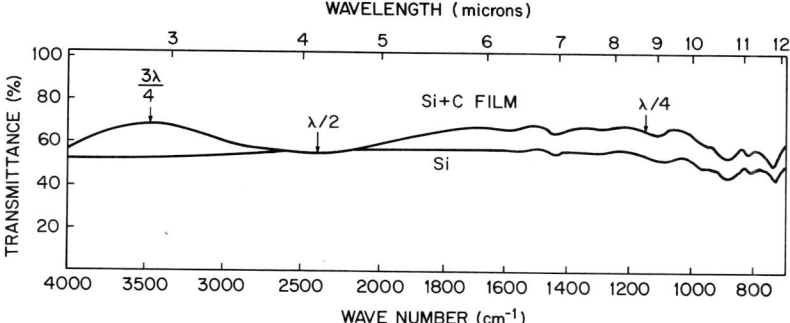

Fig. 13. The transmittance curves show the effect of coating one surface of a polished silicon slice (2 mm thick) with an amorphous carbon film free from C–H absorption bands. The wavelengths at which the optical thickness of the carbon film corresponds to the $\lambda/4$, $\lambda/2$, and $3\lambda/4$ values are marked. [After Holland and Ojha (50).]

tion conditions. Moravec (54) reported a value of 2.3 in the IR region for films deposited by the rf technique used by Holland and Ojha (47). The refractive index of a natural type-IIa diamond in the IR region 2.5–25 $\mu$m is in the range 2.379–2.375 (66). This seems to be close to the value reported for the a-C(H) films. The films can thus form a very useful optical coating on IR materials having a refractive index from 3.5 to 4. The a-C(H)-coated IR optical systems are now commercially produced. a-C(H) films are also being considered for coating interior surfaces of a fusion plasma reactor so as to prevent contamination of the plasma (67).

The films are found to be resistant to attack by acids and organic solvents (47), as is also the case for the ion-beam-grown films (32). The coatings on glass substrates have a low static coefficient of friction, e.g., 0.28 against steel and 0.2 for an a-C(H)-coated steel ball against a-C(H)-coated glass (48). Enke et al. (55) measured the effect of humidity on the coefficient of sliding friction of an a-C(H) coating on silicon wafers against a steel ball, using a pin-on-disk apparatus, and found that it varied from 0.01 to 0.19 as the percentage relative humidity was increased from $10^{-7}$ to 100 (Fig. 14). They also observed that the wear resistance of the coatings was optimum in the 0.5–5.0% relative humidity range, and that coatings with higher electical conductivity ($10^{12}$ $\Omega$ cm) showed better wear resistance than relatively lower-resistivity films ($10^6$ $\Omega$ cm).

The microhardness of the films has been measured by various workers (45, 46) and has generally been found to lie in the range 2800–3500 VHN. Diamond is reported to have a hardness value of 7000 VHN (56). The hardness value is related to the electrical conductivity of the films: the higher the conductivity, the lower the hardness (51). Glassy carbons, prepared by

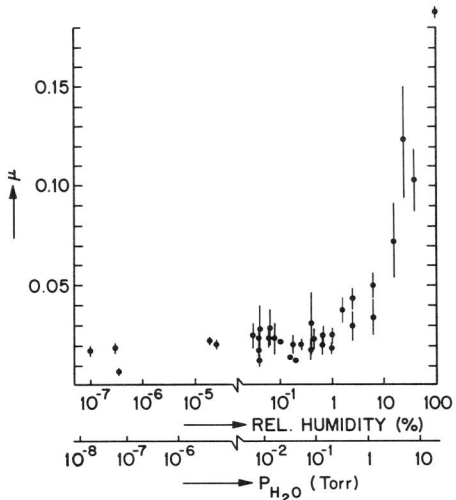

FIG. 14. Coefficient of sliding friction $\mu$ versus water vapor partial pressure $p_{H_2O}$ of a diamond-like carbon layer on a silicon substrate. [After Enke *et al.* (*55*).]

pyrolysis of resins (*23*) have a hardness value of ~6–7 Mohs, which, although appreciably higher than that of graphite, is less than the reported value for hydrogenated a-C films.

The low density of glassy carbons, ~1.4 g cm$^{-3}$ (*23*), suggests a porous structure, as compared to the hydrogenated a-C films, which have an estimated density of 2.5 g cm$^{-2}$ (*52*). The density of glassy carbon depends on the structure of the polymer from which they are derived, as the carbon atoms in the matrix will tend to retain their original positions in the polymer structure (*23*). In comparison, hydrogenated a-C films, from the very nature of the growth process, which involves deposition of films by separate carbon atoms or clusters of a few atoms, should give rise to impervious and denser film. However, on a microscopic scale, the structure and molecular weight of the starting hydrocarbon monomer, e.g., $CH_4$ or $C_6H_6$, can influence the structural and other properties of the a-C films grown under ion impact. Further investigations are necessary to elucidate this. The density of the films, estimated by measuring stopping power of the films deposited on glassy carbon substrates (*52*), is found to be around 2.5 g cm$^{-3}$, which is higher than the 1.8–2.1 g cm$^{-3}$ of amorphous carbon (*21*) and lower than the 3.15–3.53 g cm$^{-3}$ of diamonds (*56*).

The adhesion of the films is variable and depends on the nature of the substrate and its cleanliness (*46, 49*). Whitmell and Williamson (*46*) were able to deposit films up to 3–4 $\mu$m thick on metal alloy substrates, such as titanium alloy IMI 318, En 58 stainless steel, and tungsten tool steel, that were sputter-etched initially by argon ion bombardment. Films of similar thickness have also been reported to grow on silicon and germanium single crystals (50). However, films greater than 3000 Å thick deposited on glass or quartz substrates are found to peel off or spontaneously crack (49). Although thicker films can be grown on metallic substrates, on aging these either crack or their scratch resistance deteriorates. The lack of good adhesion of the films to most of the substrates has been attributed to the presence of unbonded hydrogen in the films and between the film-substrate interface which introduces a compressive stress in the films (49). The films would adhere well to substrates like silicon and titanium due to possible formation of carbides at the interface under ion impact. The presence of hydrogen in the films was confirmed by SIM measurement (49) and later estimated by stopping power measurements (52) to be as much as 25 at. %. However, as shown earlier by IR absorption studies, not all the hydrogen in the films is unbonded, and more work is required to ascertain the role of unbonded hydrogen, if it is present, in the films.

The stress in the films is not only dependent on the nature of the substrate but also on the hydrocarbon monomer from which the film is prepared. Thus films deposited on germanium single crystals in a methane discharge are more stable with regard to adhesion than are films prepared in a butane discharge under similar experimental conditions (52). Stress in a-C(H) films has been shown (57) to depend on the rf power applied for cracking hydrocarbons. Stress measurements are made by measuring the radius of curvature, as a function of the rf power, of glass slides coated with films of similar thicknesses.

The deposition rate of the films is reported to increase with increasing molecular weight of the hydrocarbon gases (45). Although various hydrocarbon gases, such as $C_2H_2$, $CH_4$, $C_2H_6$, $C_3H_8$, iso-$C_4H_{10}$, and $C_4H_{10}$, have been used for depositing a-C(H) films, no systematic work on the effect of the nature of gas and its film properties has been reported.

Amorphous C(H) films have also been deposited by dissociating acelylene in an inductively coupled rf discharge (58). Films having electrical resistivity of $\sim 10^{-12}$ $\Omega$ cm were deposited on quartz or Corning 7059 glass substrates placed on a grounded susceptor. The films were grown at substrate temperatures between 80 and 380°C. It was noted that the optical band gap of the films was a function of deposition temperature (58). The fall in band-gap values at higher deposition temperatures was attributed to the possible inclusion of graphitic phase in the films. The increase in the electrical conductivity with substrate temperature lends support to the possibility of a change from amorphous to graphitic structure. Similar values of optical band gap and its variation with substrate temperature were also obtained for a-C(H) films deposited in a dc discharge (59, 60) of acelylene. The films were grown on glass substrates placed about 1 cm below a screen cathode. It is significant that by adding phosphene or diborane to acelylene, doped n- and p-type a-C(H) films were deposited (60). The electrical conductivity of the doped films was reported to increase from $10^{-12}$ $\Omega^{-1}$ cm$^{-1}$ for an undoped film to $10^{-7}$ $\Omega^{-1}$ cm$^{-1}$. In the absence of information regarding film properties such as hydrogen content, the nature of C–H bonding, and mechanical hardness it is difficult to compare the films described in the above-mentioned work (58–60) with those deposited in a parallel-plate electrode system (52).

A deposition technique that can achieve a higher degree of ionization compared to the methods discussed earlier has been reported by Bewilogua et al. (61). In this technique benzene vapors were admitted directly into an ionization source consisting of a hot cathode, a reflector electrode, and an anode grid (Fig. 15). The substrates could be biased to a different negative voltage, thus making the ionized species extracted through the anode strike the substrates at variable energies ($\sim$ 100–250 eV) and condense on it. An electron gun (not shown in the figure) was used for neutralizing the positive charge on the substrate. The partial pressure in the chamber was in the range $10^{-3}$–$10^{-4}$ Torr. Mass spectrometric analysis of a benzene discharge (62) showed that 62% of the ions in the ionization source corresponded to such species as $C_6H_5^+$ and $C_6H_5^+$, the rest consisting mainly of fragments with two, three, or four carbon atoms. It is worth noting that since the partial pressure of the vapors is the same in the ionization source and the deposition chamber, the ions striking the substrate will collide with the benzene molecules adsorbed on it. This is likely to give rise to film growth conditions similar to those existing in the rf deposition system described earlier, except

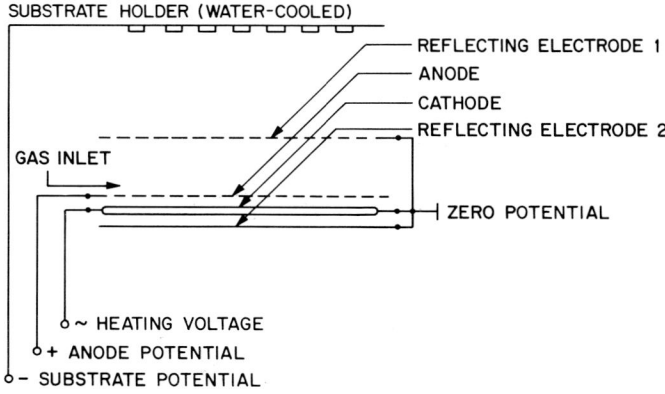

FIG. 15. Ion beam plating system for depositing carbon films. [After Bewilogua et al. (61).]

for the lower operating pressure. The ion energy of the condensing species was found to alter the properties of the films from polymer-like to "diamond-like" in agreement with the earlier reported work (51). The growth rate of the films varied from 10 to 50 nm min$^{-1}$, depending on the substrate bias voltages. The electrical resistivity of the films varied from $10^{15}$ to $10^7$ $\Omega$ cm as the ion energy was varied from 25 to 250 eV. No significant change in the conductivity was noted for the ion energies between 250 and 1000 eV. The mechanical hardness of the films deposited on hard alloy steel substrates was found to be about 2700 Kp mm$^{-2}$ and was reported to show high elastic recovery ($\sim 85\%$). Coatings less than $\sim 1$ $\mu$m thick were reported to adhere well to the substrates, but thicker coatings either cracked or peeled off.

The organic contents of the films were analyzed by electron energy loss analysis (EELA) (65). The height of the $\pi$ plasma peak at 6.8 eV was recorded for films deposited at different substrate bias voltages. The peak height was observed to decrease as the bias voltage on the substrates increased. This decrease was attributed to the reduction in the amount of hydrocarbons in the films. The main energy-loss peak ($\sim 23$ eV) for the films was observed to be lower than that for diamonds (34 eV) and graphite (27 eV). The energy-loss spectra of the freshly deposited and aged films were also different. EELA thus appears to be a sensitive technique for monitoring changes in the film structure, although the precise nature of these changes is not understood. Electron diffraction studies of the films suggested an amorphous structure. However, TEM studies in the dark field were reported to show bright spots about 0.8 mm in diameter, corresponding to clusters of about 50 atoms, thus indicating the presence of microcrystallites in the films.

*b. Boron.* The dissociation of boron trichloride in a glow discharge has been investigated by many workers (68–70). Optical emission spectra of boron trichloride in a microwave discharge (2400 to 2500 MHz) show the presence of boron monochloride, atomic chlorine, and atomic boron (70). Hultquist and Sibert (71) have reported deposition of boron on tungsten and silica filaments in an rf discharge (3 MHz) excited externally by capacitively coupling the rf source. A mixture of boron trichloride and hydrogen was introduced in a glass tube to a pressure between 5 and 20 Torr, and an rf discharge ignited to deposit boron on tungsten or silica filaments placed axially along the center of the tube. By carefully controlling the reactant gas concentration, gas flow rates, and rf input power, the authors were able to avoid fine particle formation and deposition of the films on the wall of the reactor, thus enabling deposition of thick boron films ($\sim 2$ mm), mainly on the filaments. In the absence of a detailed composition analysis of the deposits it is not certain to what degree the films contained polymers such as (BCl).

Wales (72) also reported formation of amorphous boron films grown in a corona discharge. The films were grown on a tungsten filament, which also acted as a cathode (6–13 kV; 60 Hz) surrounded by parallel wires forming the anode. Vapors of boron tribromide ($BBr_3$) mixed with hydrogen (typically around 5 Torr total pressure) were dissociated in the discharge region to produce, among other species, boron ions, which arrived at the cathode with considerable kinetic energy and were neutralized to form a boron film. Careful control of such deposition conditions as $BBr_3:H_2$ ratio, the flow rates of the two gases, the temperature of $BBr_3$, and the applied potential between cathode and anode was found to be necessary to ensure a continuous and uniform coating process, the lack of which resulted in arcing, overheating of the filament, and coarse coating. Since the cathode filament is maintained at a very high field strength in a corona discharge, any asperities on the substrate, such as dust and surface irregularities, could result in the formation of an arc. The substrates thus require careful preparation. The boron films, though amorphous, showed signs of anisotropy. Hydrogenated amorphous boron films have been deposited by dissociating diborane ($B_2H_6$) in a dc discharge (73). Attempts have also been made to dope the films with silicon and carbon (74).

## 2. Oxides

*a. Silicon Dioxide.* Silicon dioxide films used in various semiconductor devices are normally grown by thermal oxidation, in the case of silicon devices, or by thermally activated CVD (75–77). Thermal oxidation of silicon is generally carried out by heating silicon at temperatures ranging from 1000 to 1200°C in a flow of dry or wet oxygen or steam at atmospheric pressure, which produces faster growth rates and thicker oxide films. The refractive index of the oxide films in the thickness range 3000–8000 Å is 1.4618 (76) and reaches the value for fused silica (1.4601) for thicker films. The IR absorption spectra (200–1400 $cm^{-1}$) of the thermal oxide films are also similar to that of fused silica (79).

Thermally activated CVD is one of the most extensively used techniques for depositing $SiO_2$ films for semiconductor device applications. In this method silicon-carrying gases are reacted with oxygen or oxygen-containing gases (e.g., nitrous oxide) on a heated substrate, resulting in the deposition of silicon oxide films. Some of the reactions involved are (76)

$$SiH_{4(g)} + 2O_{2(g)} \xrightarrow{(T, \geq 300°C)} SiO_{2(s)} + 2H_2O_{(g)} \tag{1}$$

$$SiCl_{4(g)} + O_{2(g)} + 2H_{2(g)} \xrightarrow{800°C} SiO_{2(s)} + 4HCl_{(g)} \tag{2}$$

$$SiCl_4 + 2CO_2 + 2H_2 \xrightarrow{1100°C} 2CO + 2H_2O + SiCl_4 \rightarrow 2CO + SiO_2 + 4HCl \tag{3}$$

$$SiBr_4 + 2CO_2 + 2H_2 \xrightarrow{780°C} SiO_2 + 2CO + 4HBr \quad (4)$$

$$SiCl_4 + 2H_2 + 2NO \xrightarrow{850°C} SiO_2 + N_2 + 4HCl \quad (5)$$

$$Si(OC_2H_5)_4 \longrightarrow SiO_2 + 4C_2H_4 + 2H_2O \quad (6)$$

Reaction (1), involving $SiH_4$ and $O_2$, can take place at temperatures as low as 200°C. However, the films deposited at such low temperatures are porous and contain a large amount of $H_2O$ and SiOH. Although the water and hydroxyl content in the films can be reduced (78), the use of the films as a diffusion barrier and passivating layer requires a further baking in steam at temperatures around 800–900°C (76).

Electron beam evaporation of $SiO_2$ generally leads to films that are relatively porous, deficient in oxygen, and have high P-etch rates. They require high-temperature annealing (1000°C) in a nitrogen atmosphere to achieve density and etch rates comparable to those of thermally grown $SiO_2$ films (80). The electron-beam-evaporated films are generally susceptible to contamination by evaporation source material and desorbed gases from the vessel walls (81, 82) and can contain a large amount of trapped charges (83). Silicon dioxide films deposited by rf sputtering of fused quartz at high substrate temperatures ($\sim 450°C$) are, on the other hand, relatively denser and have lower P-etch rates, around 4.6 Å $sec^{-1}$ at 25°C (84). The sputtered films also contain trapped charges and occluded argon gas atoms in the structure, although Schwartz and Jones (85) claim that the presence of gas atoms in the films does not significantly affect the film properties. Some of the problems encountered in rf-sputtered $SiO_2$ films are reported to have been minimized by rf magnetron sputtering (86). The films deposited by this technique are claimed to contain lower trapped charge density, and are subject to lower radiation damage and a better step coverage on an unbiased substrate than are rf-sputtered films. Although bombardment of the substrate could be reduced, that due to energetic neutral particles, with energies as high as or higher than those corresponding to the target bias voltage, rebounding from the target cannot be eliminated and can cause damage to surface-sensitive devices. However, high deposition rates (700 Å $min^{-1}$), low substrate temperatures (200°C), and smooth morphological features make the technique viable for depositing secondary passivating layers on devices and for coating temperature-sensitive substrates.

The first reported attempt to deposit $SiO_2$ films in a glow discharge was by Alt et al. (87). They used an rf discharge (1–5 MHz) excited between a pair of internal electrodes inside a quartz tube to dissociate vapors of organosilicon compounds, alkoxysilanes in particular, mixed with argon gas. The films were grown on an unheated substrate, and in the absence of any detailed information related to the experimental conditions and structural analysis it is not certain whether the films were $SiO_2$ or degraded polymers. Nevertheless the films were reported to offer protection to silicon substrates against chlorine and phosphorous diffusion at temperatures of ~900–950° and ~1250°C, respectively. Ing and Davern (*88*) added oxygen gas to tetraethoxysilane (TEOS) vapors to prevent the formation of silicon-rich films. The authors used an externally excited rf discharge (0.5 MHz) and placed the substrates on a quartz pedestal, outside the main glow region. At this stage it is worth considering various reaction processes that influence the growth and properties of the films under the above-mentioned conditions. The dissociated species of the reacting gases produced in the discharge arrive at the substrate surface and give rise to a compound film. The growth characteristics, composition, structure, and properties of the films depend on the rf power dissipated in the discharge, which dissociates the reacting gases to a different degree, and on the substrate temperature, which imparts additional energy to the arriving species. Ing and Davern (*88*) have pointed out that the temperature gradient between a heated substrate and the cooler walls of the vessel can cause a convection current that can deplete the low-mass fragments and thus influence the growth rate. The authors noted that the growth rate of the oxide films deposited at 200°C increased with the TEOS vapor pressure, but the films grown at higher deposition rates (3500–5000 Å) were found to leave an organic residue when etched in HF, indicating incomplete dissociation of TEOS. The dielectric properties also reflected the influence of growth rate on the films as capacitors made from high-growth-rate films showed considerable variation on aging at 100°C with a 10.5 V bias across them. Capacitors deposited at lower growth rates (2000 Å $hr^{-1}$) were subjected to humidity and thermal aging tests and found to be much more stable, with the TCC less than 20 ppm $°C^{-1}$ between 23 and 100°C and the breakdown strength in the region $5 \times 10^6$–$10^7$ V $cm^{-1}$. The lower growth rates are expected to favor higher degrees of oxidation of the films. The dielectric constant of the silicon oxide films was reported to lie in the range 4.5–5.5. The stoichiometry of the films was not determined, but the higher value of refractive index (1.5 at $\lambda = 5.460$ μm) compared to fused quartz (1.46) suggested that the oxide films were oxygen deficient.

Secrist and Mackenzie (*89*) used a slightly different approach for depositing $SiO_2$ films. Instead of subjecting both TEOS vapors and oxygen to a glow discharge, atomic oxygen was produced in a separate low-pressure discharge system, energized by a 2450 MHz microwave generator and transported to another vessel, into where TEOS vapors were introduced. The substrates were hung on a quartz helix balance for determining the growth rate and could be heated by a furnace surrounding it. The films deposited at substrate temperatures below 290°C and at flow rates 0.15 $cm^3$ $hr^{-1}$

were found to contain silicic acid, and those deposited above 290°C were reported to be noncrystalline $SiO_2$. The deposition rate of the films was found to increase with rise in temperature above 290°C and followed the Arrhenius equation, $r = A \exp(-H/RT)$, where $r$ is the deposition rate (mg cm$^{-2}$ min$^{-1}$), $H$ the apparent heat of absorption (kcal mol$^{-1}$), $R$ the gas constant, $T$ the substrate temperature, and $A$ a constant. The films were deposited on different substrates and the apparent heat of absorption of the oxide films on silica, sodium chloride, and potassium single crystals was calculated to be 12.7 kcal mol$^{-1}$ and that for the films on aluminum and alumina was 9.5 kcal mol$^{-1}$. The difference in the two sets of values was attributed to the higher degree of absorption of hydroxyl groups on alumina and surface oxide of aluminum than on the other substrates. The refractive index ("Becke Line" method) of the higher-temperature deposits was reported to be 1.458, which was in good agreement with the value obtained for high-purity silica glass ($n = 1.458$). The IR absorption of the films deposited on NaCl showed a strong absorption peak at 1080 cm$^{-1}$ which was identified with Si–O–Si bond-stretching vibration.

Mukherjee and Evans (90), using an experimental setup similar to Ing and Davern's (88), observed that plasma dissociation of TEOS alone or in the presence of argon gas produces organo-silicon polymer films, whereas films deposited in a mixture of TEOS and $O_2$ were predominantly silicon oxide, in agreement with Secrist and Mackenzie (89). The authors noted from IR absorption studies that oxide films deposited below 500°C contained hydroxyl groups, which disappeared when the substrate temperature was raised to 600°C.

The main disadvantages in using TEOS or any other organic media for depositing silicon oxide films is, as seen above, that the films can contain polymers, and substrate temperatures as high as 600°C are required to obtain hydroxyl-free oxide films. Inclusion of carbon or silicon carbide in the films cannot be ruled out. The use of an inorganic medium such as $SiH_4$ reduces the risk, mentioned earlier, of contaminants, and they have been used by Sterling and Swann (91) and Joyce et al. (92) for depositing silicon oxide films. These workers used a rf (1 MHz, 2 kW) power supply coupled to an external coil surrounding a quartz tube to excite a discharge. Silane and nitrous oxide were separately leaked into the vessel, raising the total pressure to 0.1 Torr, with total flow rates of $\sim 10$ ml min$^{-1}$. The substrates were placed on a carbon or molybdenum susceptor. The ratio of nitrous oxide to silane had a pronounced influence on the properties of the films (92). The dielectric constant of the films deposited at a substrate temperature of 400°C varied from 3.8 for a $N_2O$ to $SiH_4$ ratio of about 8:1 to a value of 10 for the 1:1 gas ratio which is nearer to the value for silicon. The refractive index and etch rate of the oxide films were likewise dependent on the gas proportions. For example, for films deposited at 400°C, $n$ varied from 1.46, which is close to the value for fused quartz and thermally grown $SiO_2$, to 1.9 as the volume proportion of $SiH_4$ in the gas mixture was changed from 10 to 30%. The corresponding P-etch rates varied from 26 Å $sec^{-1}$ to 0.5 Å $sec^{-1}$, respectively. The P-etch rate of 26 Å $sec^{-1}$ for a stoichiometric film is still greater than that for the thermal oxide, 2.0 Å $sec^{-1}$, and rf sputtered films, 4–12 Å $sec^{-1}$ (80). Pliskin attributed the higher etch rate to the films' possibly being comparatively porous and under strain, although Joyce et al. (92) found the films to be under less stress, as evaluated by an X-ray diffraction method, than the thermally grown oxide. IR absorption studies showed that the absorption peak near 9.5 μm shifted to lower wavelengths, and the spectrum in the range 8–14 μm appeared more like that for vitreous silica and alpha quartz as the $N_2O$ content in the mixture was raised. However, as the $SiH_4$ content was raised the absorption peak at 12.4 μm disappeared, and instead one at 11.5 μm was observed. This change in the absorption wavelength was attributed to a silicon oxide film, with possibly a SiO or $Si_2O_3$ composition. The SiOH (2.8–3 μm) and Si–H (4.45 μm) bands were also observed in films deposited at substrate temperatures below 300°C, but above this temperature they were reported to be absent or weak. Pliskin (80) has given a detailed account of IR absorption characteristics and their dependence on the physical and chemical properties of silicon oxide films grown by different techniques.

One of the recent applications of plasma-deposited composite films of silica and germania has been in the field of glass fibers for optical communications. In the past, the inner wall of the quartz tubes used for drawing fibers were coated by mixed oxide films of Ge and Si prepared by thermally activated CVD (94), so as to obtain multiple coatings with a variable refractive index profile. One of the disadvantages of the high-temperature CVD methods is the contamination of oxide films by soot formed in the gas-phase reaction above the substrate. Since plasma-induced deposition requires a comparatively lower substrate temperature, the film growth is surface-reaction dominated. The discharge inside a quartz tube (8 mm diameter) was excited by a microwave cavity coupled to a 2.45 GHz generator (maximum power output 20 W) (95, 96). The reacting gases, $O_2$, $SiCl_4$, and $GeCl_4$, were separately admitted to the tube and the total gas pressure was maintained between 1 and 10 Torr. For depositing the oxide films along the length of the tube, the microwave cavity was moved along the tube axis, thus moving the discharge region and resulting in local deposition of the films. The mixed film forms in accordance with the reactions

$$SiCl_4 + O_2 \rightarrow SiO_2 + Cl_4, \qquad GeCl_4 + O_2 \rightarrow GeO_2 + Cl_4$$

Although the oxide films could be deposited at substrate temperatures below

500°C the deposits were found to contain large quantities of chlorine, and films thicker than 1000 Å tended to peel off. At elevated temperatures (800–1000°C) the films deposited at high growth rates ($\sim 50$–100 $\mu$m min$^{-1}$) were reported to be stable and to contain a very small amount of chlorine. However, it is interesting that even at such high deposition temperatures the presence of the hydroxyl radicals in the films was still observed. The deposition profile of $SiO_2$ and $GeO_2$ in a single film was found to be different for each oxide across the thickness of the films and depended on the direction of the microwave cavity movement relative to the gas flow direction. The nonuniformity in the relative concentration of the two oxides in a layer was avoided by depositing multiple layers, the microwave cavity motion allowing sufficient time for diffusion of the two oxides in each layer. The authors have reported deposition of up to 3000 layers of a desired refractive index profile. The fibers coated with $GeO_2$- and $B_2O_3$-doped $SiO_2$ film were reported to give optical losses less than 10 dB km$^{-1}$ with a minimum of 1.4 dB at a wavelength of 1.05 $\mu$m and a pulse dispersion of 0.8 ns km$^{-1}$.

With the commercial use of plasma-deposited silicon nitride films for passivating semiconductor devices, recent studies on silicon oxide films have been carried out in production-type reactors. Hollahan (97) used a parallel-plate reactor (grounded substrate holder 26 in. in diameter) for depositing silicon oxide films in a $SiH_4$ and $N_2O$ discharge. The films were deposited at substrate temperatures of 250 and 350°C and rf (50 kHz) power densities of $\sim 300$ W cm$^{-2}$. The film deposition rate (typically 600 Å min$^{-1}$) was observed to be dependent on total gas pressure and the flow rate ratio of the two reactant gases. Hollahan has also reported phosphorus doping of the films by adding $PH_3$ to $SiH_4$ and $N_2O$. A much smaller rf (13.56 MHz) power density ($10^{-2}$ W cm$^{-2}$) was used by Adam et al. (98) for depositing silicon oxide films in a radial-flow parallel-plate reactor similar in size to the one used by Hollahan (97). A detailed IR absorption study of the silicon oxide films deposited in the temperature range 100–340°C showed the presence of 2–9 at. % H in the films, bonded as $H_2O$, SiOH, and SiH. The relative concentration of these species was found to vary with deposition conditions. The films, grown typically at rates between 200 and 360 Å min$^{-1}$, were under compressive stress varying from $0.07 \times 10^9$ to $0.97 \times 10^9$ dyn cm$^{-2}$. These stress values compare favorably with those reported for CVD films deposited at high temperatures ($0.6 \times 10^9$ dyn cm$^{-2}$) (99) and thermally grown films ($3$–$4 \times 10^9$ dyn cm$^{-2}$) (104). Silicon dioxide films deposited by low-temperature (400–450°C) CVD have tensile stress of $1$–$4 \times 10^9$ dyn cm$^{-2}$ (102). The reported average density of 2.29 g cm$^{-3}$ of the plasma-deposited films is higher than the value of 2.20 g cm$^{-3}$ for fused quartz (85) and 2.07–2.27 g cm$^{-3}$ (99) for CVD silicon dioxide. The plasma-deposited films have much higher P-etch rates than the thermally grown films. The refractive index of the films is 1.47, which is slightly higher than the value for the oxide films deposited by thermal CVD (*85, 99*). Dielectric breakdown of the films occurs at fields of 4–10 MV cm$^{-1}$, which is lower than the thermal-CVD and thermally grown films. Thermal annealing of the films in air at up to 400°C has no effect on the etch rate in dilute hydrofluoric acid, but a reduction in the density of the film is observed. Films deposited on vertical steps are not conformal (*98*), in contrast to the film deposited by Hollahan (*97*). It is worth noting that much higher power density and frequency were used in the latter work. Fabrication of high-voltage and high-frequency semiconductor devices with plasma-deposited silicon oxide as an interlayer dielectric has been reported (*102*). The oxide films were deposited in a production-size system similar to the one used by Hollahan (*97*). Besides the usual advantages of good step coverage and low defect density, the plasma–oxide films were claimed to offer other distinct advantages over CVD $SiO_2$, such as low compressive stresses, which permits deposition of crack-free films up to 5 $\mu$m thick, and controlled variation in the properties of the films, which allows fabrication of an oxide step with a slope of 20–70°. Plasma-deposited oxide films also contain much less hydrogen than plasma-deposited silicon nitride films and thus offer better high-temperature stability.

The effect of oxygen content on the photoluminescence properties of plasma-deposited Si–O–H system has been reported (*103*).

*b. Aluminum Oxide.* Aluminum oxide films have been prepared by various methods, including pyrolytic CVD (*104*), reactive evaporation (*105*), reactive sputtering (*106*), rf sputtering (*107*), and plasma anodization (*110*). Kalto and Koga (*108*) utilized a glow discharge excited in a fused silica tube by an external coil (400 kHz), for dissociating aluminum chloride and oxygen deposit aluminum oxide films. The substrates were placed outside the glow region on a grounded susceptor. The deposition rate of the oxide films was found to depend not only on the partial pressures of the two gases but also on the total gas pressure, due to dependence of oxygen ionization efficiency on its partial pressure. Thus the optimum deposition rate was obtained at $\sim$1–2 Torr partial pressure of oxygen. The films deposited at a substrate temperature of 480°C were amorphous and showed a characteristic IR absorption peak near 15 $\mu$m. The P-etch rate of the films varied from 10 to 100 Å sec$^{-1}$, depending on the deposition condition. The dielectric constant, dielectric loss (tan $\delta$) at 1 MHz, and dielectric strength were reported to be 8.3, 0.01, and $5 \times 10^6$ V cm$^{-1}$, respectively. The flat-band voltage determined from $C$–$V$ measurements of the MIS structure had positive values and showed a large shift toward positive values under negative bias at 250°C. Field-effect mobility of the electrons in a metal–oxide–semiconductor FET with aluminum oxide as the gate was found to vary from 70 cm$^2$ V$^{-1}$

sec$^{-1}$ for films deposited at 480°C to 140 cm$^2$ V$^{-1}$ sec$^{-1}$ when the films were annealed at 700°C for 1 hr. Inclusion of a thin ($\sim$500 Å) thermally oxidized SiO$_2$ film between the silicon and aluminum oxide films greatly enhanced the mobility (to $\sim$530 cm$^2$ V$^{-1}$ sec$^{-1}$). Sokolowski et al. (109), using a pulse-discharge (high-current) technique, reported growth of aluminum oxide film having a metastable cubic ($\gamma$-Al$_2$O$_3$) phase.

## 3. Nitrides

*a. Silicon Nitride.* Silicon nitride films are known to offer many advantages over thermally grown silicon dioxide films when used as a passivating layer in silicon devices (76). The dense and impervious nature of the silicon nitride films, compared to the relatively porous structure of thermally grown silicon dioxide films, offers much better protection to silicon devices against moisture and alkali ions and metal-atom diffusion through the films. Amorphous silicon nitride films have a lower-energy band gap ($\sim$5 eV) than silicon dioxide films ($\sim$8 eV), and thus form a low-energy barrier with metal electrodes or silicon. This gives rise to charge injection through the insulating silicon nitride layers. To avoid this charge injection the thermally grown silicon dioxide films are retained in the device structure, and an overcoat of silicon nitride layers is generally used for passivating the devices.

Silicon nitride for semiconductor devices have usually been prepared by one of the following techniques:

(1) Pyrolytic dissociation of silicon-carrying vapors in a nitrogen atmosphere (*111, 112*).

(2) Direct thermal nitridation of silicon by nitrogen or ammonia (*113–116*). Although earlier workers (*113–115*) found the second technique to be unsatisfactory, as it yields films with mixed crystalline phases and patchy and nonuniform surfaces, Ito et al. (*116*) have perfected the technique by carefully minimizing water and oxygen impurities down to 1 ppm in the nitrogen gas used for nitriding. Uniform Si$_3$N$_4$ films with properties as good as those prepared by CVD are reported. However, the nitriding temperature in this technique is quite high ($\sim$1200–1300°C).

(3) Reactive sputtering of silicon or rf sputtering of silicon nitride (*124*).

(4) Ion nitriding of silicon. Formation of Si$_3$N$_4$ on p- and n-type silicon, placed on the cathode of a dc sputtering system, by ion bombardment in an atmosphere of nitrogen and hydrogen has also been reported (*125*).

(5) Plasma CVD.

Synthesis of silicon nitride films by pyrolytic dissociation can be brought about by reacting silicon-carrying gases with nitrogen, nitrogen and hydrogen mixtures, or ammonia at high temperatures. The standard free-energy changes are different for different sets of reactions (2), and thus the growth of silicon nitride will proceed at different rates. The reactions involving ammonia are expected to be more favorable than those with nitrogen, because ammonia has a higher free energy of formation than nitrogen. This is also the case for films grown by plasma-induced deposition (130), in which the growth rate of the silicon nitride films is much lower. In this method each turn of the coil forms a capacitive coupling with the discharge, with the glass wall as the dielectric. Thus the discharge conditions could vary radically along the axis, giving rise to nonuniform films. Careful adjustment of flow rates, gas pressure, rf power, sample holder geometry, and reaction vessel wall temperature are necessary for optimization of the film uniformity. A great deal of improvement in the film uniformity has been reported by Gereth and Scherber (130), who used two parallel disk electrodes to excite the discharge. Reinberg (136) and Rosler et al. (137) had a similar electrode arrangement, but took special care about the radial gas-flow uniformity across the silicon wafers, either by introducing the gas mixtures from the side of the grounded substrate holder plate and pumping them from a centrally located port (136), or by reversing this arrangement by admitting the gas centrally (137). Sinha et al. (138) modified this arrangement by confining the discharge between the electrodes (16 in. diameter, 1 in. apart) with the help of a metal shield around the substrate holder. This shielding arrangement confines the film deposition essentially between the aluminum electrodes as it prevents the spread of the discharge to other parts of the chamber.

The suitability of silicon nitride film as a passivating layer is determined by its various properties, such as density, etch rate, resistivity, flat-band surface charge concentration, stresses, and Si:N ratio. These properties in turn are essentially dependent on preparation conditions. Since the early experiments of Sterling and Swan in 1965) (126), the major efforts have been directed towards achieving film properties most suitable for silicon devices. Swann et al. (127) reported the formation of silicon nitride films with varying properties when different proportions of $SiH_4$ and $NH_3$ were introduced in an externally excited rf discharge (1 MHz, 500 W, $T_s = 300°C$, 11 cm$^3$ min$^{-1}$, $10^{-1}$ Torr). The resistivity, dielectric constant, and dielectric breakdown field were found to vary from $8 \times 10^{16}$ Ω cm to $5 \times 10^{12}$ Ω cm, 8 to 11, and $6 \times 10^6$ to $10^6$ V cm$^{-1}$, respectively, as the $SiH_4$ concentration was varied from 9 to 50%. These changes in the film properties were attributed to the increase in unreacted silicon content in the films at higher $SiH_4$ concentrations. This was also evidenced by a shift in the main IR absorption peak from 11.6 to 12.2 μm for silicon-rich films. The $C-V$ measurements on NaOH (0.1% $N$)-treated films of silicon nitride (1000 Å) and thermally grown $SiO_2$ on Si showed a drift of 9 V in the case of $SiO_2$ films as compared to 0.3 V for the silicon nitride films, thus indicating good resistance to sodium ion diffusion.

Kuwano (128) used a similar technique (27 MHz, 0.5 Torr) but replaced $NH_3$ with $N_2$ and studied the effect of substrate temperature (25–500°C) and silane concentration (0.1–2.5 mol %) on the properties of silicon nitride films. Structural examination of the films by transmission and reflection electron microscopy suggested that they were amorphous in the entire temperature range investigated. The dielectric constant, however, was sensitive to the deposition temperature, varying from 5 to 8 between 25 and 500°C for a fixed silane concentration of 0.5 mol %. The main IR absorption peak also showed a shift from 12.1 to 11.4 μm for the same set of films as the deposition temperature was raised. The author ascribed this increase in the dielectric constant and the IR shift to higher silicon content in the films resulting from an increased $SiH_4$ decomposition rate at higher temperatures. The presence of excess silicon in the films was also detected by measuring the etch rates in a mixture of Hf and $HNO_3$. Films prepared from the gas mixture with 0.5 mol % $SiH_4$ were reported to have a minimum etch rate.

Gareth and Scherber (130) employed the same gas constituents, $SiH_4$ and $N_2$, and a similar deposition technique, except for a lower-frequency rf source (500 kHz), and reported growth of stoichiometric $Si_3N_4$ films ($T_s \sim 350°C$). A glow discharge treatment of the silicon substrate by $H_2$ prior to silicon nitride deposition appeared to eliminate completely the hysteresis in the $C-V$ characteristics of the subsequently deposited $Si_3N_4$ films. The hydrogen treatment would strip the surface oxide layer from the substrate, resulting in a direct $Si-Si_3N_4$ interface. A small amount of hydrogen gas, when mixed with $SiH_4$, had another interesting effect in that it raised the growth rate of the silicon nitride films considerably even at very low powers. Taft (129) used a very low rf power input (10 W, 0.5–30 MHz) and high substrate temperature (400°C) and claimed the formation of stoichiometric $Si_3N_4$ films from $SiH_4-NH_3$ and $SiH_4-N_2$ gas mixtures. The films had a refractive index of 2.03 at a wavelength of 4500 Å, a dielectric constant of 7.4, a resistivity of $10^{16}$ Ω cm at $4 \times 10^6$ V cm$^{-1}$, and an etch rate of 150 Å min$^{-1}$ at room temperature in 48% hydrofluoric acid. The author also showed that excess silicon or the presence of oxygen in the silicon nitride films shifts the UV absorption edge. This effect, in conjunction with the IR absorption, can be effectively used for characterizing the films.

Since the development of the parallel-plate type reactor by Reinberg and its commercialization, most of the subsequent studies of silicon nitride films have been done in such a reactor. Rosler et al. (137) used a relatively low frequency (50 kHz) and high power input (500 W, 1 A) to deposit silicon nitride films on 3 in. silicon wafers, placed flat on a grounded plate which was separated from the rf electrode by 5 cm. The growth rate of the films, deposited from a mixture of $SiH_4$, $NH_3$, and $H_2$ at 0.2 Torr and 240°C substrate temperature, was reported to vary linearly up to about 500 W rf input, but at higher power levels it either became constant or increased further, depending on the flow rates and gas composition. The flow rate of $SiH_4$ was reported to have considerable influence on the growth rate and hence on the film thickness uniformity radially across the substrate holder; the authors claimed a 15% uniformity across wavers and from wafer to wafer. In agreement with the earlier workers (*138*), the authors also noted the sensitivity of the refractive index (measured by ellipsometry) to silane concentration as the index changed from 1.9 for 2% $SiH_4$ to about 2.4 for 12% $SiH_4$, the higher index indicating silicon enrichment of the silicon nitride films. The refractive index also increased with rise in the substrate (Si wafer) temperature, but above 250°C the variation was rather small, from 2.05 to about 2.1 between 250 and 400°C. Pliskin (*141*) noted that when a $Si_3N_4$ film plasma-deposited at 300°C was heated in an argon atmosphere to 900°C, the refractive index of the films increased from 2.177 to 2.322, accompanied by a decrease in the film thickness. The IR absorption peak corresponding to $Si_3N_4$ also shifted towards higher wavelengths, indicating silicon enrichment of the films after high-temperature treatment. The decrease in the etch rate from about 130 to 6–8 Å $min^{-1}$ in a 7:1 buffered HF solution after the high-temperature treatment was ascribed to the relatively porous structure of the film prior to annealing.

Sinha *et al.* (*138*) used a modified radial flow parallel-plate reactor (cathode diameter $\sim 40$ cm) with discharge conditions different from those of Rosler *et al.* (13.56 MHz, 1–10 V rf electrode self-bias to ground, rf power $\sim 300$ W), and studied the suitability of the plasma-grown silicon nitride films for MOS–LSI passivation. The dependence of several film properties on $SiH_4/NH_3$ ratio, gas flow rates, and substrate temperature was similar to that observed by Rosler *et al.* (*137*) and Kern and Rosler (*142*). It was significant that the stress in the films was reported to change from tensile ($0.5 \times 10^9$ dyn $cm^{-2}$) for lower rf power input (250 W) to compressive ($1–2 \times 10^9$ dyn $cm^{-2}$) for higher power levels (300 W) thus emphasizing the need for proper choice of rf power input along with other parameters for depositing films with minimum stress. The Si:N ratio in the films was also noted to depend on the rf power input, higher power levels favoring stoichiometric $Si_3N_4$ films. Higher power input to the system would raise the production of atomic nitrogen, thus reducing the unreacted silicon content in the films. Thick silicon nitride films with excessive stress ($8 \times 10^9$ dyn $cm^{-2}$) or films deposited at temperatures much lower than 275°C were found to crack when heated to about 450°C. However, the films deposited under optimized experimental conditions (e.g., 300 W rf power input, $T_s = 275$°C, pressure 0.95 Torr) were reported to withstand temperatures up to 550°C for 1/2 hr without showing any signs of cracks. The films were shown to have good step coverage and were also reported to be compatible with such Si devices as MOS, PMOS, and CMOS.

The hydrogen content in silicon nitride films deposited in a production-scale parallel-plate reactor was determined by using the 80° resonant nuclear reaction: $^{15}N + H \rightarrow {}^{12}C + {}^{4}He + \gamma$ (*139*). The films, deposited under different experimental conditions, such as rf power and frequency, silane to ammonia proportion, and pressure, were found to contain up to 20–25 at. % H. The IR absorption measurements showed that 7.5% of the hydrogen in the films was bonded to silicon and the rest to nitrogen. Similar amounts of hydrogen were also reported to be present in a-C(H) films (*52*). The etch rate of the silicon nitride films was found to be sensitive to the variation, though small, of total hydrogen content in different films. Annealing of the silicon nitride films above the deposition temperature resulted in the loss of hydrogen and changes in the physical properties: films deposited at 300°C and annealed at 900°C were found to crack, resulting in a 7% increase in the etch rates and a decrease in thickness corresponding to a 15% decrease in the density of the films (*144*). Similar cracking of the films was also noted at much lower temperatures ($\sim 500$°C) (*145*).

Nitrogen instead of ammonia has been used for reducing hydrogen content in silicon nitride films (*146*). The major difficulty in using nitrogen is the much lower dissociation rate compared to ammonia, which results in much lower deposition rates. One of the methods (Fig. 4) reported for overcoming this difficulty was to dissociate nitrogen gas in small discharge cells and transport it to the substrate in the downstream mode, where it reacted with $SiH_4$ to form silicon nitride films.

*b. Aluminum Nitride.* Aluminum nitride is a chemically and thermally stable dielectric material with a band gap of about 5.9 eV (*147*). It shows strong piezoelectricity and has been investigated for constructing acousto-electronic devices (*148, 149*). The growth of aluminum nitride coatings in crystalline or amorphous form has generally been carried out by pyrolytic dissociation of aluminum trichloride and ammonia in the temperature range 1000–1300°C (*150*).

Formation of aluminum nitride was observed when an aluminum surface was bombarded by a 20–50 keV nitrogen ion beam (*151*). A similar observation was made when a freshly evaporated aluminum surface was exposed to a nitrogen glow discharge (*152*). Stoichiometric aluminum nitride in the powder and crystalline form was reported to grow at temperatures of $\sim 1000$°C when the reaction products of an aluminum surface and a nitrogen and chloride discharge were transported to the hot zone (*153*). A comparative study of the physical properties of aluminum nitride films deposited by thermal and plasma CVD has been made (*154, 155*). The thermally activated deposition was done either by mixing AlCl$_3$ vapors and NH$_3$ very near a hot substrate (700–1300°C) or by premixing the two gases well before they reached the substrate. The latter procedure resulted in the formation of AlCl$_3$NH$_3$. The activation energy of AlN formation was estimated to be much lower ($\sim$3–7 kcal/mol) in the former case than in the latter (24 kcal/mol). In the two cases the compound formation takes place according to the following reactions:

$$AlCl_{3(g)} + NH_{3(g)} \rightarrow AlN_{(s)} + 3HCl_{(g)}$$
$$AlCl_3NH_{3(g)} \rightarrow AlN_{(s)} + 3HCl_{(g)}$$

For depositing AlN films by plasma CVD the authors introduced a mixture of aluminum trichloride and nitrogen in a 3 MHz externally excited discharge (total gas pressure 0.6 Torr). The substrates, as in the case of pyrolytic CVD, were heated to between 800 and 1200°C. The deposition rate was between 4 and 130 Å min$^{-1}$, as compared to 100–10,000 Å min$^{-1}$ for CVD-deposited AlN. Aluminum nitride films deposited pyrolytically on (100) spinel MgO·1.4Al$_2$O$_3$ in the temperature range 1050–1250°C were epitaxial with (0001) orientation, and those deposited on (111) silicon and fused quartz had (0001) fiber texture. The films grown by plasma CVD were reported to have (1123) fiber texture in addition to (0001) texture, depending on the deposition conditions. The dielectric conductivity of the films deposited by the two methods was on the order of $10^{-13}$–$10^{-17}$ $\Omega^{-1}$ cm$^{-1}$, the films deposited by plasma CVD favoring a lower value. The effective density of charges, determined by high-frequency C–V characteristics of Al–AlN–Si structures, was observed to be lower for plasma-deposited films ($3 \times 10^{11}$–$1.3 \times 10^{12}$ cm$^{-2}$) than for pyrolytically deposited films ($2$–$8 \times 10^{12}$ cm$^{-2}$), the charges being invariably positive for plasma-deposited films but either sign in the pyrolytic films. The optical band gap was reported to be $\sim$5.9–6.0 eV for films deposited by either technique. Sokolowski *et al.* (*109*) have synthesized AlN in a pulsed high-current discharge between aluminum electrodes in the presence of nitrogen gas.

*c. Phosphorus Nitride.* Amorphous phosphorous nitride (P$_3$N$_5$) is reported to be a chemically inert and transparent dielectric having an optical band gap of about 5.1 eV (*156*). It is also reported to be stable in air up to 500°C. Two different methods have been used for depositing phosphorous nitride compound in a glow discharge (*156*).

(1) By chemical transport of phosphorous by plasma-activated nitrogen gas and dissociation and deposition of phosphorous–nitrogen species on substrates. The substrates were electrically floating and heated to temperatures of $\sim$260–300°C. Typical deposition rates of the compound were $\sim$2000 Å min$^{-1}$.

(2) Rf plasma decomposition of phosphine and nitrogen. The films deposited on substrates heated to 330°C and deposition rates between 180–480° Å min$^{-1}$ were found to contain up to 13% hydrogen but the ratio of phosphorous to nitrogen was the same as that for the films deposited by method (1). The hydrogenated phosphorous nitride films were found to have a dielectric constant of $\sim 4$–5 in the frequency range 1 kHz–2 MHz, dielectric loss (tan $\delta$) $10^{-2}$ and dielectric strength $10^7$ V cm$^{-1}$. The room-temperature conductivity of the films was in the region of $10^{-17}\,\Omega^{-1}$ cm$^{-1}$. Optical absorption in the UV, visible, and IR regions, Raman spectroscopy, and X-ray photoelectron spectroscopy studies of the films were done to determine the structure of the phosphorus nitride.

*d. Boron Nitride.* Boron nitride films deposited by the high-temperature CVD technique have useful mechanical (easily machinable and extremely hard) and electrical (insulator, high breakdown field, low losses) properties. The films are not only chemically inert, like pyrolytic graphite and hydrogenated amorphous carbon films, but are also reported to be resistant to oxidation up to 2000°C (*157*). The structure and properties of the films, which depend strongly on such preparation conditions as deposition temperature, growth rate, and substrate, have been reviewed by Feist *et al.* (*76*).

The use of a glow discharge for synthesizing boron nitride films was first attempted by Sterling *et al.* (*158*). Vapors of boron tribromide were introduced into a 1 MHz rf discharge (external rf electrodes), resulting in a deposition of boron nitride films at substrate temperatures higher than 500°C. The films showed characteristic IR adsorption peaks at 7.25 and 12.7 $\mu$m and were reported to be chemically inert. Hyder and Yep (*159*) were able to grow crystalline boron nitride films in a $B_2H_6$ (diluted in hydrogen) and $NH_3$ discharge (13 MHz; total pressure 0.3–0.5 Torr) at substrate temperatures higher than 1000°C. The films were found to be nitrogen- or nitrogen–boron-rich, depending on the $NH_3B_2H_6$ ratio. The authors were able to obtain films with large single crystallites with a $NH_3:B_2H_6$ ratio of 8:1 and substrate temperature in the range 700–1000°C, the B:N ratio of the films being 41:58. The grain size and crystallite perfection of the films deposited by the glow-discharge technique were found to be better than those deposited by the high-temperature CVD technique under the same conditions of substrate temperature and gas composition. Under the discharge conditions used by the authors the growing film undergoes a weak ion bombardment which can provide extra energy for crystallization. It is worth noting here that Weissmantel *et al.* (*34*) were able to grow epitaxial films of silicon at lower substrate temperatures by glow-discharge cracking of silane, compared to the higher temperatures required in the CVD technique. The IR absorption characteristics of the plasma-grown films were similar to those reported by Sterling *et al.* The electrical resistivity of the films was reported to be typically $10^9$ Ω cm. Sokolowski (16) has reported growth of crystalline boron nitride films in a pulsed discharge.

### 4. Carbides

*a. Si–C–H.* Crystalline silicon carbide films have in the past been grown by pyrolytic decomposition (1200–2500°C) of silicon- and carbon-carrying gases on a heated substrate or by transporting carbon and silicon in a suitable carrier gas onto a heated silicon, carbon, or silicon carbide substrate. The films thus grown at high temperatures are reported to have cubic $\beta$-SiC ($\leq 1700$–2000°C) or hexagonal $\alpha$-SiC ($>2000$°C) structures (76). The use of crystalline silicon carbide as a hard and corrosion-resistant refractory is well established. In its amorphous form silicon carbide has attracted attention because, as a tetrahedrally coordinated binary compound with both carbon and silicon belonging to group IV of the periodic table, it can help in understanding tetrahedrally coordinated amorphous semiconductors (160).

Although pyrolytic decomposition has been reported as a technique for depositing amorphous silicon carbide (177), sputtering (160, 168, 176) and glow-discharge-assisted CVD (161–167, 169–171) have been considered more suitable because of the flexibility afforded by the ability to vary the substrate temperature from low (near room temperature) to high values. The formation of silicon carbide in an rf glow discharge of silane and methane (or ethylene) was reported in 1965 (161, 162), but a detailed study of its properties have only recently been studied in detail. Anderson and Spear (164), reported deposition of amorphous silicon carbide films on substrates placed on a grounded electrode by passing silane and ethylene through an rf (5 MHz) discharge. The substrates were placed just outside the main discharge, region. The composition of the films ($\sim 2$ μm thick, 50 Å min$^{-1}$ growth rate, 0.4–0.8 Torr total gas pressure), determined by electron microprobe measurements, was found to depend on the volume ratio of $SiH_4$ and $C_2H_4$ but was independent of temperature between 227 and 527°C. A significant feature of the measured physical properties of the films was the occurrence of a maximum in the optical and electrical band gaps of films of composition $Si_{0.32}C_{0.68}$. The size of the band gap depended on the substrate temperature, varying from 2.8 eV (optical band gap) to about 2.4 eV as the temperature was raised from 227 to 527°C. The dc electrical conductivity showed a minimum for $Si_{0.32}C_{0.68}$ films, its value being $\sim 10^{-13}$ Ω$^{-1}$ cm$^{-1}$ (substrate temperature 227°C) compared to $\sim 10^{-11}$ Ω$^{-1}$ cm$^{-1}$ for films composed of equal parts of silicon and carbon. In the absence of any characteristic features of the plasma-deposited films with equal silicon and carbon proportions, which are comparable to amorphous silicon carbide films prepared by other techniques with a possible stoichiometric proportion of silicon and carbon, Anderson and Spear (164) suggested that plasma-grown films could be considered an alloy of silicon and carbon with a predominance of homonuclear bondings between silicon (Si–Si) and carbon (C–C) atoms. This hypothesis was, however, rejected by Catherine and Turnan (165), who reported that up to 40 at. % hydrogen, as measured by the nuclear reaction $^1H(^{11}B,\alpha)2\alpha$, was present in the films, and that IR absorption studies showed that it was bonded to carbon and hydrogen. These authors suggested that the films should be considered a polymer consisting of a mixture of $SiH_x$, $CH_y$, and $SiCH_z$ compounds. The hydrogen content of the films did not change significantly with substrate temperature (200–500°C) and decreased only slightly, e.g. from 41 to 38 at. %, as the applied rf power (5 MHz) was raised from 100 W to 270 W. In agreement with Anderson and Spear (164), Catherine and Turnan (165) observed a variation of the optical band gap as a function of silicon content in the films, with an optimum value (~2.8 eV) for films containing silicon up to 38% of the total silicon and carbon present. The refractive index was found to increase with high silicon concentration in the films and was typically ~2.1–2.2 for films with silicon content around 40 and 50%, respectively, of the total silicon and carbon in the films. A comparison of the properties of reactivity rf-sputtered films of the type $Si_xC_yH_z$ with those deposited by glow-discharge decomposition was made by Contellec et al. (168). The authors rf sputtered silicon in a partial pressure of $CH_4$ and $H_2$. The optical band gap of films thus prepared showed a maximum value of about 2.5 eV for films with a silicon content of ~40% of the total silicon and carbon in the films, in agreement with glow-discharge-deposited films (164). The band gap was also found to vary with the hydrogen content of the films; thus an increase in hydrogen content from 25 to 38% for films containing about 50% silicon raised the band gap from 2.25 eV to 2.55 eV. A higher hydrogen content presumably leads to saturation of a larger number of dangling bonds. Contellec et al. (168) have also reported the formation of a Schottky barrier between gold and a-SiC films. Substrate bias potential during the film growth was observed to influence the properties of the a-Si–C–H films grown by rf (13.56 MHz) decomposition of $SiH_4$ and $CH_4$. The hardness of the films deposited at a substrate temperature of 300°C increased from 1200 to 1600 kgf mm$^{-2}$ as the bias voltage was changed from 0 to about $-1.5$ kV. The stresses in the films (compressive) decreased with a rise in the bias voltage. It is interesting to note that the IR absorption spectra showed a reduction in the absorption peak for C–H bonds in the films deposited at higher bias voltages, thus indicating a reduction in the hydrocarbon polymers included in the coatings. This may also explain the increased hardness of the films at higher bias voltages. The coatings were reported to contain columnar grains and were observed to be insoluble when left in a 30% KOH solution, heated to 100°C for 8 hr.

One of the attractive features of plasma-deposited a-Si–C–H films is the ability of the technique to deposit this hard and corrosion-resistant material on temperature-sensitive substrates, thus extending its usefulness. While other vapor-phase deposition techniques, such as reactive ion plating and sputtering, can also deposit silicon carbide films at relatively low substrate temperatures (ion plating having the added advantage of high deposition rates), plasma CVD has two distinct advantages: (1) It produces coatings that have fewer defects, such as pinholes, which is especially desirable for thin coatings and permits coating of complex three-dimensional shapes; (2) it is also possible to adjust the composition of the films at the substrate-coating interface so as to produce adherent and low-stress films. Mechanical properties can be varied over a wide range by altering the deposition conditions. Thus Linger (169) found that the hardness of the films, deposited on a grounded electrode in an externally excited rf (0.3–5 MHz) discharge of $SiH_4$ and $C_2H_4$, increased sharply from about 1000 kg mm$^{-2}$ to 2700 kg mm$^{-2}$ as the substrate temperature was raised from 200 to 1400°C. It is worth noting that Wasa et al. (184) have reported a hardness figure of 4000 kg mm$^{-2}$ (25 g load) for rf-sputtered silicon carbide films. In a similar temperature range the density of the films also increased from about 2.6 to 3 g cm$^{-3}$. Nodular features were observed on films deposited at low temperatures, but the films deposited at ~1370°C appeared to be featureless. Similar nodular growth was also observed by Katz et al. (170, 171) for films deposited by rf (0.4 MHz) plasma cracking of tetramethylsilane or dichlorodimethylsilane in a mixture of hydrogen and argon. As mentioned earlier, silicon carbide films contain an enormous amount of hydrogen, which can give rise to mechanical stresses in the film and film–substrate interface. These stresses would become more pronounced for thicker coatings. A detailed study of the stresses in films and ways of reducing them is necessary before plasma-deposited silicon carbide films can be of use as hard and wear-resistant coatings.

Photoluminescence has been observed in a-Si–C–H films (172) and found to depend on its composition and preparation technique. For example, the films deposited on quartz substrates heated to 200°C by plasma decomposition of $SiH_4$ and $C_2H_2$ were reported to show a maximum in photoluminescence activity at 2.1 eV for films of composition $Si_{0.4}C_{0.6}$. The films deposited by plasma decomposition of TMS (185), however, showed a different behavior. The composition of the coatings (~2 μm thick) deposited at substrate temperatures ranging from 40 to 600°C was found to be $Si_{0.1}C_{0.9}$. The optical band gap of the films deposited at temperatures below 200°C was 2.8 eV. The photoluminescence spectra for such films was observed to shift towards higher energies compared to those found in the earlier work (*172*), resulting in white luminescence.

Phosphorus- and boron-doped a-Si–C–H films have been used for making a-Si–C–H/a-Si(H) heterojunction solar cells (*174*). The conversion efficiency of such a cell is claimed to be 7.14% under AM1 illumination. s-Si–C–H films have also been used as masks in X-ray lithography (*175*).

*b. Titanium Carbide.* Titanium carbide is a hard and wear-resistant material widely used for enhancing the lifetime of cutting tools and bearings. Thick films of titanium carbide have been deposited by CVD (*178*), reactive physical vapor deposition (*179*, *180*), and rf sputtering (*181*). Deposition of TiC films has also been attempted by plasma-induced CVD (*182*, *183*). The films were deposited on steel substrates placed on the cathode in a dc discharge in a titanium tetrachloride, hydrogen, acetylene, and argon gas mixture. The films thus grew under constant ion bombardment. The cathode voltage was $\sim 3$–5 kV and the current density 16–49 A m$^{-2}$. The growth rates were typically 8000 Å min$^{-1}$. X-ray and electron microprobe analysis showed that the films contained crystalline titanium carbide with excess carbon and about 4% chlorine. An interfacial film deficient in carbon was observed to form between the titanium carbide and the steel substrate. The microhardness of the films varied from 800 to 2900 kg mm$^{-2}$ as a function of its carbon content. The kinetic friction coefficient against stainless steel (13 N; 100 mm sec$^{-1}$) had a value of 0.23 and was independent of film composition. The hardness and friction coefficients were comparable to those of the bulk TiC. The film surface, however, grew nodular features, presumably due to preferential sputter-etching of the constituents in the films.

## IV. Plasma Oxidation

The growth of oxides on metals and semiconductors immersed in an oxygen plasma under appropriate discharge conditions is a simple and controlled method of producing thin insulating films ($<1$ μm) at ambient temperatures (*186*). In this method the substrates to be oxidized are typically biased to a small positive potential with respect to the plasmas. The low-pressure ($10^{-1}$–$10^{-4}$ Torr) discharge is excited by a dc, rf, or microwave field. The applied positive potential on the specimen surface attracts electrons and negative ions from the plasma, and the oxide growth appears to take place by diffusion of negative oxygen ions towards the substrate surface and simultaneous out-diffusion of the metal ions towards the oxide surface under the influence of the applied electric field. The experimental evidence of the direction of the diffusing species has been obtained by two separate methods, one involving the use of $^{18}O_2$ as a marker during the growth process (*187*) and the other using a thin aluminum film deposited on the substrate prior to oxidation, and subsequent location of the film in the oxide (*215*). Besides the externally applied substrate bias potential, the oxide growth also depends on the electron density, the partial pressure of oxygen gas, and the substrate temperature. A large external positive potential on the substrate considerably enhances the growth rate but does not appear to be essential for oxide growth. Films have been grown on electrically isolated GaAs having a thickness of ~1500 Å at moderate growth rates (~400 Å min$^{-1}$) using a magnetically confined discharge to raise the concentration of the charged species at the oxide surface. Similar results were obtained by Kraitchman (*188*) and Bardos *et al.* (*189*) with electrically floating silicon wafers in a microwave-excited discharge, although the oxide growth rate was rather low (25–50 Å min$^{-1}$), possibly due to a low electron density.

That the oxide growth rate and the oxide film thickness increase on application of a positive potential to the substrate led earlier workers (*190*) to believe that negative oxygen ions in the discharge were essential for the oxide growth. This claim has been disputed on the basis of two separate experimental observations. Oliver *et al.* (*191*), assuming a Maxwellian distribution of negative ions in the plasma, measured oxide growth rate by applying a combination of dc and rf potentials to a tantalum substrate while maintaining a constant anode current (thus applying a constant field strength across the oxide). They concluded that negative oxygen ions are not essential in plasma oxidation. Chang and Sinha (*207*) noted that in spite of a positive bias potential on a GaAs substrate immersed in a magnetically confined rf plasma, the oxide growth was insignificant when the substrate surface was parallel to the magnetic field, but was much higher when the sample surface was perpendicular to the magnetic lines of force. The authors concluded that it was a supply of electrons rather than negative ions in the plasma that was necessary for the oxide growth. The electrons can produce negative oxygen ions at the sample surface by ionizing adsorbed oxygen atoms, and in the electron space-charge sheath by such processes as dissociative attachment.

Earlier experiments on plasma-oxidation were mostly performed in dc discharges (hot or cold cathodes) because it is a simple method of creating a low-pressure plasma. Some of the obvious disadvantages of using a dc plasma are: low ionization efficiency for a given applied power density, resulting in a lower electron density; and a high voltage on the (cold) cathode, leading to sputtering effects and contamination of the films by the cathode material. The general practice for preventing the sputtered particles from reaching a specimen is to place it at a sufficient distance from the cathode that the sputtered material is backscattered before reaching the specimen and at the same time to turn the sample to be oxidized away from the cathode. However, composition analysis of the plasma-grown oxide by Auger electron spectroscopy (AES) has shown that despite the above-mentioned precautions, the cathode material does get transferred to the anode surface (*192*). It is interesting to note that the authors observed that the transfer of the cathode material takes place only when the anode draws current, and not when it is in an electrically floating state.

An rf discharge can be excited by either internal or external electrodes and can be sustained at much lower pressures ($\sim 10^{-3}$ Torr) than can a dc (cold cathode) discharge ($<10^{-2}$ Torr). The use of an external magnetic field can further lower the partial pressure at which an rf discharge can be sustained in addition to raising the electron density. The electron density in an rf discharge is also much higher than that in a dc discharge for a given power density and pressure, thus giving rise to a higher concentration of negative oxygen ions and consequent faster oxidation rates. Although detailed AES analysis of GaAs oxide films grown in an rf discharge excited by internal electrodes (*215*) has been made, no contamination of the oxide film by the cathode material, of the type noted by Leslie *et al.* (*192*), has been reported.

### 1. GaAs Oxide

One of the major obstacles in the fabrication of GaAs devices, such as MOSFET and CCD's, and their surface passivation has so far been the growth of dense and uniform native oxide films having good insulating and oxide-interface properties. GaAs oxide films have chiefly been grown by "wet" anodic oxidation (*193–197*), thermal oxidation (*198–200*), plasma oxidation (*202–218*), and by *in situ* oxidation of epitaxial GaAs layers grown by MBE in the presence of oxygen and arsenic vapors (*201*). The thermal oxidation of GaAs at high temperatures, which is essential for reasonable growth rates, results in oxide films that are deficient in arsenic due to evaporation of volatile $As_2O_3$ (1 atm vapor pressure at 450°C). The nonstoichiometry of the oxide films makes them porous and electrically leaky. Anodic oxidation in aqueous solution, although it occurs at room temperature and at fast growth rates, produces films that have high water content and that require postoxidation annealing (*193–195*). Oxidation in nonaqueous solutions (*196*) has also been reported. Plasma oxidation of GaAs has been widely investigated as an alternative method for growing device-quality oxide films, because of the simplicity of the technique and the good control of film properties based on the properties of the discharges.

The early work on plasma oxidation was done by Weinreich (*202*), who used a microwave discharge (2.5 GHz, 0.2–0.6 Torr oxygen partial pressure).

Amorphous oxide films (300–4000 Å thick) were thus grown on n-type GaAs held at a positive potential with respect to ground. The refractive index of the oxide film was reported to be 1.78 and dielectric breakdown strength $\sim 5 \times 10^6$ V cm$^{-1}$. Different techniques and discharge conditions have since been used for growing GaAs oxide films. Chester and Robinson (203) placed GaAs wafers (p- and n-type (100) orientation) in the negative glow region of a cold-cathode oxygen discharge. The oxide ($\sim 2500$ Å thick) was grown by first using a constant current of 1 mA cm$^{-2}$, after which the applied voltage to the sample was held constant until the current dropped to a constant minimum value. The films thus grown were rich in Ga and had a wide oxide–semiconductor interface (e.g., 500 Å wide for 1740-Å-thick oxide films).

Koshiga and Sugano (204) reported growth of oxide films on n- and p-type GaAs in a dc glow discharge. The oxide films were amorphous, nonstoichiometric (Ga-rich), and had a high density of states ($10^{12}$ eV$^{-1}$ cm$^{-2}$) at the oxide–GaAs interface. Insulated gate field-effect transistors (IGFET) incorporating GaAs oxide were fabricated; the mobility of the devices, estimated from measured transconductance values, was 3750 cm$^{-2}$ V$^{-1}$ sec$^{-1}$. The transconductance, however, showed a large frequency dispersion, apparently due to the high density of interface states.

Yokoyama et al., (206) were able to grow oxide films at varying rates ($\sim 50$–250 Å min$^{-1}$) by changing the magnetic field (150–600 G) (Fig. 16). The substrates were placed on a grounded electrode and no external bias potential was applied. The increased oxidation rates at higher magnetic

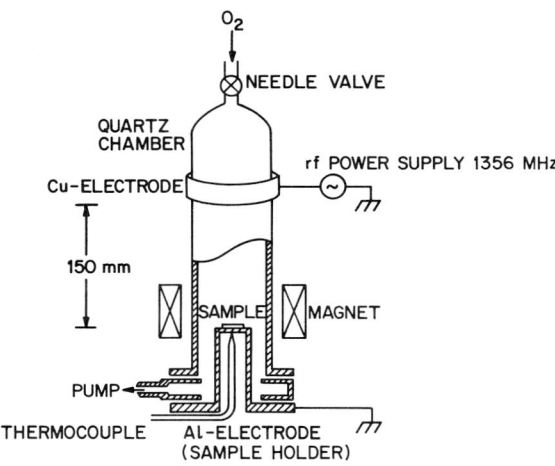

Fig. 16. Plasma oxidation in a magnetic field. [after Yokoyama et al. (206).]

fields could have resulted from a rise in the plasma density. Another significant factor that can influence the growth rate under the discharge conditions (0.05 Torr $O_2$ pressure; 150 W input power at 13.56 MHz) employed by Yokoyama et al., is the mildly energetic positive ion bombardment of the wafer surface due to a negative floating potential (with respect to the plasma) on the wafer surface. Capacitive coupling of the substrate/oxide to the applied rf field could also give rise to surface potentials up to 20–30 V with respect to ground. The surface potential thus developed on the substrate surface will drop at higher applied magnetic fields because of the increased ionization and concomitant higher current density, leading to reduced sputtering effects and higher growth rates. The oxide films were found to be uniform but deficient in $As_2O_3$. The oxide GaAs (p-tyde) interface state density, obtained from $C$–$V$ measurements (1 MHz), was reported to be in the region of $10^{10}$ cm$^{-2}$ eV$^{-1}$ at $\sim 0.5$ eV below the valence band. The $C$–$V$ characteristics, however, showed hysteresis and frequency dispersion in the accumulation region.

A detailed study of the composition and electrical characteristics of plasma-grown GaAs oxide has been made by Chang and co-workers (207–215). The experimental arrangement (Fig. 17) employed by them consisted of a 27 MHz rf generator that excited a discharge at an oxygen partial pressure of about $2 \times 10^{-3}$ Torr between two electrodes situated inside the chamber (207). An external magnetic field (500 G) was used to confine the discharge between rf electrodes and the substrate, which were about 50 cm apart. The wafers (Si-doped GaAs, $N_d \sim 1.7 \times 10^{17}$ cm$^{-3}$, (100) orientation) were biased at a positive dc potential (20–100 V) with respect to the plasma potential. Under constant discharge conditions (input rf power 300 W) the oxide growth rate was found to increase both with bias potential and substrate temperature (0–100°C). Rutherford backscattering and ion-induced X-ray measurements of the oxide films showed that they

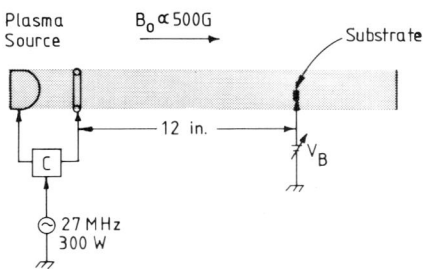

FIG. 17. Plasma oxidation system of Chang (215). The rf signal is fed to the aluminum electrodes via a matching coupler. [After Chang (215).]

consisted of $Ga_2O_3$ and $As_2O_3$. Depth profiling of the oxide by ion-mill AES revealed that the bulk of the oxide composition was uniform but was deficient in $As_2O_3$. By choosing a fast oxide growth rate the authors were able to minimize the ratio of $Ga_2O_3$ to $As_2O_3$ in the oxide to about 1.2. The growth rate also had a significant influence on the width of the oxide–GaAs interface. The oxide films grown at higher rates ($\sim 50$ Å $sec^{-1}$), achieved by raising the substrate bias, had a narrower interface with the GaAs substrate ($\sim 20$ Å for a 100 Å film) than the ones grown at slower rates (25 Å sec; 40 Å width for a 1000 Å film). It is interesting to note that similar observations have also been made for films grown by anodic oxidation (wet), in which the interface width for a given oxide thickness decreased for low current densities (197).

The As deficiency in plasma-grown GaAs oxide films was also accompanied by accumulation of arsenic at the oxide–GaAs interface (212). The amount of arsenic was found to be larger for thicker oxide films but decreased at higher oxide growth rates. TEM micrographs showed the presence of amorphous arsenic, which crystallized when the oxide was annealed at 400–450°C. The oxide composition remained uniform after the films were annealed in nitrogen for 1/2 hr at a temperature below 450°C, but became nonuniform above 450°C due to volatilization of $As_2O_3$. Annealing the oxide films in hydrogen at similar temperatures caused phase separation of the oxide, which made MOS devices leaky. The density of the interface states, however, decreased for hydrogen-annealed films.

The problem of perferential out-diffusion of arsenic during the oxidation process is reported to have been curtailed by depositing a thin layer of aluminum film ($\sim 100$ Å) prior to oxidation' The presence of the aluminum film, besides improving the oxide properties, also gave some insight into the oxidation mechanism. A simultaneous depth profiling and TEM study of the oxide films on GaAs (Te doped, $N_d = 2 \times 10^{17}$ $cm^{-3}$) revealed that

(1) The aluminum film was completely oxidized to an amorphous $Al_2O_3$ film.

(2) The Ga:As ratio in the oxide film between the GaAs and the $Al_2O_3$ film was nearly 1, in contrast with oxidation of GaAs without Al film, in which case the oxide was always deficient in As. The $Al_2O_3$ layer thus apparently acted as a preferential filter, controlling the migration of ionized species.

(3) Excess arsenic at the oxide–GaAs interface was not detected.

(4) The reduction in arsenic accumulation at the interface and improved stoichiometry were considered to be responsible for the lower MOS fixed-charged density, $\sim 5 \times 10^{10}$ $cm^{-2}$, after the oxide was annealed in hydrogen for 1/2 hr at 550°C, as compared to $\sim 5 \times 10^{11}$ $cm^{-2}$ in the case of oxide

films grown without an aluminum film. A breakdown voltage of $\sim \pm 4 \times 10^{-6}$ V cm$^{-1}$ was reported for the MOS diodes. A marked improvement in the $C-V$ characteristics (at 1 MHz) was also noted, as the curve (Fig. 18) went into inversion at negative voltages.

Shinoda and Yamaguchi (216) claim to have improved the oxide properties by growing an initial Ga-rich oxide layer. An initial positive oxygen ion bombardment of the wafer by biasing it at a negative potential (400–500 V) resulted in a thin layer of Ga-rich oxide. Subsequent oxidation of the wafer by biasing it at a positive potential yielded an oxide film having a better stoichiometry than the oxide films grown without the initial positive ion bombardment. However, Shinoda and Yamaguchi did not report any measurements of the oxide composition or the oxide–GaAs interface width. According to them, the pinhole density of the oxide films grown after the initial ion bombardment showed a large decrease (0–5 cm$^{-2}$) compared to the oxides on unbombarded wafers (500–1500 cm$^{-2}$). The low pinhole density resulted in a higher dielectric breakdown voltage, $\sim 3 \times 10^6$ V cm$^{-1}$. Chang (215), however, has reported growth of oxide films (1500 Å) on GaAs (without an Al layer) in a discharge of CF$_4$ and O$_2$ gas mixtures. The films were annealed in an atmosphere of N$_2$ and H$_2$ (N$_2$:H$_2$ = 1) for 1/2 hr at 450°C. The high-frequency $C-V$ curves for these films show inversion at negative bias and a minimal hysteresis effect. The improvement in the electrical characteristics was attributed to a reduction in excess arsenic at the oxide–GaAs interface, the exact mechanism for which is not well understood.

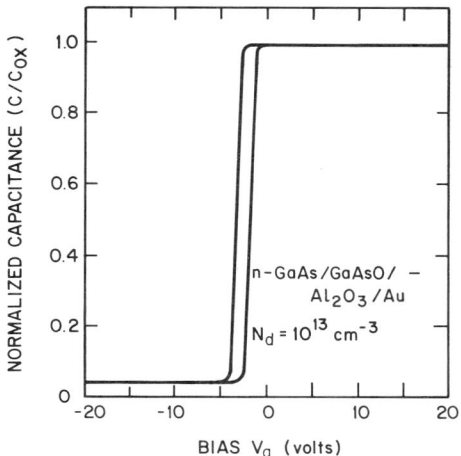

FIG. 18. Typical $C-V$ curves for a MOS device, where $C_{ox}$ is the capacitance per unit area of oxide layer and $V_g$ is the gate voltage. [After Chang and Coleman (212).]

Chang et al. (*210*) have also reported the growth of an oxide film on the ternary compound $Ga_{0.64}Al_{0.36}As$ by plasma oxidation. AES analysis of the oxide showed it to consist of $Ga_2O_3$, $As_2O_3$, and $Al_2O_3$. The fixed-charged density of the oxide was found to be $\sim 6.8 \times 10^{11}$ cm$^{-2}$.

Plasma oxidation of InP and $Hg_{1-x}Cd_xTe$ has also been reported (*218, 219*).

## 2. Silicon Oxide

In silicon device technology, silicon oxide films are normally grown by high-temperature ($\sim 1100°C$) oxidation of silicon. Such high-temperature processes have many drawbacks, especially for submicron device structures: the dopant profile is changed, and defects such as stacking faults, dislocations, and so-called bird's-beak-shaped defects are generated. High-pressure oxidation processes have succeeded in bringing the oxidation temperature down to about 700°C (*221*), but are limited to very low growth rates. Oxidation of silicon at much lower temperatures ($\sim 270°C$) and higher growth rates ($\sim 60$ Å min$^{-1}$) by using a low-pressure oxygen discharge was reported nearly 20 years ago (*220*). The current interest in plasma oxidation results from the advantage the low-temperature process offers in VLSI technology.

Earlier work on plasma oxidation was done in microwave and dc arc discharges (*222–224*). Recently reported investigations have mainly been carried out in inductively excited rf plasmas (*225–229*). High oxide growth rates ($\sim 300$ Å min$^{-1}$, bias voltage $\sim +80$ V to ground) as substrate temperatures of $\sim 600°C$ were claimed without the occurrence of defects such as stacking faults and impurity redistribution in the silicon substrate (*225*). Ray and Reisman (*226–228*) and Ray (*229*) have made a detailed investigation of the growth-kinetics properties and behavior of the films in FET device structures. The authors noted that at low pressures (<10 mTorr), oxidation of the silicon substrate (56 mm diameter) was negligible and only sputtering of quartz from the reactor wall contributed to the oxide growth. The substrates were electrically floating in the plasma and were heated to temperatures of $\sim 600°C$ and higher. At higher pressures (>10 mTorr), oxide growth was recorded. However, at these pressures it was observed that the growth rate of oxide on the silicon surface facing away from the discharge was significantly higher ($\sim 0.5$ Å min$^{-1}$); the reason for this was not known. The properties of the plasma-grown oxide are summarized and compared with those of thermally grown films in Table III.

## 3. rf Oxidation

Metal–oxide–metal structures are used in such devices as Josephson junction switches (*230–237*) and IR detectors and mixers (*238*). A very

TABLE III

Properties of Plasma $SiO_2$ Grown at 500°C Compared to Thermal $SiO_2$ Grown at 1100°C[a]

| Properties | Plasma $SiO_2$, 500°C growth temperature | Thermal $SiO_2$, 1100°C growth temperature |
|---|---|---|
| Etch rate in 1:9 BHF (nm min$^{-1}$) | 74–76 | 75 |
| Refractive index | 1.461–1.465 | 1.462 |
| Stress (dyn cm$^{-2}$) | 1.5–1.6 × 10$^9$ | 3.1–3.4 × 10$^9$ |
| Fixed charge (cm$^{-2}$) | 2.6 × 10$^{10}$ | 2 × 10$^{10}$ |
| Interface states (cm$^{-2}$ eV$^{-1}$) | 2.6 × 10$^{10}$ | 2 × 10$^{10}$ |
| Retention time (sec) | ~100 | >500 |
| Breakdown strength (MW cm$^{-1}$) | 4.8 | 10 |
| Boron depletion | Absent | Present |
| Bird's beak effect | Absent | Present |
| Oxidation-induced defects | Absent | Present |

[a] After Ray and Reisman (228). Reprinted by permission of the publisher, The Electrochemical Society, Inc.

thin (5–25 Å), uniform, and defect-free oxide film is essential in these structures. Plasma anodization, as discussed earlier, has been found to be inadequate in producing pinhole-free thin oxide films at controlled rates. Greimer (230–231) was able to overcome these problems by placing the substrate to be oxidized on the cathode of an rf system and adjusting the cathode potential to a value that resulted in simultaneous etching and oxidation by oxygen ions during the oxide growth. The net growth rate is a function of oxide film thickness and becomes zero at a thickness of ~5–100 Å thick. The steady-state oxide film thickness depends on discharge parameters, such as cathode potential and oxygen pressure. Several mechanisms have been proposed for the growth of a steady-state oxide thickness (230–233). This rf oxidation technique, however, suffers from such drawbacks as contamination of the oxide film due to sputtering of the walls, deposition of back-sputtered material, and excessive heating of the sample during oxidation (237). Oxidation by a low-energy ion beam (30–180 eV) is claimed to offer better control and less contamination of the oxide film (237).

## V. Plasma Carburizing

Carburizing of steels involves transport of carbon in the form of a hydrocarbon gas to a work piece that is heated to temperatures of ~1000°C. The carbon thus deposited diffuses into the steel and forms carbides. Conventionally the carburizing process has been carried out either at atmospheric pressure (*239*) or in vacuum (*240*). Vacuum carburizing is reported to be a more controlled and economical process, as it utilizes only 1% of the total gas required for atmospheric carburizing to produce a similar case depth. An even more economical and rapid carburizing of steels has been reported which makes use of a low-pressure glow discharge (*241*). In principle the method is similar to plasma nitriding. The steel to be carburized is made a cathode (Fig. 19) in a dc cold-cathode glow discharge of a hydrocarbon gas (partial pressure $\sim 1$–20 Torr) and heated by an external furnace to temperatures of $\sim 1050°C$. The discharge current is varied only for controlling the temperature. As mentioned earlier, in plasma nitriding the discharge current is the only medium of heating the cathode.

The hydrocarbon gas used in plasma carburizing dissociates not only pyrolytically but also by electron bombardment in the discharge region. The presence of discharge thus enhances the deposition rate of carbon on the metal surface. The deposited carbon diffuses into the metal at much faster rates than in vacuum carburizing. A case depth of 1 mm is reported to have been achieved by plasma carburizing in half the time required by vacuum carburizing carried out at the same temperature and partial pressure of

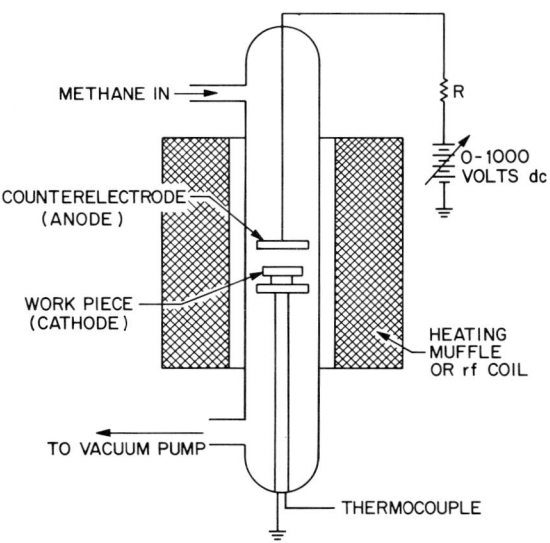

FIG. 19. Plasma carburizing apparatus. [After Grube and Gay (*242*). Copyright © 1978, American Society for Metals and the Metallurgical Society of AIME.]

methane (Fig. 20). Theoretical and experimental consideration of the diffusion process appear to suggest that high infusion rates of carbon into the surface, rather than high diffusion rates, are responsible for the high carburizing rates achieved in plasma carburizing (241). As in the case of plasma nitriding, energetic ion bombardment of the cathode during carburizing would also influence the growth process.

The case hardness achieved by plasma carburizing has been found to be greater than that achieved by vacuum carburizing. The case depth and hardness uniformity in different regions of an irregularly shaped component depend on the ion-sheath uniformity, which in turn determines the nature of the carbonaceous species arriving at the surface and also the flux and energy of the ionized species bombarding it.

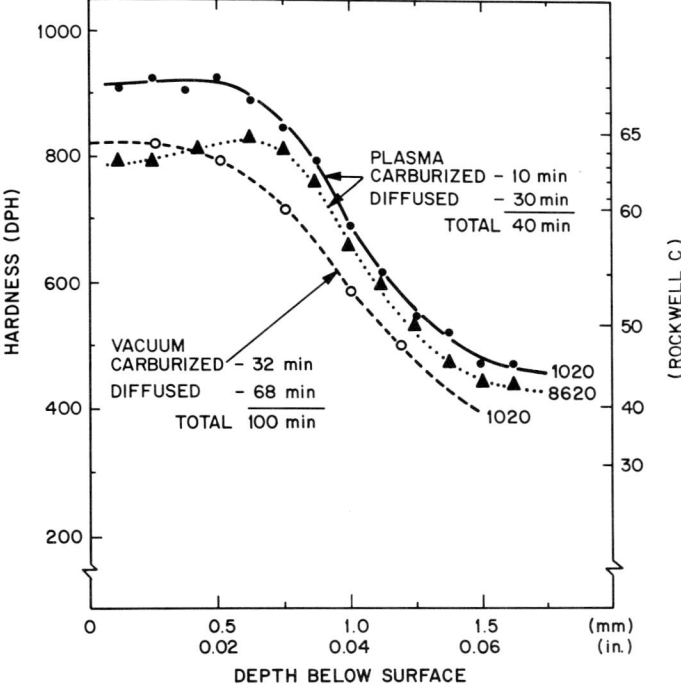

FIG. 20. Hardness profile through carburized cases on AlSi 1020 and 8620 steels after plasma carburizing for 10 min and diffusing for 30 min. Ratio of diffusing time to carburizing time is 3:1. [After Grube and Gay (242). Copyright © 1978, American Society for Metals and the Metallurgical Society of AIME.]

## VI. Glow-Discharge Nitriding

Nitriding of steels in a low-pressure nitrogen discharge is an established industrial process of forming hard and wear-resistant surfaces. The surface hardening of metals immersed in a high-pressure ammonia discharge was observed by Egan (*242*) as early as 1930. The use of a low-pressure glow discharge for nitriding was introduced by Bosse *et al.* and Berghouse (*243, 244*), and Berghouse later (*245*) patented the process in which the articles to be nitrided were made a cathode in the glow discharge of a nitrogen and hydrogen gas mixture. By this technique the work piece receives energetic gas ions and can also be heated to temperatures of $\sim 500°C$, both conditions being essential for forming hard and deep nitride layers.

The gas discharge is maintained in the abnormal region of the glow discharge to obtain the high current and power densities that have been found essential for a rapid nitride growth. Typically the current density and cathode voltages during nitriding vary in the region $0.1-10$ mA cm$^{-2}$ and $0.5-1$ kV, respectively. The maintenance of constant discharge conditions in the abnormal mode, and especially preventing it from turning into an arc, requires power supplies that can be rapidly interrupted to quench arcs that can damage the cathode surface (*246, 247*). A localized arc can develop when either contaminant particles on the cathode surface emit gas due to ion bombardment and heating, or because of electrical breakdown resulting from the build-up of surface charge on insulating particles on the cathode.

Nitrogen gas introduced into the discharge forms ionized ($N^+$, $N_2^+$) atomic and excited species. The cathode thus undergoes energetic ion bombardment and at the same time adsorbs neutral gas species such as atomic nitrogen. The individual role of ionized and neutral species in the growth mechanism is still not precisely understood. Edenhofer (*248, 249*) has contended that reactive sputtering of the cathode and back-diffusion of the sputtered species are responsible for the nitriding process. According to him the sputtered Fe atoms react with nitrogen to form unstable FeN, which is backscattered onto the cathode and decomposes to lower nitrides, such as $Fe_2N$, $Fe_3N$, and $Fe_4N$. A fraction of the nitrogen atoms released due to such a decomposition are believed to diffuse into the steel. Hudis (*250*) has confirmed that ion bombardment of the specimen to be nitrided is essential for achieving rapid nitride growth and large case depth. No significant nitriding of a steel specimen placed in a floating position was observed, thus indicating that adsorbed gas species are insufficient for nitriding. However, the author has not indicated the exact mechanism that leads to an enhanced nitriding rate for a specimen undergoing ion bombardment. Hudis has also observed that the addition of hydrogen to the nitrogen gas enhances the nitriding rate, case depth, and hardness of steel specimens (Fig. 21). Mass spectrometric analysis of the ionized species bombarding the cathode showed that the presence of hydrogen in the discharge considerably reduces ionized nitrogen ($NH^+$, $N_2^+$) as compared to the pure nitrogen discharge and instead creates predominantly $H^+$ and $NH_3^+$ species. The role of these species in improving the nitriding process is once again not well understood. A nitriding mechanism based on the formation of vacancy and nitrogen ion pairs and their subsequent diffusion inside the metal was proposed by Brokman and Tuler (*251*). The authors used crossed electric and magnetic fields to enhance the ion current density and studied their effect on the hardness of AISI 304 stainless steel. They contended that ion bombardment of the surface causes the formation of vacancy ion pairs, which have a relatively high local diffusion coefficient. This diffusion coefficient was found to be proportional to the current density (100–1000 A m$^{-2}$).

Besides steel, refractory metals such as Zr and Ti (*252, 253*) and carbides such as TiC (*254*) have been nitrided in a glow discharge. Zr and Ti were nitrided in an externally excited discharge. The specimens were kept electrically floating with respect to the ground and heated to temperatures in the range 800–900°C. The time dependence of the nitriding rate was found to

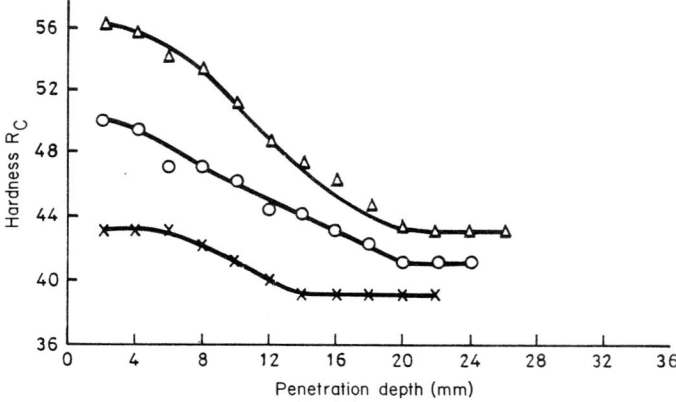

FIG. 21. Hardness profile for three samples of 4340 heat-treated steel ion nitrided with different gas compositions but the same total gas pressure [after Hudis (*251*)]:

| Symbol | $p_{N_2}$ (Torr) | $p_A$ (Torr) | $p_{H_2}$ (Torr) |
|---|---|---|---|
| △ | 1.2 | 0 | 4.8 |
| ○ | 1.2 | 4.0 | 0.8 |
| × | 1.2 | 4.8 | 0 |

follow the parabolic relation $\Delta w^2 = kt + C$, where $w$ is the weight gain per unit area and $k$ is the rate constant. In the case of Zr, the nitrided layer was found to consist of ZrN with small traces of α-Zr. The maximum hardness of several-micron-thick ZrN layers was ~1000 kg mm$^{-2}$ (25 g load). In agreement with earlier workers, addition of hydrogen to the nitrogen gas was observed to increase the case depth. Similarly, nitrided titanium metal consisted of TiN and Ti$_2$N phases' The nitrided layers, typically a few microns thick, were reported have hardnesses of ~1200 kg mm$^{-2}$ in the diffusion region of the layer, and the extrapolated surface hardness was ~2000 kg mm$^{-2}$.

The recent increase in applications involving hard and wear-resistant alloy steel surfaces has resulted their plasma nitriding (254). It was noted that the hardness of a high-chromium alloy steel (composition in wt. %: 56.5 Fe, 27.5 Ti, 7.4 C, and 20 Mo) containing distributed grains of titanium carbide was increased to about 1500 Knoops on plasma hardening as compared to untreated steel. X-ray diffraction analysis showed that nitriding resulted in the formation of iron components such as Fe$_3$N and Fe$_2$N at low current densities and low temperatures (~475°C) and $\gamma$-Fe$_4$N and Fe$_3$N at high current densities and high temperatures (~750°C). However, no nitrides of Ti were detected. The hardness of the nitrided steel also depends on the treatment temperature; the maximum value of approximately 1500 Knoops occurs at temperatures around 600°C. Conventional hardening techniques comprising multiple-cycle heat-quenching at ~1090°C yielded a much lower hardness figure of 900 Knoops. Plasma nitriding of Inconel 625 (255) in a dc glow discharge has also been shown to result in a surface that is about 4.5 times harder than that of the untreated specimens.

The kinetics of nitriding of different metals, such as Ti, Mo, and Mb, in low-pressure glow discharges has been studied by Wirz et al. (256) by means of in situ thermogravimetry using a quartz spring microbalance. The authors found that the dependence of the nitriding rate with time followed a $t^{1/2}$ dependence only during the initial period, after which the rate was controlled by the successive formation of crystallographic phases of different stoichiometry. The different crystallographic phases gave rise to varying nitrogen diffusion rates into the metal. The nitriding rate of molybdenum placed at a floating potential in a nitrogen discharge was found to be orders of magnitude faster than the nitriding rates in molecular nitrogen at atmospheric pressures. The nitride layer on molybdenum underwent transformation from the hexagonal $\delta$-MoN phase to $\delta'$-MoN, leading to a large increase in the nitriding rate. Niobium, however, was found to nitride in a nitrogen glow discharge at rates similar to those in molecular nitrogen at atmospheric pressures. Plasma-nitrided Ti and stainless steel (No. 304) were subjected to further glow-discharge treatment in hydrogen gas (257). The titanium nitride layer was unaffected by the hydrogen plasma, but the nitrided stainless steel was considerably depleted in nitrogen after hydrogen plasma treatment.

## VII. Conclusions

From the above discussions it is seen that a wide range of new* and useful materials can be deposited by plasma-CVD methods. The low deposition temperatures allow one to deposit materials on temperature-sensitive substrates and devices. Plasma cleaning and/or etching in a reactive discharge followed by plasma deposition can in principle produce films with good adhesion to substrates. It is also possible to produce a graded material interface by changing the gases used for deposition in order to produce good adhesion and low stress. The properties of the films can be varied by changing deposition conditions, such as gas flow pressure and substrate temperature, and discharge conditions, such as electron density, electron temperature, and substrate bias potential, which influences the energy of the ion striking a growing film. Electrical properties of the films can be modified by doping. Plasma-deposited films have good conformity and low defect density (such as pinholes). The films, however, are generally highly stressed and further studies of the factors controlling stress are necessary. The processes occurring in a reactive plasma are complex and need better understanding to permit deposition of films of desired properties. Diagnosis of the reactive species in the discharge (neutral, excited, and ionized) by such techniques as optical emission spectroscopy (*264, 265*) and ion sampling by mass spectrometry (*266, 267*), and improved understanding of their role in the deposition process should lead to better control over film properties (*268, 269*). Research in these areas is leading to plasma-CVD techniques of even greater versatility and usefulness.

## References

1. D. E. Carlson, in "Polycrystalline and Amorphous Thin Film Devices," (L. L. Kazmerski, ed.), p. 175. Academic Press, New York, 1980.
2. S. Veprek, Z. Iqbal, H. R. Oswald, and A. P. Webb, *J. Phys. C* **14**, 295 (1981).
3. S. C. Brown, "Introduction to Electrical Discharges in Gases." Wiley, New York, 1966.
4. F. K. McTaggard, "Plasma Chemistry in Electrical Discharges." Elsevier, Amsterdam, 1967.

* Some new materials whose deposition has been reported are a-P (*18*), a-As (*259*), a-Ge (*260*), a-Ge–Si (*261*), Ge–C (*262*), Ga–N (*263*), As–Se and As–S (*264*).

5. M. Venugopalan (ed.), "Reactions under Plasma Conditions," Vols. 1 and 2. Wiley (Interscience), New York, 1971.
6. J. R. Hollohan and A. T. Bell (eds.), "Techniques and Applications of Plasma Chemistry." Wiley, New York, 1974.
7. T. M. Shaw, in "Formation and Trapping of Free Radicals" (A. M. Bass and H. P. Broida, eds.). Academic Press, New York, 1960.
8. F. Kaufman, *Adv. Chem. Ser.* **50** (1969).
9. G. Carter and J. S. Colligan, "Ion Bombardment of Solids." Elsevier, Amsterdam, 1969.
10. V. N. Kondratiev, "Chemical Kinetics of Gas Reactions." Addison Wesley, Reading, Massachusetts, 1964.
11. L. Holland and S. M. Ojha, *Vacuum* **26** (2), 53 (1976).
12. J. R. Hollahan and R. S. Rosler, in "Thin Film Processes" (J. L. Vossen, ed.), p. 335. Academic Press, New York, 1978.
13. J. L. Vossen, *J. Electrochem. Soc.* **126** (2), 319 (1979).
14. R. S. Rosler and G. M. Engle, *Solid State Technol.* December, p. 88 (1979).
15. "Ionic Systems." San Jose, California.
16. M. Sokolowski, *J. Cryst. Growth* **46**, 136 (1979).
17. M. Sokolowski, A. Sokolowski, A. Michalski, Z. Romanowski, A. Rusek-Mazurek, and M. Wronikowski, *Thin Solid Films* **80**, 249 (1981).
18. S. Veprek, *Curr. Top. Mater. Sci.* **4** (1980).
19. S. Fujimori and K. Nagai, *Jpn. J. Appl. Phys.* **20**(3), L194 (1981).
20. I. S. McLintock and J. C. Orr, in "Chemistry and Physics of Carbon" (P. L. Walker, Jr. and P. A. Thrower, eds.), Vol. 11, p. 242. Dekker, New York, 1973.
21. J. J. Hauser, *J. Non-Cryst. Solids* **23**, 21 (1977).
22. N. Hosokawa, A. Konishi, H. Hiratsuka, and K. Annoh, *Thin Solid Films* **73**, 115 (1981).
23. G. M. Jenkins and K. Kawamura, "Polymeric Carbons—Carbon Fibre, Glass and Char." Cambridge Univ. Press, London and New York, 1976.
24. M. L. Kaplan, P. H. Schmidt, C. H. Chen, and W. M. Walsh, Jr., *Appl. Phys. Lett.* **36**(10), 867 (1980).
25. J. A. Brinkman, U. S. Patent No. 3,142,539, July 28, 1964.
26. W. G. Eversole, Canadian Patent 628,567, October 3, 1961.
27. J. C. Angus, H. A. Will, and W. S. Stanko, *J. Appl. Phys.* **39**(6), 2915 (1968).
28. D. J. Poferl, N. C. Gardner, and J. C. Angus, *J. Appl. Phys.* **44**(4), 1428 (1973).
29. S. P. Chauhan, J. C. Angus, and N. C. Gardner, *J. Appl. Phys.* **47**(11), 4746 (1976).
30. D. G. Teer and M. Salama, *Proc. Conf. Ion Plating Allied Tech.*, Edinburgh, June (1977).
31. S. Aisenberg, U.S. Patent 3,961,103, June 1976.
32. S. Aisenberg and R. Chabot, *J. Vac. Sci. Technol.* **10**, 104 (1974).
33. E. G. Spencer, P. H. Schmidt, D. C. Joy, and F. J. Sansalone, *Appl. Phys. Lett.* **29**, 118 (1976).
34. Chr. Weissmantel, *Proc. Int. Vac. Congr., 7th, Int. Conf., 3rd, Solid Surf.*, Vienna p. 1533 (1977).
35. Chr. Weissmantel, K. Bewilogua, H. Erler, and G. Reiss, *Proc. Conf. Ion Plating Allied Tech.*, London p. 272 (1979).
36. J. H. Freeman, W. Temple, and G. A. Gard, *Nature (London)* **275**, 634 (1978).
37. M. Millard, in "Techniques and Applications of Plasma Chemistry" (J. R. Hollahan and A. T. Bell, eds.). Wiley, New York, 1974.
38. M. Shen, ed., "Plasma Chemistry of Polymers." Dekker, New York, 1976.
39. M. Duval and A. Theoret, *J. Electrochem. Soc.* **122**(4), 581 (1975).
40. H. Hiratsuka, G. Akovali, M. Shen, and A. T. Bell, *J. Appl. Polym. Sci.* **22**, 917 (1978).
41. H. F. Winters, *J. Chem. Phys.* **63**(8), 3462 (1975).

42. S. M. Ojha and L. Holland, *Proc. Int. Vac. Congr., 7th, Int. Conf. Solid Surf., 3rd, Vienna* p. 1667 (1977).
43. L. P. Anderson and S. Berg, *Vacuum* **28**(10/11), 449 (1978).
44. S. Berg and L. P. Anderson, *Proc. Int. Thin Film Congr., 4th, Loughborough, 1978, Thin Solid Films* **58**, 117 (1979).
45. L. P. Anderson, S. Berg, H. Norström, R. Olaison, and S. Towa, *Thin Solid Films* **63**, 155 (1979).
46. D. S. Whitmell and R. Williamson, *Thin Solid Films* **35**, 255 (1979).
47. L. Holland and S. M. Ojha, *Thin Solid Films* **38**, L17 (1976).
48. S. M. Ojha and L. Holland, *Thin Solid Films* **40**, L31 (1977).
49. S. M. Ojha and L. Holland, *Proc. Conf. Ion Plating Allied Tech., Edinburgh, June* p. 238 (1977).
50. L. Holland and S. M. Ojha, *Thin Solid Films* **48**, L21 (1978).
51. L. Holland and S. M. Ojha, *Thin Solid Films* **58**, 107 (1979).
52. S. M. Ojha, H. Norstrom, and D. McCulluch, *Thin Solid Films* **60**, 213 (1979).
53. E. I. Zorin, V. V. Sukhorukov, and D. I. Tetel'baum, *Sov. Phys. Tech. Phys.* **25**(1), 103 (1980).
54. T. J. Moravec, *Thin Solid Films* **70**, L9 (1980).
55. K. Enke, H. Dimigen, and H. Hubsch, *Appl. Phys. Lett.* **36**(4), 291 (1980).
56. R. C. Weast, ed., "Handbook of Chemistry and Physics," p. F22. CRC Press, Cleveland, Ohio, 1975.
57. D. J. McCulluch, personal communication.
58. D. A. Anderson, *Philos. Mag.* **35**(1), 17 (1977).
59. B. Meyerson and F. W. Smith, *J. Non-Cryst. Solids* **35–36**, 435 (1980).
60. B. Meyerson and F. W. Smith, *Solid State Commun.* **34**, 531 (1980).
61. K. Bewilogua, D. Dietrich, L. Pagel, C. Schürer, and C. Weissmantel, *Surf. Sci.* **86**, 308 (1979).
62. Chr. Weissmantel, K. Bewilogua, C. Schürer, K. Breuer, and H. Zscheile, *Thin Solid Films* **61**, L1 (1979).
63. Chr. Weissmantel, *Thin Solid Films* **58**, 101 (1979).
64. U. Ebersbach, C. Scurer, and C. Weissmantel, *Z. Phys. Chem.* **260**(5), 938 (1979).
65. C. Weissmantel, K. Bewilogua, D. Dietrich, H. J. Erler, H. J. Hinneberg, S. Klose, W. Nowick, and G. Reisse, *Thin Solid Films* **72**, 19 (1980).
66. D. F. Edwards and E. Ochoa, *J. Opt. Soc. Am.* **71**(5), 607 (1981).
67. S. K. Das, M. Kaminsky, L. H. Rovner, J. Chin, and K. Y. Chen, *Thin Solid Films* **63**, 227 (1979).
68. J. W. Fraser and R. T. Holzmann, *J. Am. Chem. Soc.* **80**, 2907 (1958).
69. R. T. Holzmann and W. F. Morris, *J. Chem. Phys.* **29**, 677 (1958).
70. T. Wartic, R. Moore, and H. I. Schlesinger, *J. Am. Chem. Soc.* **71**, 3265 (1949).
71. A. E. Hultquist and M. E. Sibert, *Adv. Chem. Ser.* **80**, 182 (1969).
72. R. D. Wales, *Adv. Chem. Ser.* **80**, 198 (1969).
73. F. H. Cocks, P. L. Jones, and L. J. Dimmey, *Appl. Phys. Lett.* **36**, 970 (1980).
74. B. L. Zalph, L. J. Dimmey, H. Park, P. L. Jones, and F. H. Cocks, *Phys. Status Solidi A* **62**, K185 (1980).
75. W. A. Pliskin, D. R. Kerr, and J. A. Perri, in "Physics of Thin Films" (G. Hass and R. E. Thun, eds.), Vol. 4, p. 257. Academic Press, New York, 1967.
76. W. M. Feist, S. R. Steele, and D. W. Readey, in "Physics of Thin Films" (G. Hass and R. E. Thun, eds.), Vol. 5, p. 237. Academic Press, New York, 1969.
77. G. L. Scnable, W. Kern, and R. B. Comizzoli, *J. Electrochem. Soc.* **122**, 1092 (1975).
78. W. A. Pliskin and H. S. Lehman, *J. Electrochem. Soc.* **112**, 1013 (1965).

79. J. Wong, *J. Electron. Mater.* **5**, 113 (1976).
80. W. A. Pliskin, *J. Vac. Sci. Technol.* **14**, 1064 (1977).
81. F. G. Allen, T. M. Buck, and J. T. Law, *J. Appl. Phys.* **31**, 979 (1960).
82. E. G. Baylander, J. R. Piedmont, L. D. Shubin, and R. C. Smith, *J. Appl. Phys.* **34**, 3407 (1963).
83. F. L. Shuermeyer, W. R. Chase, and E. L. King, *J. Appl. Phys.* **42**, 5856 (1971).
84. W. A. Pliskin and E. E. Conrad, *IBM J. Res. Dev.* **8**, 42 (1964).
85. G. C. Schwartz and R. E. Jones, *IBM J. Res. Dev.* **14**, 52 (1970).
86. K. Urbanck, *Solid State Technol.* **87**, April (1977).
87. L. L. Alt, S. W. Ing, Jr., and K. W. Laendle, *J. Electrochem. Soc.* **110**, 465 (1963).
88. S. W. Ing, Jr., and W. Davern, *J. Electrochem. Soc.* **112**, 285 (1965).
89. D. R. Secrist and J. D. Mackenzie, *J. Electrochem. Soc.* **113**, 914 (1966).
90. S. P. Mukherjee and P. E. Evans, *Thin Solid Films* **14**, 105 (1972).
91. H. F. Sterling and R. C. G. Swann, *Solid State Electron.* **8**, 653 (1965).
92. R. J. Joyce, H. F. Sterling, and J. H. Alexander, *Thin Solid Films* **1**, 481 (1967/1968).
93. P. C. Schultz, *Bull. Am. Ceram. Soc.* **52**, 383 (1973).
94. J. Koenings, D. Kuppers, H. Lydtin, and H. Wilson, *Proc. Int. Conf. Vapor Deposit.*, 5th p. 270 (1970).
95. P. Geittner, D. Kuppers, and H. Lydtin, *Appl. Phys. Lett.* **28**, 645 (1976).
96. D. Kuppers and H. Lydtin, *in* "Topics in Current Chemistry" (F. L. Boschke, ed.), Vol. 89. Springer-Verlag, Berlin and New York, 1980.
97. J. R. Hollahan, *J. Electrochem. Soc.* **126**(6), 931 (1979).
98. A. C. Adams, F. B. Alexander, C. D. Capio, and T. E. Smith, *J. Electrochem. Soc.* **128**(7), 1545 (1981).
99. R. Lathlaen and D. A. Diehl, *J. Electrochem. Soc.* **116**, 620 (1969).
100. W. Kern, G. L. Schnable, and A. W. Fisher, *RCA Rev.* **37**, 3 (1976).
101. A. C. Adams and C. D. Capio, *J. Electrochem. Soc.* **126**, 1042 (1979).
102. E. P. G. T. Van de Ven, *Solid State Technol.* April, p. 167 (1981).
103. R. A. Street and J. C. Knights, *Philos. Mag.* **B 42**(4), 551 (1980).
104. E. Ferrieu and B. Pruniaux, *J. Electrochem. Soc.* **116**, 1008 (1969).
105. T. Tanaka and S. Iwauchi, *Jpn. J. Appl. Phys.* **7**, 1420 (1968).
106. C. A. T. Salama, *J. Electrochem. Soc.* **117**, 913 (1970).
107. K. Ando and K. Matsumura, *Thin Solid Films* **52**, 153 (1978).
108. K. Katto and Y. Koga, *J. Electrochem. Soc.* **118**(10), 1619, (1971).
109. M. Sokolowski, A. Sokowska, A. Micalski, B. Gokieli, Z. Ramanowski, and A. Rusek, *Cryst. Growth* **42**, 507 (1977).
110. B. Micheletti, P. E. Norris, and K. H. Zaininger, *Solid State Technol.* April, p. 27 (1971).
111. V. Y. Doo, D. R. Nichols, and G. A. Silvey, *J. Electrochem. Soc.* **113**(12), 1279 (1966).
112. T. L. Chu, C. H. Lee, and G. A. Gruber, *J. Electrochem. Soc.* **114**(7), 717 (1967).
113. M. D. Brown, P. V. Gray, F. K. Heumann, H. R. Philipp, and E. A. Taft, *J. Electrochem. Soc.* **115**, 311 (1968).
114. C. J. Mogab, P. M. Petroff, and T. T. Sheng, *J. Electrochem. Soc.* **122**, 815 (1975).
115. G. J. Kominiak, *J. Electrochem. Soc.* **122**, 1271 (1975).
116. C. J. Mogab and E. Lugijjo, *J. Appl. Phys.* **47**, 1302 (1976).
117. S. Rigo, G. Amsel, and M. Croset, *J. Appl. Phys.* **47**, 2800 (1976).
118. W. Kaiser and C. D. Thermond, *J. Appl. Phys.* **30**, 427 (1959).
119. A. N. Knoop and R. Sticker, *Electrochem. Technol.* **3**, 84 (1965).
120. S. M. Hu, *J. Electrochem. Soc.* **113**, 693 (1966).
121. R. G. Frieser, *J. Electrochem. Soc.* **115**(10), 1092 (1968).
122. T. Ito, S. Hijiya, T. Nozaki, H. Arakawa, M. Shinoda, and Y. Fukukawa, *J. Electrochem. Soc.* **125**(3), 448 (1978).

123. T. Ito, S. Hijya, H. Ishikawa, and M. Shinoda, *Int. Electron Dev. Meet., Washington, D.C.* p. 284 (1977).
124. S. M. Hu and L. V. Gregor, *J. Electrochem. Soc.* **114**(8), 826 (1976).
125. M. Sokoewski, A. Sokoewska, E. Rolinski, and A. Michalski, *Thin Solid Films* **30**, 29 (1975).
126. H. F. Stirling and R. C. G. Swann, *Solid State Electron.* **8**, 653 (1965).
127. R. C. G. Swann, R. R. Mehta, and T. P. Cauge, *J. Electrochem. Soc.* **114**(7), 713 (1967).
128. Y. Kuwano, *Jpn. J. Appl. Phys.* **8**(7), 876 (1969).
129. E. A. Taft, *J. Electrochem. Soc.* **118**(8), 1341 (1971).
130. R. A. Gereth and W. E. Scherber, *J. Electrochem. Soc.* **119**(1), 1248 (1972).
131. D. A. Anderson and W. E. Spear, *Philos. Mag.* **35**, 1 (1976).
132. W. Kern and R. S. Rosler, *J. Vac. Sci. Technol.* **14**(5), 1082 (1977).
133. Y. Catherine and G. Turban, *Int. Round Table Plasma Polym. Surf. Treat., IUPAC, Limoges* July 11–12 (1977).
134. A. K. Sinha and T. E. Smith, *J. Appl. Phys.* **49**(5), 2756 (1978).
135. A. K. Sinha and E. Lugujjo, *Appl. Phys. Lett.* **32**(4), 245 (1978).
136. A. R. Reinberg, U.S. Patent No. 3,757,733, Sept. 11, 1973.
137. R. S. Rosler, W. C. Benzig, and J. Baldo, *Solid State Technol.* June, p. 45 (1976).
138. A. K. Sinha, H. J. Levinstein, T. E. Smith, G. Quintana, and S. E. Haszko, *J. Electrochem. Soc.* **125**(4), 601 (1978).
139. W. A. Lanford and M. J. Rand, *J. Appl. Phys.* **49**(4), 2473 (1978).
140. M. J. Rand and D. R. Wonsidler, *J. Electrochem. Soc.* **125**(1), 99 (1978).
141. W. A. Pliskin, *J. Vac. Sci. Technol.* **14**(5), 1064 (1977).
142. W. Kern and R. S. Rosler, *J. Vac. Sci. Technol.* **14**(5), 1083 (1977).
143. W. A. Lanford and M. J. Rand, *J. Appl. Phys.* **49**, 2473 (1978).
144. H. J. Stein, V. A. Wells, and R. E. Hampy, *J. Electrochem. Soc.* **126**(10), 1750 (1979).
145. M. Vandenberg, *Electrochem. Soc. Spring Meet., Boston* May 6–11, p. 234 (1979).
146. M. Shiloh, B. Gayer, and F. E. Brinckman, *J. Electrochem. Soc.* **128**(7), 1555 (1981).
147. W. M. Yim and R. J. Praff, *J. Appl. Phys.* **45**, 1456 (1974).
148. M. T. Wouk and D. K. Winslow, *Appl. Phys. Lett.* **13**, 5578 (1974).
149. J. H. Collins, P. J. Hagan, and G. R. Pulliam, *Ultrasonics* October, p. 218 (1970).
150. A. J. Norika and D. W. Ing, *J. Appl. Phys.* **39**, 5578 (1968).
151. K. W. Ehler, U.S. Patent No. 3,341,352, 1967.
152. G. Lewicki and C. A. Mead, *Phys. Rev. Lett.* **16**(21), 266 (1971).
153. S. Veprek, C. Brendel, and H. Schafer, *J. Cryst. Growth* **9**, 266 (1971).
154. H. Arnold, L. Biste, D. Bolze, and G. Eichhorn, *Krist. Tech.* **11**(1), 17 (1976).
155. J. Bauer, L. Biste, and D. Bolze, *Phys. Status Solidi A* **39**, 173 (1977).
156. S. Veprek, Z. Iqbal, J. Bunner, and M. Scharli, *Philos. Mag. B* **43**(3), 527 (1981).
157. J. M. Blocher, Jr., in "Vapor Deposition" (C. F. Powell, J. H. Oxley, and J. M. Blocher, Jr., eds.), p. 650. Wiley, New York, 1966.
158. H. F. Stirling, J. H. Alexander, and R. J. Joyce, *Special Ceram.* **4**, 139 (1968).
159. S. B. Hyder and T. O. Yep, *J. Electrochem. Soc.* **123**(11), 1721 (1976).
160. E. A. Fagan, in "Amorphous and Liquid Semiconductors" (J. Skuke and W. Brunig, eds.), p. 601. Taylor & Francis, London, 1974.
161. H. F. Stirling and R. C. G. Swann, *Solid State Electron.* **8**, 653 (1965).
162. H. F. Stirling and R. C. G. Swann, Brit. Patent No. 1,104,935, 1965.
163. Y. Catherine and G. Turban, *Int. Round Table Plasma Polym. Surf. Treat. IUPAC, Limoges* July (1977).
164. D. A. Anderson and W. E. Spear, *Philos. Mag.* **35**, 1 (1977).
165. Y. Catherine and G. Turban, *Thin Solid Films* **60**, 193 (1979).
166. Y. Catherine and G. Turban, *Thin Solid Films* **70**, 101 (1980).

167. O. A. Weinreich and A. Ribner, *J. Electrochem. Soc.* **115**, 1090 (1968).
168. M. Le. Contellec, J. Richard, A. Guivarc'h, E. Ligeon, and J. Fontenille, *Thin Solid Films* **58**, 407 (1979).
169. K. R. Linger, *Proc. Conf. Ion Plating Allied Tech.*, Edinburgh June, p. 223 (1977).
170. M. Katz, A. Grill, D. Itzhak, and R. Arni, *Int. Symp. Plasma Chem.* (IUPAC), 4th Zurich Sept., p. 444 (1979).
171. M. Katz, D. Itzhak, A. Grill, and R. Arni, *Thin Solid Films* **72**, 497 (1980).
172. D. Engemann, R. Fischer, and J. Knecht, *Appl. Phys. Lett.* **32**(9), 567 (1978).
173. H. Munekata, S. Murasato, and H. Kukimoto, *Appl. Phys. Lett.* **37**(6), 536 (1980).
174. Y. Tawada, H. Okamoto, and Y. Hanakawa, *Appl. Phys. Lett.* **39**(3), 237 (1981).
175. H. Yoshihara, H. Mori, M. Kiuchi, and T. Kadota, *Jpn. J. Appl. Phys.* **17**(9), 1693 (1978).
176. C. J. Mogab and W. D. Kingery, *J. Appl. Phys.* **39**, 3640 (1968).
177. T. E. Hartman, I. C. Blair, and C. A. Mead, *Thin Solid Films* **2**, 79 (1968).
178. G. Perssons, *Met. Prog.* **97**, 81 (1970).
179. A. C. Rahuram, *J. Vac. Sci. Technol.* **9**, 1389 (1972).
180. H. Mori and H. Yoshihara, *Jpn. J. Appl. Phys.* **17**(3), 575 (1978).
181. J. E. Greene and M. Pestes, *Thin Solid Films* **37**, 373 (1976).
182. F. J. Hazelwood and P. C. Iordanis, *Proc. Conf. Ion Plating Allied Tech.*, Edinburgh p. 248 (1977).
183. F. J. Hazelwood, *Int. Conf. Adv. Surf. Coat. Technol.* London p. 29 (1978).
184. K. Wasa, T. Nagai, and S. Hayakawa, *Thin Solid Films* **31**, 235 (1976).
185. H. Yoshihara, H. Mori, and M. Kiuchi, *Thin Solid Films* **76**, 1 (1981).
186. C. J. Dell'Oca, D. L. Pulfrey, and L. Young, in "Physics of Thin Films" (G. Hass and R. E. Thun, eds.), Vol. 6, Academic Press, New York, 1971.
187. J. F. O'Hanlon and M. Sampogna, *J. Vac. Sci. Technol.* **10**, 450 (1973).
188. J. Kreitchman, *J. Appl. Phys.* **38**, 4323 (1967).
189. L. Bardos, G. Loncar, I. Stoll, J. Musil, and F. Zacek, *J. Phys. D* **8**, L195 (1965).
190. J. F. O'Hanlon and W. B. Pennebaker, *Appl. Phys. Lett.* **18**, 554 (1971).
191. G. Olive, D. L. Pulfrey, and L. Young, *Thin Solid Films* **12**, 427 (1972).
192. J. D. Leslie, V. Keeth, and K. Knorr, *J. Electrochem. Soc.* **125**(1), 45 (1971).
193. H. Hasegawa and H. L. Hartnagel, *J. Electrochem. Soc.* **123**, 713 (1976).
194. R. A. Logan, B. Schwartz, and W. J. Sandburg, *J. Electrochem. Soc.* **120**, 1385 (1973).
195. D. Law and C. A. Lee, *J. Electrochem. Soc.* **123**, 168 (1976).
196. E. B. Stoneham, *J. Electrochem. Soc.* **121**, 1382 (1974).
197. M. Croset, J. Diaz, D. Dicumegard, and L. M. Mercandall, *J. Electrochem. Soc.* **126**, 1543 (1979).
198. F. Kishiga and T. Sugano, *Jpn. J. Appl. Phys.* **16** (Suppl. 17), 465 (1977).
199. D. N. Butcher and B. J. Sealy, *Electron. Lett.* **13**, 558 (1977).
200. I. Shiota, N. Miyamoto, and J. Nishizawa, *J. Electrochem. Soc.* **124**, 1405 (1977).
201. K. Ploog, A. Fischer, R. Trommer, and M. Hirose, *J. Vac. Sci. Tech.* **16**(2), 290 (1979).
202. O. A. Weinreich, *J. Appl. Phys.* **37**, 2924 (1966).
203. L. A. Chester and G. Y. Robinson, *Appl. Phys. Lett.* **32**(1), 60 (1978).
204. T. Sugano and Y. Mori, *J. Electrochem. Soc.* **121**, 113 (1974).
205. F. Koshiga and T. Sugano, *Thin Solid Films* **59**, 39 (1979).
206. M. Yakoyama, T. Mimura, K. Odani, and M. Fukuta, *Appl. Phys. Lett.* **32**(1), 58 (1978).
207. R. P. H. Chang and A. K. Sinha, *Appl. Phys. Lett.* **29**(1), 58 (1976).
208. R. L. Kaufman, L. C. Feldman, J. M. Poate, and R. P. H. Chang, *Appl. Phys. Lett.* **30**(7), 319 (1977).
209. R. P. H. Chang, C. C. Chang, and T. T. Shang, *Appl. Phys. Lett.* **30**(12), 657 (1977).
210. R. P. H. Chang, C. C. Chang, J. J. Coleman, R. L. Kaufman, W. R. Wagner, and L. C. Feldman, *J. Appl. Phys.* **48**(12), 5384 (1977).

211. R. P. H. Chang and J. J. Coleman, *Appl. Phys. Lett.* **32**(5), 332 (1978).
212. R. P. H. Chang, A. J. Polak, D. L. Allara, C. C. Chang, and W. A. Lanford, *J. Vac. Sci. Technol.* **16**(3), 888 (1979).
213. C. C. Chang, R. P. H. Chang, and S. P. Murarka, *J. Electrochem. Soc.* **125**(3), 481 (1978).
214. R. P. H. Chang, *Thin Solid Films* **56**, 89 (1979).
215. R. P. H. Chang, to be published.
216. Y. Shinoda and M. Yamaguehi, *Appl. Phys. Lett.* **34**(8), 484 (1979).
217. K. Watanabe, M. Hashiba, Y. Hirohava, M. Nisbino, and T. Yamashina, *Thin Solid Films* **56**, 63 (1979).
218. K. Kanazama and H. Matsunami, *Jpn. J. Appl. Phys.* **20**(3), L221 (1981).
219. Y. Nemirovsky and R. Goshen, *Appl. Phys. Lett.* **37**(9), 813 (1980).
220. R. I. Nazarova, *Russ. J. Phys. Chem.* **36**, 522 (1962).
221. L. E. Katz and L. C. Kimerling, *J. Electrochem. Soc.* **125**(10), 1680 (1978).
222. J. R. Ligenza, *J. Appl. Phys.* **36**(9), 2703 (1965).
223. E. R. Skelt and G. M. Howell, *Surf. Sci.* **7**, 490 (1967).
224. B. J. R. Ligenza and M. Kuhn, *Solid State Technol.* Dec., 33 (1970).
225. V. Q. Ho and T. Sugano, *IEEE Trans. Electron. Dev.* **ED-27**(8), 1436 (1980).
226. A. K. Ray and A. Reisman, *J. Electrochem. Soc.* **128**(11), 2461 (1981).
227. A. K. Ray and A. Reisman, *J. Electrochem. Soc.* **128**(11), 2466 (1981).
228. A. K. Ray and A. Reisman, *J. Electrochem. Soc.* **128**(11), 2424 (1981).
229. A. K. Ray, *Thin Solid Films* **84**, 389 (1981).
230. J. H. Greiner, *J. Appl. Phys.* **42**, 5151 (1971).
231. J. H. Greiner, *J. Appl. Phys.* **45**, 32 (1974).
232. See for example, articles in *IBM J. Res. Dev.* **24**(2) (1980).
233. A. J. Fromhold, Jr. and M. Baker, *J. Appl. Phys.* **51**(12), 6377 (1980).
234. T. Imamura, H. Suzuki, and H. Hasuo, *Fujitsu Sci. Tech. J.* Dec., p. 21 (1979).
235. T. Waho, K. Kuroda, and A. Ishida, *J. Appl. Phys.* **51**(8), 4508 (1980).
236. K. Kuroda, T. Waho, and A. Ishida, *J. Appl. Phys.* **51**(8), 4513 (1980).
237. J. M. E. Harper, M. Heiblum, I. L. Speidell, and J. J. Cuomo, *J. Appl. Phys.* **52**(6), 4118 (1981).
238. K. Gustafson and J. R. Whinnery, *J. Quantum Electron.* **QE-14**, 159 (1978).
239. Anom., *Met. Prog.* May, p. 45 (1977).
240. P. C. Jindal, *Met. Prog.* April, 78 (1973).
241. W. I. Grube and J. G. Gay, *Metall. Trans. A* **9A**, 1421 (1978).
242. J. J. Egan, U.S. Patent No. 1,837,256, May 1930.
243. J. Von Bosse, K. Richter, and E. F. Kruppa, Swiss Patent, 172436, July 1932.
244. B. Berghouse, German Patent 669639, July 1932.
245. B. Berghouse and H. Bucek, U.S. Patent 2,946,708, Sept. 1955.
246. C. K. Jones, Marblehead, and S. W. Martin, U.S. Patent No. 3,437,784, April 1966.
247. L. Holland, Br. Patent 1,134,562, Nov. 1968.
248. B. Edenhofer, *Heat Treat. Met.* **1**, 23 (1974).
249. B. Edenhofer, *Heat Treat. Met.* **2**, 59 (1974).
250. M. Hudis, *J. Appl. Phys.* **44**(4), 1489 (1973).
251. A. Brokman and F. R. Tuler, *J. Appl. Phys.* **52**(1), 468 (1981).
252. M. Komuma and O. Matsumoto, *J. Less Common Met.* **55**, 97 (1977).
253. M. Konuma and O. Matsumoto, *J. Less Common Met.* **52**, 145 (1977).
254. F. E. Gifford and K. W. Cooley, **11**(2), 511 (1974).
255. A. S. Rizk and D. J. McCulloch, *Surf. Technol.* **9**, 303 (1979).
256. E. Winz, H. R. Oswald, and S. Veprek. *Int. Symp. Plasma Chem.*, 4th, Zurich p. 492 (1979).
257. C. Braganza, S. Veprek, E. Winz, H. Sussi, and M. Textor, *Int. Symp. Plasma Chem.*, 4th, Zurich p. 100 (1979).

258. J. C. Knight and J. E. Mahan, *Solid State Commun.* **21,** 983 (1977).
259. D. I. Jones, W. E. Spear, P. G. LeComber, S. Li, and R. Martins, *Philos. Mag.* **339,** 147 (1979).
260. D. Hauschildt, R. Fischer, and W. Fuhs, *Phys. Status Solidi B* **102,** 563 (1980).
261. D. A. Anderson and W. E. Spear, *Philos. Mag.* **35,** 1 (1977).
262. K. Matsushita, Y. Matsuno, T. Hairu, and Y. Shibata, *Thin Solid Films* **80,** 243 (1981).
263. V. Smid and H. Fritzsche, *Solid State Commun.* **33,** 735 (1980).
264. F. J. Iampas and R. W. Griffith, *Solar Cell* **2,** 385 (1980).
265. M. A. Maesler, T. Okumura, and W. Paul, *J. Vac. Sci. Technol.* **17**(6), 1332 (1980).
266. G. Turban, Y. Catherine, and B. Grolleau, *Proc. Int. Symp. Plasma Chem.*, *4th*, Zurich p. 164 (1979).
267. I. Haller, *Appl. Phys. Lett.* **37**(3), 282 (1980).
268. G. Turban and Y. Catherine, *Thin Solid Films* **48,** 57 (1978).
269. G. H. Bauer and G. Bilger, *Thin Solid Films* **83,** 223 (1981).

# Author Index

Numbers in parentheses are reference numbers and indicate that an author's work is referred to although the name is not cited in the text. Numbers in italics show the page on which the complete reference is listed.

## A

Abarenkov, I., 105(178), *162*
Abeles, B., 72(73), 79(73), *160*
Abraham, W. G., 143(313), *166*
Adams, A. C., 263, 264(98), *292*
Airey, R. W., 147(318, 319), *166*
Aisenberg, S., 246(31, 32), 247, 253(32), *290*
Aitchison, R. E., *234*
Akovali, G., 248(40), 251(40), *290*
Alexander, F. B., 263(98), 264(98), *292*
Alexander, J. H., 261(92), 262(92), 271(158), *292, 293*
Allara, D. L., 277(212), 279(212), 280(212), *295*
Allen, F. G., 259(81), *292*
Allen, G. A., 124(253), *164*
Alt, L. L., 259, *292*
Amsel, G., *292*
Anderson, D. A., 255(58), 256(58), 272(164), 273, *291, 293, 296*
Anderson, J., 115(214), *163*
Anderson, L. P., 248(43, 44, 45), 251(43, 44, 45), 253(45), *291*
Ando, H., 186(93), 197(93), 200(93), *234*, 264 (107), *292*
Angus, J. C., 246(27, 28, 29), *290*
Annoh, K., 246(22), 254(22), *290*
Antypas, G. A., 114(191, 192, 194), 123(251), 124(252), 125(256), 126(263, 264), 128 (256, 264), 132, *162, 165*
Apker, L., 60(34), 101(34), 108(34), *159, 162*
Arakawa, H., *292*
Aranovich, J., 176(54), 177(54), 178(54), 179 (54), *233*
Arnaud D'Avitoya, F., 116(223), 118(227), *163*
Arni, R., 272(170, 171), 274(170, 171), *294*

Arnold, H., 270(154), *293*
Arnon, O., 7, 36(63), *50, 51*
Aronovich-Magran, J., *235*
Asao, A., 71, 75, 76, 78, 79, *159, 160*
Augustine, F., 181(69, 70), 190(69, 70), *234*
Ashley, E. J., 8(33), 28(55), 36(55), 37(67), *50, 51*
Ashok, S., *235*
Aslam, M., 147(318), *166*
Astridge, R. A., 135(291), *165*

## B

Babalola, I. A., 114(207), *163*
Babenko, Yu. E., 178(64), *234*
Baker, M., 282(233), 283(233), *295*
Baldo, J., 266(137), 267(137), 268(137), *293*
Ball, G. J., 30, 36, *51*
Banerjee, A., 169, 170(27), 172, 173(27), 174 (27), 177(54a), 178(27), 179(27, 65), 180 (65), 183(25, 27, 28, 73), 185(30, 54a, 73), 186(30, 54a, 73), 186(30, 54a, 73), 187 (54a, 73), 189(27, 28, 29, 30), 190(27, 30), 192(30), 193(27), 194(27), 195(24), 196 (24, 27), 197(27, 73), 198(30, 73), 199(29, 30, 73), 225(28), 228(30), *233, 234*
Bardas, D., 143(313), *166*
Bardos, L., 276, *294*
Bartelink, D. J., 60(32), 75(32), *159*
Bass, S. J., 130(275), *165*
Bassani, F., 57(20), 58(20), 68(20), *159*
Bates, C. W., Jr., 75(87), *160*
Baud, C., 121(236), *164*
Bauer, G. H., 289(269), *296*
Bauer, J., 270(155), *293*
Baumeister, P., 7, 36(63), *50, 51*

Baylander, E. G., 259(82), *292*
Beaumont, D., 6(26), *50*
Beck, A. H., 88, 89(128), 99(128), *161*
Behrndt, K. H., 12, *50*
Bell, A. T., 238(6), 248(40), 251(40), *290*
Bell, R. L., 113(187), 114(191, 194), 123(251), 125, 126(263, 264), 128(264), 132(258), 135(297), 137, 138, *162*, *163*, *164*, *165*
Belyaeva, N. N., 204(97), 217(97), *234*
Bennett, H. E., 6, 8(33), 28(55), 36, 38(64), *50*, *51*
Bennett, J. M., 6, 28(55), 36(64), 37(67), 38(64), *50*, *51*
Benzig, W. C., 266(137), 267(137), 268(137), *293*
Berg, S., 248(43, 44, 45), 251(43, 44, 45), 253(45), *291*
Berghouse, B., 286, *295*
Berglund, C. N., 81, 82, *160*
Berning, P. H., 10(37), 48(37), *50*
Bessonova, E. S., 176(52), *233*
Bewilogua, K., 246(35), 248(35), 251(62, 63), 256, 257(65), *290*, *291*
Bhat, B. M., 71(64), 72(64), 85(101), *160*
Bhatia, T. B., 148(320), *166*
Bhide, G. K., 85(101), 148(320), *160*, *166*
Bibik, V. F., 71(60), 74(60), 78(60), 80(60), 81(60), 84(60), 119(60), *159*
Bilger, G., 289(269), *296*
Biste, L., 270(154, 155), *293*
Blair, I. C., 272(177), *294*
Blocher, J. M., 271(157), *293*
Bode, D. E., 169, 201, 230(33), *233*
Boeckner, C., 29(59), *51*
Boerio, A. H., 157(335), *166*
Boileau, A. R., 91(140), *161*
Borzyak, P. G., 71(60), 74, 78, 80, 81, 84, 94, 95, 101(172), 103, 119(60), *159*, *161*, *162*
Borisova, I. I., 176(55), *233*
Botvinkin, O. K., 176(55), *233*
Bougnot, J., 169(19), *232*
Bovey, E., 71(65), *160*
Bower, K., 14, *50*
Bradford, A. P., 2(5, 7), 3(14), 10, 12(39), 29(57, 58), *49*, *50*, *51*
Braganza, C., 288(257), *295*
Brauer, G., 88(132), *161*
Brendel, C., 269(153), *293*
Bresse, J. F., 176(50), 179(50), 180(50), 183(50), 186(50), 187(50), 190(50), 197(50), 200(50), *233*
Breuer, K., 251(62), 256(62), *291*
Briggs, G. S., 157(336), *166*
Brinckman, F. E., 269(146), *293*
Brinkman, J. A., 246(25), *290*
Brokman, A., 287, *295*
Brown, M. D., 265(113), *292*
Brown, S. C., 238(3), *289*
Brust, D., 57, 58, 68(20), *159*
Bube, R. H., 169(12, 13, 14, 15, 16), 176(54), 177(54), 178(54), 179(54), 183(16, 18), 189(14), 190(14), 193(14), 194(13, 14), 196(12, 14), *232*, *233*, *235*
Bubulac, L. O., 137(309), *165*
Bucek, H., 286, *295*
Buch, F., 169(15, 17), *232*
Buck, T. M., 259(81), *292*
Bunner, J., 270(156), *293*
Burton, J. A., 101(171), 102(171), 137(304), *162*, *165*
Burge, D. K., 36, 38(64), *51*
Burt, M. G., 126(265, 266), 128(265, 266), *164*
Butcher, D. N., 277(199), *294*

# C

Campbell, N. R., 54(3), 71, *158*
Canfield, L. R., 5, 10(38), *50*
Capio, C. D., 263(98), 264(98), *292*
Carlson, D. E., 238(2), *289*
Carter, G., 238(9), *290*
Carve, G. E., 176(56), *233*
Cashman, R. J., 168(5), *232*
Catchpole, C. E., 147(318), *166*
Catherine, Y., 272(163, 165, 166), 273(266, 268), *293*, *296*
Cauge, T. P., 266(127), *293*
Ceren, E., 39(68), *51*
Chabot, R., 246(32), 247, 253(32), *290*
Chadi, D. J., 115(215), *163*
Chaldyshev, V. A., 105(176, 177), 106(176), *162*
Chamberlin, R. R., 168, 169(2, 3, 4, 10, 11), 176(4), 183(4), 190(4), 194(11), *232*
Chang, C. C., 277(209, 210, 213), 279, 280(212), 282(210), *294*, *295*

Chang, R. P. H., 276(207, 215), 277(206, 208, 209, 210, 213), 279, 280(212), 281, 282, *294*, *295*
Chase, W. R., 259(83), *292*
Chau, Y. C., 196(94), 197(94), *234*
Chauhan, S. P., 246(29), *290*
Chckowski, D. H., 146(317), 147(317), *166*
Chelikowsky, J. R., 115(216), *163*
Chen, C. H., 246(24), *290*
Chen, K. Y., 253(67), *291*
Chernayev, V. N., 177(61), *234*
Chester, L. A., 277(203), 278, *294*
Chiarotti, G., 56(17), *158*
Chidambaram, K., 176(53), *233*
Chikawa, J., 88(131), 96(131), *161*
Chikazumi, S., 94(148), *161*
Chin, J., 253(67), *291*
Chopra, K. L., 71(67), *160*, 169, 172, 176(26, 49, 51, 53), 177(54a), 179(65), 180(65), 183(25, 28, 73), 185(30, 49, 54a, 73), 186 (26, 30, 49, 54a, 73), 187(54a, 73, 87, 88), 189(28, 29, 30, 88, 89), 190(30, 87, 88), 191(88), 192(30), 194, 195(24), 196(24), 197(26, 73), 198(30, 49, 73), 199(29, 30, 49, 73), 200(30, 54a, 73), 203(42), 204(43), 205(42, 43), 207(42), 208(42), 210(42), 211(42), 214(42), 215(42), 217(40, 43), 217 (104), 218(99, 111), 219(106, 111, 112), 220(112), 223(106, 113), 224(120), 225 (28), 226(122, 124, 125), 227(124), 228 (30), 229(106, 111), 230(99, 106), *233*, *234*, *235*
Choyke, W. J., 76, *160*
Christman, S. B., 121(238), *164*
Chu, J. C., 137(309), *165*
Chu, T. L., 265(112), *292*
Chye, P. W., 74(85), 78, *160*, 114(206, 207), 115(212, 217), 116(212, 222), 117(212), *163*
Chynoweth, A. G., 60(29, 31), *159*
Clemens, H. J., 114(210), *163*
Cocks, F. H., 258(73, 74), *291*
Cody, G. D., 71, 79(72), *160*
Cohen, M. L., 114(203), 115(216), *163*
Cohen, R. W., 72(73), 79(73), *160*
Coleman, J. J., 277(210, 211), 279(210, 211), 282(210), *295*
Colligan, J. S., 238(9), *290*
Collins, J. H., 269(149), *293*

Comizzoli, R. B., 258(77), *291*
Condas, G. A., 85(102), *160*
Conrad, E. E., 259(84), *292*
Cooley, K. W., 287(254), 288(254), *295*
Cougulin, J. P., 73(80), *160*
Coulter, J. K., 8(35), 15(44), 40(44), *50*
Courtney Pratt, J. S., 148(321), *166*
Coutts, M. D., 72(73), 79(73), *160*
Cowache, P., 226(113), *235*
Cox, J. T., 2(8, 9, 10), 3(11, 12), 26(52), 28(8), 29(9), 31(8, 10), 33(11), *49*, *50*
Crawshaw, D. D., 142, *165*
Crombeen, J. E., 122(247), *164*
Croset, M., 277(197), 280(197), *292*, *294*
Crowe, K. R., 70(55), *159*
Crowell, C. R., 84, *160*
Cuomo, J. J., 282(237), 283(237), *295*

**D**

Dachraoui, M., 226(113), *235*
Dalal, V., 137(307), *165*
Dankov, P. D., 73, *160*
Das, S. K., 253(67), *291*
Das, S. R., 169(25, 28), 172(25), 183(25, 28), 189(28), 225(28), *233*
Davern, W., 260(88), 261, *292*
Davey, J. E., 76, *160*
Dekker, A. J., 64, 101(170), *159*, *162*
Dell'Oca, C. J., 275(186), *294*
Deshotels, W. J., 181, 190(69, 70), *234*
Derrien, J., 116(223), 118(227), *163*
Deutscher, K., 86(114), *161*
Diaye, A. N., 169(16), 183(16), *232*
Diaz, J., 277(197), 280(197), *294*
Dickey, J., 60(34), 101(34), 108(34), *159*
Dicumegard, D., 277(197), 280(197), *294*
Diehl, D. A., 263(99), 264(99), *292*
Dietrich, D., 256(61), 257(65), *291*
Dimigen, H., 253(55), 254(55), *291*
Dimitrova, P. P., 204(98), *234*
Dimmey, L. J., 258(73, 74), *291*
Djoglev, D. H., 204(98), *234*
Doo, V. Y., 265(111), *292*
Doremus, R. H., 72(74), 79(74), *160*
Dorn, F. W., 87, *161*
Dowman, A. A., 88, 89, 99, *161*
Dragieva, I. D., 204(98), *234*

Drummeter, L. F., 42(71), 44(71), *51*
Drummeter, L. R., 29(56), *51*
Duke, C. B., 115(218), *163*
Dutault, F., 174(46, 47), 175(46), 180(46), 231(46), *233*
Dutta, V., 226(124, 125), 227(124), *235*
Duval, M., 248(39), 251(39), *290*
Dvoinin, V. I., 204(97), 217(97), *234*
Dyatlovitskaya, B. I., 101(172), *162*

# E

Eastman, D. E., 114(196), *163*
Ebersbach, U., *291*
Ebina, A., 103, 104, 105, *162*
Eckart, F., 82, 101(97), 102, *160*
Eden, R. C., 59(22), 137(309), *159*, *165*
Edenhofer, B., 286, *295*
Edgecumbe, J., 114(191, 194), 125(256), 128(256), 132(281), *162*, *163*, *164*, *165*
Edwards, D., 121(244), *164*
Edwards, D. F., 253(66), *291*
Egan, J. J., 286, *295*
Ehler, K. W., 269(151), *293*
Ehrenreich, H., 79, 80, *160*
Eichhorn, G., 270(154), *293*
Elliott, D., 190(91), *234*
Enck, R. S., 132(281), 143(313), *165*, *166*
Engemann, D., 274(172), 275(172), *294*
Engle, G. M., 242, *290*
Engstrom, R. W., 150(322), *166*
Enke, K., 253, 254, *291*
Enstrom, R. E., 118(229), 120(229), 126(229), 127(229), *163*
Entine, G., 176(48a), *233*
Erler, H., 246(35), 248(35), *290*
Erler, H. J., 257(65), *291*
Esaki, L., 115(213), *163*
Escher, J., 122(249), 123(249, 251), 126(249), 129(249), 131(249), 139, *164*, *165*
Escher, J. S., 54(9), 118(229), 120(229), 126(229), 127(229), 130, 132(281), 136(9), 138(9), 139(9), 158, *158*, *163*, *165*
Ettenberg, M., 113(190), 129(272), *162*, *165*
Excher, J. S., *165*
Evans, G. B., 112(184), 118(184), *162*
Evans, P. E., 261, *292*

Eversole, W. G., 246(26), *290*

# F

Fagan, E. A., 272(160), *298*
Fahrenbruch, A. L., 169(15, 17, 18), 183(18), *232*
Fano, U., 134(288), *165*
Farnsworth, P. T., 152(325), *166*
Faulkner, K. R., 135(291), *165*
Faust, R. C., 71(69), *160*
Feigelson, R. S., 169(12, 16), 182(72), 183, 193(72), 196(12), *232*, *234*
Feist, W. M., 258(76), 259(76), 265(76), 271, 272(76), *291*
Feldman, L. C., 277(208), 279(208), *294*
Feng, T., 226(117), *235*
Ferrieu, E., 263(104), 264(104), *292*
Feuerbacher, B., 8(34), *50*
Fischer, A., 187(75), *234*, *294*
Fisher, A. W., *292*
Fisher, G. B., 62, *159*
Fisher, D. G., 100(166), 113, 118(229), 120(229), 125, 126, 127, 128(185), 129(185, 270, 271), 131, 133, 134(185), *162*
Fischer, R., 274(172), 275(172), *294*, *296*
Fishman, C., 226(117), *235*
Fitts, R. W., 150(322), *166*
Fiyimori, S., 246(19), *290*
Foex, M., 168, *232*
Fofanov, G. M., 169(37), 201(37), *233*
Folks, J. R., 126(261), 128(261), 132(261), *164*
Fonash, S. J., *235*
Fontenille, J., 272(168), 273(168), *294*
Fowler, G. W., 129(271), *164*
Fowler, P., 6, *50*
Fowler, R. H., 54(10), *158*
Frank, G., 124(254), *164*
Fraser, J. W., 257(68), *291*
Freeman, C. F., 145(316), 152(316), 153, *166*
Freeman, J. H., 248, 271, *290*
Frieser, R. G., *292*
Fritzsche, H., *296*
Frohlich, H., 54(14), *158*
Fromhold, A. J., Jr., 282(233), 283(233), *295*
Fuhs, W., *296*
Fukuhara, S., 92(141), *161*
Fukukawa, Y., *292*

Fukuta, M., 278(206), *294*

## G

Garbe, S., 114(193), 121(193), 124(254), 126(260), *162, 164*
Garbuny, M., 91(139), *161*
Gard, G. A., 248(36), 271(36), *290*
Gardner, N. C., 246(28, 29), *290*
Garfield, B. R. C., 86(116), 87, 89, 97(116), 98, 99, *161*
Garner, C. M., 115(217), 116(222), *163*
Garner, P. W., 114(206), *163*
Gates, D. M., 6, *50*
Gay, J. G., 284(241), 285(241), *295*
Gayer, B., 269(146), *293*
Geittner, P., 262(95), *292*
Gereth, R. A., 266, 267, *293*
Ghosh, C., 59(25), 60(39), 64(47), 68(52), 70(54), 74(82), 85(104), 86(52, 112, 115), 89(52), 90(52), 91, 96, 97, 98, 99(52), 100, 101, 103(39, 115), 105, 107, 108(39), 109(104, 175, 180), 110(180), 111, 112, 145(315), 148(320), 158, *159, 160, 161, 162, 166*, 226(117), *235*
Gifford, F. E., 287(254), 288(254), *295*
Gobrecht, R., 96(160), 97(160), *162*
Goetz, G. W., 157(335), *166*
Goisa, S. N., 121(243), *164*
Gokieli, B., 265(109), *292*
Goldstein, B., 114(201), 118(226), 119, 121(201, 240), *163, 164*
Golovcenco, I., 187(77), *234*
Gorlich, P., 54(4), 85(4), 92(4), *158*
Goshen, R., 282(219), *295*
Govindarajan, N., 71(64), 72(64), 75(89), *160*
Gray, P. V., 265(113), *292*
Greene, J. E., 275(181), *294*
Gregor, L. V., 265(124), *293*
Gregory, P. E., 54(9), 74(85), 78, 114(206), 115(217), 117(221), 118, 120(235), 121(230), 130(276), 136(9), 138, 139, 158, *163, 164, 165*
Greiner, J. H., 282(230, 231), 283, *295*
Griffith, R. W., 289(264), *296*
Grill, A., 272(170, 171), 274(170, 171), *294*
Grill, C., 169(21), *232*
Grolleau, B., 289(266), *296*

Groth, R., 187(82), *234*
Grube, W. I., 284(241), 285(241), *295*
Gruber, G. A., 265(112), *292*
Guichar, G. M., 114(211), *163*
Guivarc'h, A., 272(168), 273(168), *294*
Gumnick, J. L., 70(55), *159*
Gustafson, K., 282(238), *295*
Gundry, P. M., 114(202), 121(241, 242), *163, 164*
Guntzmann, G., 87, 88, *161*
Gwilliams, G. F., 143(313), *166*

## H

Haacke, G., 169(7), 186(93), 197(93), 200(93), *232, 234*
Haaymaü, P. W., 187(78), *234*
Hacman, D., 48(77), *51*
Hagino, M., 87, 95, 135(295), *161, 162, 165*
Hahn, R. B., 202(95), *234*
Hahn, R. E., 42(72), 45(72), *51*
Hairu, T., *296*
Hale, G. M., 46(75), *51*
Hall, J. A., 151(323), 152, 154(327), *166*
Haller, I., 289(267), *296*
Hampy, R. E., 269(144), *293*
Hanakawa, Y., 275(174), *294*
Haneman, D., 122(246), *164*
Hansen, J. R., 91(139), *161*
Harmer, A. L., 70(56), 121(245), 131(245), *159, 164*
Harper, J. M. E., 282(237), 283(237), *295*
Harper, W. J., 76, *160*
Harris, J. S., 137(309), *165*
Harris, L., 6, 31, *50, 51*
Hartman, T. E., 272(177), *294*
Hartnagel, H. L., 277(193), *294*
Hasegawa, H., 277(193), *294*
Hashiba, M., 277(217), *295*
Hass, G., 2(1, 3, 4, 5, 7, 8, 9, 10, 11, 12), 3(11, 12, 14, 15), 4(16), 8(16, 31, 32), 10(37, 38, 39), 12(39), 15(44, 45), 18(3, 47), 16, 22(48), 23(48), 24(49), 25(50), 26(51, 52), 28(54), 29(56, 57, 58), 31(8, 10, 48, 60), 33(11), 35(12), 37(48), 38(48), 40(44, 69), 41(47), 42, 44(47, 71), 48(37, 49), *49, 50, 51*
Hasuo, H., 282(234), *295*

Haszko, S. E., 266(138), 268(138), *293*
Hauschildt, D., *296*
Hauser, J. J., 246(21), *290*
Hayakawa, S., 274(184), *294*
Hazelwood, F. J., 275(182, 183), *294*
Heaney, J. B., 28(54), 29(57, 58), 31(60), *51*
Hebb, H. M., 64, *159*
Heiblum, M., 282(237), 283(237), *295*
Heiman, W., 119(234), *164*
Heine, V., 105(178), *162*
Heitmann, W., 28, 35, 36(62), 48(78), *50*, *51*, 73, 74, 76(81), 77, 83(81), 84, *160*
Helms, C. R., 74(84), 75(84), 81(84), *160*
Henry, W. M., 157(336), *166*
Hernandez, J. P., 59(24), 109(179), 111(179), 112(179), *159*, *162*
Herriott, D. R., 7, *50*
Hertz, H., 54, *158*
Herzig, H., 14(41), *50*
Heumann, F. K., 265(113), *292*
Hijiya, S., *292*, *293*
Hill, J. E., 168(3), 169(3), *232*
Himpsel, F. J., 114(196), *163*
Hinneberg, H. J., 257(65), *291*
Hiratsuka, H., 246(22), 248(40), 251(40), 254(22), *290*
Hirohava, Y., 277(217), *295*
Hirose, M., *294*
Ho, V. Q., 282(245), *295*
Hoene, E. L., 72, 73(75, 81), 74(81), 76(81), 77(81), 83(81), 84(81), 89, 99, 119(234), *160*, *161*, *164*
Hoffmann, R., 48(76), *51*
Hofman, H. H., 86(114), *161*
Hollahan, J. R., 263, 264, *292*
Holland, L., 16, *50*, 238(11), 241(11), 247(50), 248(42, 47), 251, 252, 253, 254(49), 255(49), 257(51), 286(247), *290*, *291*, *295*
Hollohan, J. R., 238(6), 239(12), 241(12), *290*
Holtom, R., 121(242, 245), 126(261), 128(261), 132(261), 135(245, 292), *164*, *165*
Holton, R., 70(56), 114(202), 121(241), *159*, *163*, *164*
Holzmann, R. T., 257(68, 69), *291*
Hopkins, G. P., 133(283), *165*
Horvath, E., 177(63), *234*
Hosokawa, N., 246(22), 254(22), *290*
Hovel, H. J., 176(57), *234*
Howarth, L. E., 84(99), *160*
Howell, G. M., 282(223), *295*

Howorth, J. R., 70(56), 121(245), 126(261), 128(261), 132(261, 282), 133(283), 135(291, 292), *159*, *164*, *165*
Hu, S. M., 265(124), *292*, *293*
Hubsch, H., 253(55), 254(55), *291*
Huchital, D. A., 118(225), 131(225), *163*
Hudis, M., 286, 287, *295*
Huijser, A., 114(205, 208), *163*
Hultquist, A. E., 257, *291*
Hunter, W. R., 2(3, 9), 4(16), 5, 8(16), 18(3), 29(9), *49*, *50*
Hutcheson, E. T., 8, *50*
Hyder, S. B., 54(9), 130(276, 280), 136(9), 138(9), 139(9), 158, *158*, *165*, 271, *293*

I

Iampas, F. J., 289(264), *296*
Iams, H. A., 154(329, 330), *166*
Imamura, S., 95(154, 155), 96, *162*
Imamura, T., 282(234), *295*
Ing, D. W., 269(150), *293*
Ing, S. W., Jr., 259(87), 260, 261(88), *292*
Inkson, J. C., 126(265, 266), 128(265, 266), *164*
Iordanis, P. C., 275(182), *294*
Iqbal, Z., 238(2), 270(156), *289*, *293*
Ishida, A., 282(235, 236), *295*
Ishikawa, H., *293*
Ito, T., *292*, *293*
Itzhak, D., 272(170), 171, 274(170, 171), *294*
Iwauchi, S., 264(105), *292*

J

Jack, K. H., 87, 88, *161*
Jackson, D. A., 135(299), *165*
Jacobi, K., 115(219), 122(248), *163*, *164*
James, L. W., 76(92), 77, 81(92), 114(191, 192, 194), 118(92, 231), 123(251), 124(252), 125(257), 126(263, 264), 128, 130(277), 137(308), 138(308), *160*, *162*, *163*, *164*, *165*
Janssen, G. H., 187(78), *234*
Janssen, J. E., 45, *51*
Jenkins, G. M., 246(23), 254(23), *290*
Jensen, J. D., 218(110), 219(110), *235*
Jeric, S., 73(81), 74(81), 76(81), 77(81), 83(81), 84(81), *160*

Jindal, P. C., 284(240), *295*
Joannopoulos, J. D., 114(203), *163*
Johnson, T. H., 169, 201, *233*
Jones, C. K., 286(246), *295*
Jones, D. I., *296*
Jones, P. L., 258(73, 74), *291*
Jones, R. E., 259, 263(85), 264(85), *292*
Jones, T. H., 88(128), 89(128), 99(128), *161*
Joy, D. C., 246(33), 248(33), 252(33), *290*
Joyce, R. J., 261, 262, 271(158), *292, 293*

# K

Kadota, T., 275(175), *294*
Kagonovich, E. B., 223(114), 229(114), 230 (114), *235*
Kaintha, R. C., 169(41, 42), 179(65), 180(65), 203, 205(42), 207(42), 208(42), 210(42), 211(42, 99), 212(99, 102, 105, 106), 213 (105, 106), 214, 215(42), 216(102), 217 (102), 218(99, 102), 219(102, 106, 112), 220(102, 112), 221(102), 222(102), 223 (102, 106, 113), 226(121a), 229(41, 102, 106), 230(99, 102, 106), *233, 234, 235*
Kaiser, W., *292*
Kaminsky, M., 253(67), *291*
Kan, H., 135(295), *165*
Kanazama, K., 277(218), 282(218), *295*
Kane, E. O., 68(53), *159*
Kane, J., 177(59, 62), *234*
Kanev, V., 87(118, 119), *161*
Kangro, A., 96(159), *162*
Kansky, E., 73(81), 74(81), 77(81), 83(81), 84 (81), 99(164), 119(234), *160, 162, 164*
Kaplan, M. L., 246(24), *290*
Karavaev, G. F., 105(176), 106(176), *162*
Kariya, T., 176(48), 224(48), *233*
Kaseman, P. W., 157(336), *166*
Katayam, S., 176(48), 224(48), *233*
Katsumo, H., 135(295), *165*
Katto, K., 264, *292*
Katz, L. E., 282(221), *295*
Katz, M., 272(170, 171), 274, *294*
Kaufman, F., 238(8), *290*
Kaufman, R. L., 277(208), 279(208), *294*
Kaur, I., 169(43), 204(43), 205, 208, 212(103), 214(103, 103a), 217(43, 103), 218(103), 222(103, 103a), 229(103), 230(103), *233, 235*

Kawamura, K., 246(23), 254(23), *290*
Kein, W., 177(59), *234*
Keith, V., 277(192), *294*
Kelbanikov, N. S., 71(63), *160*
Kelkar, G. N., 148(320), *166*
kelly, M. J., 71(59), 73, *159*
Kern, W., 177(62), *234*, 258(77), 268, *291, 292, 293*
Kerr, D. R., 258(75), *291*
Kidder, M., 135(293), *165*
Kimerling, L. C., 282(221), *295*
King, E. L., 259(83), *292*
Kingdon, K. H., 71(57), 114(197), *159, 163*
Kingery, W. D., *294*
Kiseler, V. P., 105(177), *162*
Kishiga, F., 277(198), *294*
Kitaev, G. A., 169, 201, 203, 204, 205(38), 210, 217(97), *233, 234*
Kiuchi, M., 274(185), 275(175), *294*
Klein, Z., 39(68), *51*
Klemm, W., 87, *161*
Klimin, A. I., 105(176, 177), 106(176), *162*
Klose, S., 257(65), *291*
Knapp, J. A., 114(196), *163*
Knecht, J., 274(172), 275(172), *294*
Knight, J. C., *296*
Knights, J. C., 264(103), *292*
Knoop, A. N., *292*
Knorr, K., 277(192), *294*
Kobayashi, A., 94(148), *161*
Koening, J., 181(70), 190(70), *234*
Koenings, J., 262(94), *292*
Koch, H., 49(80), *51*
Koehler, W. E., 6, *50*
Koga, Y., 264, *292*
Kohn, E. S., 135, *165*
Koller, L. R., 54(2), 71, *158*
Kominiak, G. J., 265(115), *292*
Komuma, M., 287(252, 253), *295*
Kondratiev, V. N., 238(10), *290*
Konishi, A., 246(22), 254(22), *290*
Koppelmann, G., 48(78), *51*
Koriufskii, A. D., 126(262), *164*
Korotkikh, V. L., 126(262), *164*
Korzo, V. F., 177(60, 61), *234*
Koshiga, F., 277(205), *294*
Koval, I. F., 121(243), *164*
Krall, H. R., 143(314), 144, *166*
Kramerinko, G. S., 71(60), 74(60), 78(60), 80 (60), 81(60), 84(60), 119(60), *159*

Kreitchman, J., 276, *294*
Kressel, H., 125(294), *165*
Krolikowski, W. F., 60(43), *159*
Kruppa, E. F., 286(243), *295*
Kryzhanoyskii, B. P., 187(83), *234*
Kuhn, M., 282(224), *295*
Kukimoto, H., *294*
Kumar, S. N., 226(122), *235*
Kunze, C., 85(107), 102, *161*
Kuppers, D., 262(94, 95, 96), *292*
Kuroda, K., 282(235, 236), *295*
Kuwano, Y., 267, *293*
Kwok, N. L., 186(94), 197(94), *234*

## L

LaBate, E. E., 84(99), *160*
Ladwig, H., 187(76), *234*
Laendle, K. W., 259(87), *292*
Lahaye, J., 174(47), *233*
Laitiner, H. A., 190(91), *234*
Lampkin, C. M., 172, *233*
Lanford, W. A., 269(139), 277(212), 279(212), 280(212), *293, 295*
Langmuir, I., 114, 163
Lapeyre, G. J., 115(214), *163*
Laponsky, A. B., 133(287), *165*
Lathlaen, R., 263(99), 264(99), *292*
Law, D., 277(195), *294*
Law, H. B., 154(331), *166*
Law, J. T., 259(81), *292*
LeComber, P. G., *296*
LeContellec, M., 272(168), 273, *294*
Lee, B. W., 115(218), *163*
Lee, C. A., 277(195), *294*
Lee, C. H., 265(112), *292*
Lehman, H. S., *291*
Leslie, J. D., 277, *294*
Lettington, A. H., 30, 36, *51*
Leverett, V., 121(241), *164*
Levine, J. D., 121, *164*
Levinstein, H. J., 266(138), 268(138), *293*
Lewicki, G., 269(152), *293*
Li, S., *296*
Ligenza, J. R., 282(222, 224), *295*
Ligeon, E., 272(168), 273(168), *294*
Limansky, I., 151(324), *166*
Lindau, I., 114(206), 115(212, 217, 220), 116, 117(212), 128(268), *163, 164*

Linger, K. R., 272(169), 274, *294*
Lis, S., 176(48a), *233*
Lisina, G. A., 92(144), *161*
Liu, Y. Z., 130(279), *165*
Logan, R. A., 277(194), *294*
Loncar, G., 276(189), *294*
Love, R. B., 14, *50*
Lubezky, I., 39, *51*
Lubinsky, A. R., 115(218), *163*
Ludeke, R., 115(213), *163*
Lugijjo, E., *292, 293*
Lunden, A. B., 169(35), 201(35), 203(35), 204 (35), *233*
Lydtin, H., 262(94, 95, 96), *292*
Lye, R. G., 64, 101(170), *159, 162*
Lytle, W. O., 187(84), *234*

## M

Ma, Y. Y., 169(14, 15), 189(14), 190(14), 193 (14), 194(14), 196(14), *232*
McAfee, K. B., 60(27, 28, 29), *159*
McAuley, J. W., 187(80), *234*
McCarroll, W. H., 85(109), 86, 87, 88, 89, 93 (111), 94, 96, 99, 103, *161, 162*
McCaldin, J. O., 117(224), 119(224), *163*
McCulluch, D., 251(52), 252(52), 254(52), 255 (52, 57), 269(52), *291*
McCulloch, D. J., 288(255), *295*
McDonie, A. F., 100(166), *162*
McFarland, M., 2(7), 3(14), *49, 50*
McGee, J. D., 147, *166*
McGill, T. C., 117(224), 119(224), *163*
McKay, K. G., 60(27, 28), *159*
Mackenzie, J. D., 260, 261, *292*
McKinley, A., 114(209), *163*
Maclean, B. N., 169, 201, *233*
McLintock, I. S., 246(20), *290*
McTaggart, F. K., 73(79), *160*, 238(4), *289*
Madden, R. P., 2(2), 5, 10(37, 38), 18(2), 48 (37), *49, 50*
Maesler, M. A., 289(265), *296*
Mahan, J. E., *296*
Major, S., 177(54a), 185(54a), 186(54a), 187 (54a), 200(54a), *233*
Makowski, M. P., 181(70), 190(70), *234*
Malhotra, L. K., 169(23), 176(53), 194(23), 226(122), *233, 235*
Malitson, I. H., *51*

Maloney, T. J., *165*
Manasevit, H. M., 129(273), *165*
Manifacier, J. C., 176(50), 179(50), 180(50), 183(50), 183(50), 186(50), 187(50), 190, 197(50), 200(50), *233*
Marcel van der Leij, 180(68), *234*
Margaritondo, G., 121(238), *164*
Maris, O., 169(21), *232*
Marjin, M., 169(21), *232*
Mark, P., 115(218), *163*, 194(92), 195(92), 199 (92), *234*
Martin, S. W., 286(246), *295*
Martinelli, R. U., 54(8), 113, 114(8, 195), 124 (185), 125, 128(8, 185), 129(185), 131, 133, 134(185), 135(296), 141(296), *158, 162, 163, 164, 165*
Martins, R., *296*
Marucchi, J., 169(21), *232*
Matsumoto, O., 287(252, 253), *295*
Matsumura, K., 264(107), *292*
Matsunami, H., 277(218), 282(218), *295*
Matsuno, Y., *296*
Matsushita, K., *296*
Matyash, A. A., 126(262), *164*
Maxwell Garnett, J. C., 71, 79(72), *160*
Mead, C. A., 117(224), 119(224), *163*, 269 (152), 272(177), *293, 294*
Mealmaker, W. E., 186(93), 197(93), 200(93), *234*
Mee, C. H. B., 81, 100, 105, 109(105), 110 (183), *160, 162*
Mehta, B. R., 176(49, 51), 185(49), 186(49), 187(87), 190(87), 198(49), 199(49), *233, 234*
Mehta, R. R., 266(127), *293*
Melnik, P. V., 121(243), *164*
Memzes, C. A., *235*
Mercandall, L. M., 277(197), 280(197), *294*
Meyer, N. I., 60(32), 75(32), *159*
Meyerson, B., 256(59, 60), *291*
Micalski, A., 265(109), *292*
Michalski, A., 242(17), 243(17), 265(125), *290, 293*
Micheletti, B., 264(110), *292*
Micheletti, F. B., 194(92), 195(92), 199(92), *234*
Millard, M., 249(37), *290*
Miller, D. J., 122(246), *164*
Miller, F. D., 91(140), *161*
Mimura, T., 278(206), *294*

Mingazin, T. A., 88(130), *161*
Mitch, A., 135(293), *165*
Mitchell, K., 54(11, 12, 13), *158*
Mitchell, K. W., 169(15), *232*
Miyake, K., 94(147), 95(147), *161*
Miyamoto, N., 277(200), *294*
Miyazawa, H., 92(141), 94(148), *161*
Mogab, C. J., 265(114), *292, 294*
Mokrushin, S. G., 169(34, 36), 201(34), *233*
Moll, J. L., 60(32), 75(32), 84(98), 128, 130 (279), *159, 160, 164, 165*
Monch, W., 114(210), *163*
Monin, J., 60(42), 63, *159*
Moon, R. L., 114(191, 194), 123(251), 126 (263, 264), 128(264), 137(308), 138(308), *162, 163, 165*
Moravec, T. J., 252(54), 253, *291*
Moore, R., 257(70), *291*
Morgulis, N. D., 101(172), *162*
Mori, H., 274(185), 275(175, 180), *294*
Mori, Y., 277(204), 278, *294*
Morris, W. F., 257(69), *291*
Morton, G. A., 143(314), 144, 154(329), *166*
Moss, T. S., 169(8), *232*
Mostovskii, A. A., 101, 105(176, 177), 106, *162*
Mukherjee, S. P., 261, *292*
Munekata, H., *294*
Murarka, S. P., 277(213), 279(213), *295*
Murasato, S., *294*
Musatov, Al. L., 126(262), *164*
Musil, J., 276(189), *294*

# N

Nagai, K., 246(19), *290*
Nagai, T., 274(184), *294*
Nakamura, T., 135(295), *165*
Nakhodkin, N. G., 121(243), *164*
Nanev, K., 87(118, 119), *161*
Nathan, R., 100, 105, 109(105), *162*
Nathanson, H. C., 136(300, 301, 302, 303), 137(302), *165*
Nazarova, R. I., *295*
Neil, K. S., 81, *160*
Nelson, H., 135(294), *165*
Nemirovsky, Y., 282(219), *295*
Nichols, D. R., 265(111), *292*
Nicks, L. J., 287(68), 288(68), *312*

Ninomiya, T., 86(113), 99, *161*
Nisbino, M., 277(217), *295*
Nishizawa, J., 277(200), *294*
Noga, K., 94(148), *161*
Norika, A. J., 269(150), *293*
Norris, P. E., 264(110), *292*
Norström, H., 248(45), 251(45, 52), 252(52), 253(45), 254(52), 255(52), 269(52), *291*
Nowick, W., 257(65), *291*
Nozaki, T., *292*

## O

Ochoa, E., 253(66), *291*
Odani, K., 278(206), *294*
Oertel, G., 94(149), 96(149), *161*
O'Hanlon, J. F., 276(187, 190), *294*
Ojha, S. M., 238(11), 241(11), 247(50), 248(42, 47, 51), 251, 252, 253, 254(49, 52), 255(49, 52), 257(51), 269(51), *290*
Okamoto, H., 275(174), *294*
O'Katoy, M. A., 187(83), *234*
Okumura, T., 289(265), *296*
Olaison, R., 248(45), 251(45), 253(45), *291*
Olive, G., 276, *294*
Oliver, M. B., 99, *162*
Olsen, G. H., 125(255), 129(272), *164, 165*
Orr, J. C., 246(20), *290*
Ortiz, A., 176(54), 177(54), 178(54), 179(54), *233*
Osantowski, J. F., 10(39), 12(39), *50*
Oswald, H. R., 238(2), 288(256), *289, 295*
Oudeacoumar, 169(22), 174(22), 180(22), 189(22), 194(22), 196(22), *233*

## P

Pachori, R. D., 214(103a), 222(103a), *235*
Paff, R. J., 99(162), *162*
Pagel, L., 256(61), *291*
Palmer, I. C., 126(261), 128(261), 132(261), 133(283), *164, 165*
Pamplin, B. R., 182(72), 183, 193(72), *234*
Pandya, D. K., 169(26, 40, 41, 42, 43), 176(26, 49, 51), 179(65), 180(65), 185(49), 186(26, 49), 187(87, 88), 189(88), 190(87, 88), 191(88), 197(26), 198(49), 199(49), 203(42), 204(43), 205(43), 207(42), 208(42), 210(42), 211(99), 212(99, 104, 105, 106), 213(104, 105, 106), 214(42, 108), 215(42), 217(40, 41, 42, 104), 218(99, 111), 219(106, 111, 112), 220(112), 223(106, 113), 224(120), 226(124, 125), 227(124), 229(41, 106, 111), 230(99, 106), *233, 234, 235*
Pankove, J. I., 133(285), *165*
Park, H., 258(74), *291*
Paul, W., 289(265), *296*
Pavaskar, N. R., *235*
Peck, R. L., 36(64), 38(64), *51*
Pelissier, A., 135(293), *165*
Pellicori, S. F., 36(66), *51*
Pennebaker, W. B., 276(190), *294*
Pensak, L., 157(334), *166*
Peria, W. T., 121(244), *164*
Perotin, M., 169(21), 176(50), 179(50), 180(50), 183(50), 186(50), 187(50), 190(50), 197(50), 200(50), *232, 233*
Perri, J. A., 258(75), *291*
Perry, A. J., 177(63), *234*
Perry, D. L., 7, *50*
Perssons, G., *294*
Persyk, D. E., 142, *165*
Pestes, M., 275(181), *294*
Peterson, P. E., 45(74), *51*
Petroff, P. M., 265(114), *292*
Petrova, R., 87(119), *161*
Pettas, H. J., 133(283), *165*
Philip, R., 71(70), 72, *160*
Philipp, H., 60(35), 101(35), 102, 103, 108, 109(35), *159, 162*
Philipp, H. R., 265(113), *292*
Phillipp, H. R., 79, 80, *160*
Phillips, J. C., 57, 58(20), 68(20), *159*
Piaget, C., 135(298), *165*
Pianetta, P., 114(206), 115(217, 220), 116(222), 128(268), *163, 164*
Piedmont, J. R., 259(82), *292*
Pliskin, W. A., 258(75), 259(80, 84), 262, 268, *291, 292, 293*
Ploog, K., *294*
Plumb, R. C., 14, *50*
Poate, J. M., 277(208), 279(208), *294*
Poferl, D. J., 246(28), *290*
Polak, A. J., 277(212), 279(212), 280(212), *295*
Polkosky, J. J., 85(108), *161*
Pommier, R., 169(21), *232*
Ponomarenko, I. N., 105(176), 106(176), *162*
Pool, P. J., 132(282), *165*

Powar, S. H., 212(100), *235*
Powell, J. R., 147(318), *166*
Praff, R. J., 269(147), *293*
Prem Nath, 169(24, 25), 172(25), 183(25), 195(24), 196(24), *233*
Prescott, C. H., 71(59), 73, *159*
Pruniaux, B., 263(104), 264(104), *292*
Pulfrey, D. L., 275(186), 276(191), *294*
Pulliam, G. R., 269(149), *293*

## Q

Querry, M. R., 46(75), *51*
Quinn, J. J., 62, 63, *159*
Quintana, G., 266(138), 268(138), *293*

## R

Racheva, T. M., 204(98), *234*
Rahuram, A. C., 275(179), *294*
Raichoudhury, P., 129(274), *165*
Ramanowski, Z., 265(109), *292*
Ramberg, E., 101(171), 102(171), *162*
Ramsey, J. B., 2(8), 15(44), 24(49), 28(8, 54), 31(8), 40(44), 48(49), *49, 50, 51*
Ramsey, J. W., 45(74), *51*
Rand, M. J., 269(139), *293*
Randhawa, H. S., 169(28), 183(28), 189(28), 225(28), *233*
Rangarajan, L. M., 85(101), *160*
Ranke, W., 115(219), *163*
Rasigne, G., 71(71), 72(71), 79(71), *160*
Ray, A. K., 282, 283, 241(229), 282(229), *295*
Readey, D. W., 258(76), 259(76), 265(76), 271(76), 272(76), *291*
Reddington, R. W., 156(332), *166*
Reddy, G. B., 226(124, 125), 227(124), *235*
Reinberg, A. R., 266, *293*
Reisman, A., 282, 241(229), 282(229), *295*
Reiss, G., 246(35), 248(35), *290*
Reisse, G., 257(65), *291*
Ribner, A., 272(167), *294*
Richard, J., 272(168), 273(168), *294*
Richter, K., 286(243), *295*
Rigo, S., *292*
Rilu, E. W., 177(58), *234*
Ritter, E., 2(6, 7), 3(13), 15(45), 41, 42, 48(76), 49(13, 70, 81), *49, 50, 51*

Rizk, A. S., 288(255), *295*
Robbie, J. C., 88, *161*
Robinson, G. Y., 277(203), 278, *294*
Rogers, R. L., III, 157(336), *166*
Rohtagi, A., 180(66), 184(66), 187(66), 190(66), *234*
Rolinski, E., 265(125), *293*
Romanowski, Z., 242(17), 243(17), *290*
Rook, H. L., 14, *50*
Rose, A., 154(326, 330, 331), *166*
Rosler, R. S., 239(12), 241(12), 242, 266, 267, 268, *290, 293*
Rougeot, H., 121(236), *164*
Rourd, P., 71(71), 72(71), 79(71), *160*
Rovner, L. H., 253(67), *291*
Rowe, J. E., 121(238), *164*
Rowe, M., 85(103), *160*
Rusek, A., 265(109), *292*
Rusek-Mazurek, A., 242(17), 243(17), *290*
Rusu, Gh. I., 187(77), *234*
Rusu, M., 187(77), *234*
Ryabova, L. A., 177(60), *234*
Ryder, E. J., 64(44), *159*

## S

Sack, R. A., 54(14), *158*
Saget, P., 135(298), *165*
Saggan, B., 72, 73(75), *160*
Sahai, R., 137(309), *165*
Sakata, T., 95, *161, 162*
Salama, C. A. T., 264(106), *292*
Salama, M., 246(30), 248, *290*
Salzberg, C. D., 26(51), 28(51), *50*
Sampogna, M., 276(187), *294*
Sandburg, W. J., 277(194), *294*
Sankaran, R., 123(251), 130(276), 139, *164, 165*
Sansalone, F. J., 246(33), 248(33), 252(33), *290*
Saperstein, W. A., 128(268), *164*
Savelli, M., 169(21), 224(116), *232, 235*
Saxena, R. R., 54(9), 136(9), 138(9), 139(9), 158, *158*
Sayama, Y., 74, *160*
Schach, M., 29(56), *51*
Schade, H., 135(294), *165*
Schafer, H., 269(153), *293*
Schagan, P., 137(306), *165*
Scharli, M., 270(156), *293*

Scheer, J. J., 54(7), 87, 89, 112, 114(204), *158, 161, 163*
Scherber, W. E., 266, 267, *293*
Schlesinger, H. I., 257(70), *291*
Schmidt, P. H., 246(24, 33), 248(33), 252(33), *290*
Schmidt, R. N., 45, *51*
Schnable, G. L., *292*
Schoder, R. B., 218(110), 219(110), *235*
Schroeder, D. K., 136(302), 137(302), *165*
Schroeder, H. H., 3(15), 42(15), *50*
Schroder, D. K., 136(303), *165*
Schürer, C., 251(62, 63), 256(61, 62), *291*
Schulte, H. J., 7, *50*
Schultz, P. C., *292*
Schwartz, B., 277(194), *294*
Schwartz, G. C., 259, 263(85), 264(85), *292*
Schweizer, H. P., 177(59, 62), *234*
Scnable, G. L., 258(77), *291*
Scott, G. D., 71(68), *160*
Scott, N. W., 2(4), *49*
Scurer, C., *291*
Sealy, B. J., 277(199), *294*
Sebenne, C. A., 114(211), *163*
Secrist, D. R., 260, 261, *292*
Sehgal, N. K., 169(40), 212(104), 213(104), 214(108), 217(40, 104), 218(111), 219 (111), 229(111), *233, 235*
Seitz, F., 64(46), *159*
Semenov, V. N., 178(64), *234*
Senitzsky, B., 60(30), *159*
Sennett, R. S., 71(68), *160*
Seraphin, B. O., 42(72), 45(72), *51*
Serreze, H. B., 176(48a), *233*
Shang, T. T., 277(209), 279(209), *294*
Shanthi, E., 169, 176(46), 183(73), 185(30, 73), 186(26, 30, 73), 187(73), 189(29, 30), 190 (30), 192(30), 196(29), 197(26, 73), 198 (30, 73), 199, 200(30, 73), 228(30), *233, 234*
Sharma, N. C., 169(40, 41), 212(104), 213 (104), 214, 217(40, 104), 218(107, 111), 219(111), 229(41, 107, 111), 230(107), *233, 235*
Sharma, P. P., *235*
Shaw, C. C., 6(26), *50*
Shaw, T. M., 238(7), *290*
Shcherbakova, V. Ya., 204(97), 217(97), *234*
Shefov, A. S., 71(61), 92(144), *160, 161*
Shen, M., 248(38, 40), 251(38, 40), *290*

Sheng, T. T., 265(114), *292*
Sheppard, C. J. R., 70(56), 126(261), 128(261) 132(261), 135(245, 292), *159, 164, 165*
Shibata, Y., *296*
Shikalgar, A. G., 212, *235*
Shiloh, M., 269(146), *293*
Shinoda, M., *292, 293*
Shinoda, Y., 277(216), 281, *295*
Shiojiri, M., 88(131), 96(131), *161*
Shiota, I., 277(200), *294*
Shockley, W., 64(44, 45), *159*
Shubin, L. D., 259(82), *292*
Shuermeyer, F. L., 259(83), *292*
Sibert, M. E., 257, *291*
Silver, M., 8(33), *50*
Silvey, G. A., 265(111), *292*
Simon, I., 5(19), *50*
Simon, R. E., 60(33), 89(139), 113(186, 188), 135(289), 137(305), 157(336), *159, 161, 162, 165, 166*
Simpson, W. I., 129(273), *165*
Singh, V. P., 224(115), *235*
Sinha, A. K., 266, 268, 276, 277(206), 279 (206), *293, 294*
Sinha, A. P. B., *235*
Skaiman, J. S., 168(4), 169(4), 176(4), 183(4), 190(4), *232*
Skarman, J. S., 169(9, 10), 194(9, 11), *232*
Skeath, P., 128(268), *164*
Skeath, P. R., 115(212), 116(212), 117(212), *163*
Skelt, E. R., 282(223), *295*
Skovlin, D. O., 203(96), *234*
Slack, L. H., 177(58), 180(66), 184(66), 187 (66), 190(66), *234*
Smid, V., *296*
Smirnov, V. A., 88(130), *161*
Smith, D. L., 118(225), 131(225), *163*
Smith, F. W., 256(59, 60), *291*
Smith, H. M., 143(314), 144, *166*
Smith, N. V., 62, *159*
Smith, R. A., 67(51), *159*
Smith, R. C., 259(82), *292*
Smith, T. E., 263(98), 264(98), 266(138), 268 (138), *292, 293*
Soboleva, N. A., 71(61), *160*
Sokoewska, A., 265(125), *293*
Sokoewski, M., 265(125), *293*
Sokolowski, M., 242(16, 17), 243(16, 17), 244 (16), 265, *290, 292*

Sokolwski, A., 242(17), 243(17), *290*
Sokowska, A., 265(109), *292*
Sokolova, T. P., 169(39, 41), 201(39), 229(41), *233*
Sommer, A. H., 54(5, 6, 15), 60(15), 61, 67(15), 71(62, 66), 73, 74(62, 66), 75(86), 76(66), 79, 80, 81, 82(66), 84, 85(66), 86, 87(117), 88, 92(66, 142, 143), 93, 94, 96, 99(162), 100(166), 101, 102, 103, 105(106), 113 (188), 114(100), 118(228), 140(66), *158, 160, 161, 162, 163*
Sonnenberg, H., 113(189), 118(233), *162, 163*
Spear, W. E., 272(164), 273, *293, 296*
Speidell, I. L., 282(237), 283(237), *295*
Spencer, E. G., 246(33), 248, 252(33), *290*
Spicer, W. E., 54(15), 55(16), 59, 60(15, 33, 38, 43), 60(15, 33, 38, 43), 61, 65(23), 67 (15), 69(16), 71(62), 74(62, 84, 85), 75 (84), 78(85), 81, 82, 90(16), 91(16, 138), 92(143), 95(16), 96, 97, 101(16), 102, 103, 105, 108, 109(179), 111(179), 112(179), 114(100, 206, 207), 115, 116(212, 222), 117, 120(235), 121(237), 123, 128(268), 130(279), 137(305), *158, 159, 160, 161, 162, 163, 164, 165*
Spitz, J., 231(126), *235*
Spitzer, W. G., 84(99), *160*
Squillante, M. R., 176(48a), *233*
Srinivasan, M., 71(64), 72(64), 75(89), 148 (320), *160, 166*
Stanko, W. S., 246(27), *290*
Steele, S. R., 258(76), 259(76), 265(76), 271 (76), 272(76), *291*
Stefan, V., 187(77), *234*
Stein, H. J., 269(144), *293*
Steinmann, W., 8(34), *50*
Stephen, R. E., *51*
Sterling, H. F., 261, 262(92), 266, *292, 293*
Sticker, R., *292*
Stirling, H. F., 271, 272(161, 162), *293*
Stocker, B. J., 128(269), 129(269), *164*
Stoll, I., 276(189), *294*
Stoneham, E. B., 277(196), *294*
Street, R. A., 264(103), *292*
Strl'chenko, S. S., 126(262), *164*
Strong, J., 5(22), *50*
Struchinskii, G. B., 101(168), *162*
Stuck, R., 176(50), 179(50), 180(50), 183(50), 186(50), 187(50), 190(50), 197(50), 200 (50), *233*

Stupp, E., 135(293), *165*
Su, C. Y., 115(212), 116(212, 222), 117(212), *163*
Sugano, T., 84(98), *160*, 277(198, 204, 205), *294, 295*
Suhrmann, R., 96(159), *162*
Sukegawa, T., 114(207), 135(295), *163, 165*
Sukhorukov, V. V., *291*
Sunami, H., 74(85), 78(85), *160*
Surridge, R. K., 135(291), *165*
Sussi, H., 288(257), *295*
Suzuki, H., 282(234), *295*
Suzuki, M., 71, 75, *159*
Suzuki, T., 73, *160*
Svechnikov, S. V., 223(114), 229(114), 230 (114), *235*
Swann, C. G., 261, *292*
Swann, R. C. G., 266, 272(161, 162), *293*
Syms, C. H. A., 130(278), *165*
Sze, S. M., 84, *160*
Szepessy, L., 176(50), 179(50), 180(50), 183 (50), 186(50), 190(50), 197(50), 200(50), *233*
Szostak, D., 118(226), 119, *163*
Szostak, D. J., 129(272), *165*

**T**

Tachiya, H., 86(113), 99(113), *161*
Taft, E., 60(35), 101(35), 102, 103, 108, 109 (35), *159*
Taft, E. A., *162*, 265(113), 267, *292*
Takahashi, T., 87, 95(156), 103, 104, 105, *161, 162*
Taketoshi, K., 86(113), 99(113), *161*
Talbot, M., 176(48a), *233*
Tanaka, T., 264(105), *292*
Tarnopol, M. S., 187(85), *234*
Tauc, J., 59(26), *159*
Tawada, Y., 275(174), *294*
Taylor, J. B., 114(198), *163*
Tech, M., 226(121), *235*
Teer, D. G., 246(30), 248, *290*
Temple, W., 248(36), 271(36), *290*
Terekhova, T. S., 169(38), 201(38), 203(38), 205(38), 210(38), *233*
Tetel'baum, D. I., *291*
Textor, M., 288(257), *295*
Thakoor, A. P., 169(28), 176(49, 51), 183(28),

185(49), 186(49), 187(87), 189(28), 190(87), 198(49), 199(49), 225(28), *233*, *234*
Theoret, A., 248(39), 251(39), *290*
Thermond, C. D., *292*
Thomas, R. N., 136(300, 301, 302, 303), 137, *165*
Thuault, C. D., 114(211), *163*
Thumwood, R. F., 87, *161*
Tietjan, J. J., 113(188), *162*
Titov, V. A., 126(262), *164*
Tjapkina, V. V., 73, *160*
Toft, A. R., 10(39), 12(39), *50*
Tolmasova, V. N., 71(61), *160*
Tousey, R., 2(1), 4(16), 5(18), 8(16), *49*, *50*
Towa, S., 248(45), 251(45), 253(45), *291*
Trawny, E. W. L., 70(56), 126(261), 128(261), 132(261), 135(292), *159*, *164*, *165*
Triolo, J. J., 28(54), 29(57, 58), 31(60), *51*
Trommer, R., *294*
Tuler, F. R., 287, *295*
Turban, G., 272(163, 165, 166), 273, 289(266, 268), *293*, *296*
Turcolte, R., 176(48a), *233*
Turkevich, J., 225(254), 227(254), *241*
Turnbull, A. A., 112(184), 118(184), 137(306), *162*, *165*
Turner, A. F., 3(15), 42(15), *50*

## U

Uebbing, J. J., 76(92), 77, 81(92), 113(187), 114(200), 118(92, 231), *160*, *162*, *163*
Unger, P., 71(65), *160*
Urbank, K., 259(86), *292*
Uritskaya, A. A., 169(34, 36), 203(34, 35), 204(35), *233*

## V

Van Asselt, R. L., 157(336), *166*
Van Bommel, A. J., 122(247), *164*
Vandenberg, M., 269(145), *293*
Vander Lindea, P. C., 187(78), *234*
Van de Ven, E. P. G. T., 263(102), 264(102), *292*
Van Hove, L., 57, *159*

Vankar, V. D., 169(24), 195(24), 196(24), *233*
Van Laar, J., 54(7), 112, 114(204, 205, 208), 129(7), *158*, *163*
Vannimenns, J., 135(298), *165*
Van Rooy, T. L., 114(208), *163*
Van Vechten, J. A., 114(196), *163*
Varma, B. P., 68(52), 70(54), 74(82), 85(104), 86(52, 112, 115), 89(52), 90(52), 92, 97, 98, 99(52), 100, 101, 103(115), 105, 109 (104, 175, 180), 110(180), 111, 145(315), 147, 148(320), *159*, *160*, *161*, *162*, *166*
Varma, R. R., 114(209), *163*
Varob'eva, O. V., 176(52), *233*
Vavilov, V., 60(37), *159*
Vecht, A., 217(109), *235*
Vedel, J., 226(113), *235*
Veeneman, D., 187(78), *234*
Venugopalan, M., 238(5), *290*
Veprek, S., 238(2), 242(18), 243(18), 245(18), 269(153), 270(156), 288(256, 257), *289*, *290*, *293*, *295*
Verma, R. L., 148(320), *166*
Vincent, C. A., 180(67), *234*
Vine, J., 136(303), *165*
Viverito, T. R., 177(58), 180(66), 184(66), 187(66), 190(66), *234*
Vogel, T. P., 91(139), *161*
Von Bosse, J., 286, *295*
Vorobeva, O. B., 101(168), *162*
Vossen, J. L., 169(6), *232*, 240(13), 241(13), *290*

## W

Wachtel, M. M., 87, 88, *161*
Wada, M., 95(156), *162*
Wagner, A. E., 187(84), *234*
Wagner, L. F., 121(237), *164*
Waho, T., 282(235, 236), *295*
Wales, R. D., 258, *291*
Walkenhorst, W., 10(36), *50*
Wallis, G., 95(150), 102, 103, *161*
Walsh, W. M., Jr., 246(24), *290*
Wartic, T., 257(70), *291*
Wasa, K., 274, *294*
Watanabe, K., 277(217), *295*
Waylonis, J., 8(32), *50*

Weast, R. C., 253(56), 254(56), *291*
Webb, A. P., 238(2), *289*
Weimer, P. K., 154(331), 156(333), *166*
Weinreich, O. A., 272(167), 277, *294*
Weissmantel, C., 256(61), 257(65), *291*
Weissmantel, Chr., 246(34, 35), 248, 251(62, 63), 256(62), *290*, *291*
Welcher, F. J., 202(95), *234*
Wells, V. A., 269(144), *293*
Whetten, N. R., 133(287), *165*
Whinney, J. R., 282(238), *295*
Whitaker, H. H., 118(228), *163*
Whitmell, D. S., 248(46), 250, 251(46), 253(46), 254, *291*
Wickstrom, R. A., 136(302), 137(302), *165*
Wilcox, D. A., 143, *166*
Will, H. A., 246(27), *290*
Williams, B. F., 113(186, 188), 118(228, 229), 120(229), 126(229), 127(229), 135(289), *162*, *163*, *165*
Williams, R. H., 114(209), *163*
Williamson, R., 248(46), 250, 251(46), 253(46), 254, *291*
Wilman, H., 73, *160*
Wilson, H., 262(94), *292*
Winslow, D. K., 269(148), *293*
Winters, H. F., 248(41), *290*
Winz, E., 288, *295*
Wolff, P. A., 60, 84(36), *159*
Wong, J., 258(79), *292*
Wonsidler, D. R., *293*
Wooten, F., 59(24), 85(110), 91(138), 109, 111, 112(179), *159*, *161*, *162*
Wouk, M. T., 269(148), *293*
Wronikowski, M., 242(17), 243(17), *290*
Wu, C. S., 169(12, 13), 194(13), 196(12), *232*

# Y

Yadav, M. S., 226(121), *235*
Yakoyama, M., 278, *294*
Yamaguehi, M., 277(216), 281, *295*
Yamashina, T., 277(217), *295*
Yang, S. J., 75(87), *160*
Yee, E. M., 135(299), *165*
Yep, T. O., 271, *293*
Yim, W. M., 269(147), *293*
Yin, S., 169(16, 18), 183(16, 18), *232*
Yokoto, K., 176(48), 224(48), *233*
Yoshihara, H., 274(185), 275(175, 180), *294*
Yoshimatsu, Y., 176(48), 224(48), *233*
Young, L., 275(186), 276(191), *294*

# Z

Zacek, F., 276(189), *294*
Zaininger, K. H., 264(110), *292*
Zalm, P., 87, 89, *161*
Zalph, B. L., 258(74), *291*
Zamerowski, T. J., 129(272), *165*
Zeitsov, N. S., 71(63), *160*
Zellmer, D. L., 190(91), *234*
Zemel, J. N., 218(110), 219(110), *235*
Zelmer, D. L., 190(90), *234*
Zhuralev, N. N., 88(130), *161*
Zingaro, R. A., 203(96), *234*
Zintl, E., 88(132), *161*
Zorin, E. I., *291*
Zscheile, H., 251(62), 256(62), *291*
Zworykin, V. K., 101(171), 102(171), 154(328, 329), *162*, *166*

# Subject Index

## A

Absorption coefficient
  for negative electron affinity materials, 70
  in optical absorption, 55
Adherent silver mirrors, with protective coatings, 36–39, *see also* Silver; Silver mirrors
Affinity materials, negative electron, *see* Negative electron affinity materials
Ag–O–Cs photocathode, *see* Silver–oxygen–cesium photocathode
Alkali antimonides, 84–112
  band structures, 105
  bi- and trialkali, 86–87
  cathode preparation in, 85–87
  chemical composition and crystal structure, 87–90
  electrical properties, 94–101
  high-temperature instability, 91
  monoalkali, 85–86
  optical properties, 101–112
  photocathode superficial oxidation and, 92
  photoelectron spectroscopy, 108
  photoemission properties, 90–94
  as secondary emitters, 101
Alloying effects, in spray pyrolytic process, 181–187
Aluminum
  evaporated, *see* Evaporated aluminum
  nonabsorbing surface films on, 24
  reflectance, 9, 17–19
  reflectance as function of incidence angle for, 21
  as reflective coating, 8–10
Aluminum evaporator
  filaments and baffle plates for, 12
  film deposition in, 12–14
  for large mirrors, 10–11
Aluminum film deposition, in evaporator, 12–14
Aluminum films, purity, 14

Aluminum mirror coatings, optical constants and reflectance, 17–19
Aluminum mirrors, *see also* Evaporated aluminum mirrors
  with high reflectance in 8–12 $\mu$m region for normal-to-high angles of incidence, 35–36
  larger than 50 cm, 11
Aluminum nitride films, plasma chemical vapor deposition for, 269–270
Aluminum oxide, 10
Aluminum oxide coatings
  deposition, 28
  reduced reflectance of at higher incidence angles, 34
Aluminum oxide films
  as overcoatings, 25
  plasma chemical vapor deposition for, 264–265
Antimonides, alkali, *see* Alkali antimonides
Arsenic phosphorus alloys, smaller band gap in, 122–123
Auger electron spectroscopy, of plasma-grown oxides, 277
Avalanche breakdown, band shape and, 60

## B

Band-bending
  heterojunction model, 119–120
  surface escape probability and, 127
  surface dipole model, 120
  vs. lattice constant for III–V compounds, 131
Band gap energy
  negative vs. positive electron affinity in, 113
  vs. surface escape probability, 127
Band shape, effective mass ratio and, 60
Beckman spectrophotometer, 7
Bialkali antimonides

## SUBJECT INDEX

as mutlialkali photocathode, 86
optical properties, 103
electrical properties, 97
Bias-assisted photocathode, energy band diagram for, 139
Bismuth–silver–oxygen–cesium photocathode, 84
Bloch wave functions, 59
Boron nitride films, plasma chemical vapor deposition for, 271–272
Boron trichloride, dissociation, 257

## C

Cadmium selenide films
solution growth process for, 205
thickness vs. time for, 207
Cadmium sulfide films
doping with indium and gallium, 181
properties, 193–194
spray deposited, 169, 172–173
spray pyrolysis, 174–176, 180
transmission spectra, 194
zinc addition to, 196
Carbon dioxide laser emission line for, 32–34
Carbon films, dc and rf systems for deposition of, 250
Carrier transport, of negative electron affinity materials, 123–125
Case hardening, by plasma carburizing, 285
Cesiated silver films, photoemission from, 81
Cesium, role of in multialkali photocathodes, 97–98
Cesium activation, from silver oxide films, 74–75
Cesium antimonide
absorption constant, 102
crystal structure, 88
electrical behavior, 96
energy distribution spectra, 108
Cesium–bismuth compounds, energy distribution spectra, 108
Cesium monolayer coverage, Fermi level position and, 118
Cesium/oxygen activation, electron affinity lowering by, 114–123
Chalcogenide films, *see also* Chemical solution deposition; Solution-grown films; Spray pyrolysis
deposition conditions and, 217
optical and electrical properties, 218–222
photoconductivity characteristics, 230
polymorphic structure, 217
pyrolytic reaction and, 168–169, 201, 206
Chemical deposition process, *see also* Chemical solution deposition
selective coatings from, 226–227
transparent conducting coatings and, 228
Chemical solution deposition, 167–232, *see also* Solution growth process
complexing agents in, 203, 208
electrical and optical properties of films in, 192–201
impurity and dopant effects in, 211–212
large-area applications of, 223–230
for multicomponent films, 212–213
oxide films and, 213–214
properties of commonly used transparent conducting oxides and, 200
solution growth process and, 201–211
structural properties of films in, 187–191
substrate in, 210–211
transport properties in, 217–223
Chemical vapor deposition
in negative electron affinity photocathode fabrication, 129
plasma, *see* Plasma chemical vapor deposition
pyrolytic, 238, *see also* Spray pyrolysis
Chromium steel, plasma nitriding of, 288
Copper mirror coatings, 17
Copyrolysis, in spray pyrolytic process, 183–187
CVD, *see* Chemical vapor deposition

## D

Dark current emission, of negative electron affinity materials, 135, 139
Dark mirror coatings, low visible reflectance, 45
Dielectric constant, and joint density of states, 57–58
Dielectric films, evaporated, *see* Evaporated dielectric films

## SUBJECT INDEX

Dopant effects, in chemical solution deposition, 211–212
Doping, multicomponent, in spray pyrolysis, 181–187
Dynode materials, for photomultipliers, 141–144
Dynode systems, focusing designs for, 142

## E

EELA, see Electron energy loss analysis
Electrical properties
  of alkali antimonides, 94–101
  of spray pyrolysis films, 192–201
Electron affinity, see also Negative electron affinity
  lowering of by Cs/O activation, 114–123
  negative vs. positive, 112–113
  reduction of by p-type doping, 70
  surface barrier and, 66
Electron beam evaporation, in $SiO_2$ and $Al_2O_3$ coating deposition, 28
Electron–electron scattering
  energy loss in, 59–60, 110
  secondary carriers in, 60
Electron energy loss analysis, in plasma chemical vapor deposition, 257
Electron escape, surface barrier and, 66–67
Electron gun, in aluminum film deposition, 2, 14, 28
Electron–hole pairs, generation of, 60
Electron–phonon interaction, pair production and, 60
Electron transport, in photoemission, 59–66
Equivalent screen background input, for image intensifiers, 150–151
ESBI, see Equivalent screen background input
Evaporated aluminum, reflectance as function of incidence angle for, 21
Evaporated aluminum films
  purity, 14
  silicon monoxide effects on UV irradiation in, 27–28
Evaporated aluminum mirrors
  germanium and silicon monoxide coatings for, 42–43
  hafnium dioxide and yttria coatings for, 35
  infrared reflectance data for, 29–30
  magnesium fluoride coating for, 48–49
  with overcoatings, 25–32
Evaporated dielectric films, water absorption in, 46–49
Evaporated rhodium films, reflectance-enhancing coatings for, 39–42
Evaporated rhodium mirrors, overcoated vs. bare film, 41
Evaporated silver, reflectance of vs. incidence angle for, 20, see also Silver; Silver mirror coatings
Evaporated silver films
  high reflectance and visibility, 24
  minimum polarization introduced by, 24
Evaporator, aluminum, see Aluminum evaporator

## F

Fermi age theory, 64
Fermi level, in electron escape, 67
Fermi level pinning position
  during submonolayer cesium coverage, 118
  surface states and, 115–117
Film composition, in spray pyrolytic process, 179–181
Films
  solution grown, see Solution-grown films
  thin, see Thin films
Film structural properties, in chemical solution deposition, 187–191
Four-mirror polarizer, 33–34
Free-electron metal, plasma scattering in, 62–63
Front-surface mirrors, see also Mirror coatings; Reflection-type mirrors
  reflectance-enhancing coatings for, 39–42
  for UV–IR applications, 1–49

## G

Gallium arsenide
  forward-biased p–n junction, 135
  heteropolar nature, 128

lattice mismatch with GaP, 103–131
plasma oxidation, 277–282
Gallium arsenide films, Ga:As ratio in, 280
Gallium phosphide, lattice mismatch with GaAs, 130–131
Gallium phosphide dynodes, photomultiplier applications, 141–144
Gating-type image tube, 149–150
Germanium
 band structure, 57
 joint density of state spectra for, 58
Germanium blank, transmittance, 251–252
Germanium coatings, for metal-dielectric mirrors, 42–43
Glow-discharge nitriding, 286–289
Gold, evaporation, 15
Gold mirror coatings
 reflectance, 17, 19
 optical constants for, 19
Goniometer–reflectometer, in reflectance measurements, 4

## H

Hafnium dioxide, for aluminum mirror coatings, 35
Heat mirrors, chemical deposition process for, 228
High-sensitivity photocathodes, chemical analysis, 99
High-vacuum evaporator, in reflecting coating deposition, 8
Hydrocarbon gas, in plasma carburizing, 284–285
Hydrogen sulfide
 on silver film, 37–38
 in spray pyrolysis, 202

## I

Image converters, 150
Image intensifiers
 characteristics, 150–152
 electrostatic focusing, 147–148
 magnetic focusing, 146–147
 mechanical transfer function for, 151–152
 microchannel plate type, 148–149
 photoemissive materials for, 145–149
 proximity focusing, 145–146
Image isocon, 156–157
Image orthicon, 156
Image tubes
 gating tubes as, 149–150
 image converters and, 150
 image intensifiers and, 150–152
 photoemissive materials for, 145–157
 signal-generating tubes and, 152–157
Incidence angle
 polarization values for, 23
 reflectance as function of, 18–23
Inconel, plasma nitriding of, 288
Indium–gallium–arsenic alloys, smaller band gap of, 122–123
Indium–gallium–arsenic–phosphorus quaternary systems, 123
Indium–GaAs alloy cathodes, thermalized electron distribution in, 126
Infrared reflectance data, for SiO-coated aluminum mirrors, 29–30
Inorganic film deposition, 162–232, *see also* Chemical solution deposition; Spray pyrolysis
Ion-beam deposition, in plasma chemical vapor deposition, 246–248
Ion plating, in plasma chemical vapor deposition, 248–249
IR, *see* Infrared reflectance data

## J

Joint density of states, in optical absorption, 57

## L

Large-area thin-film devices, chemical solution deposition processes in, 223–230
Lead selenide films, optical and electrical properties, 219–221
LEED (low-energy electron deposition), in reconstruction of surface atoms, 115

Liquid-phase epitaxy, in negative electron affinity photocathode fabrication, 129

## M

Magnesium fluoride coatings, 48–49
Maxwell–Garnett absorption
  in photocathode materials, 71–72, 79
  for thin silver films, 71–72
MCP, *see* Microchannel plate image intensifier
Mechanical transfer function, for image converters, 151
Metal–dielectric mirrors, germanium coatings for, 42–43
Metallic front surface mirrors, reflectance of without overcoatings, 16–21
Metals, optical properties, 18
Microchannel plate, image intensifier, 148–149
Mirror coatings
  aluminum, 17–19
  copper, 17
  gold, 17
  metal–dielectric, 42–45
  optical constants and calculated reflectance, 19
  platinum, 17
  with protective layers and reflectance-enhancing surface films, 22–42
  reflectance measurements for, 3–8, 19
  rhodium, 15–19, 39–42
  silver, 17–19
Modulation transfer function, calculation of, 125
Molybdenum, nitriding rate for, 288
Monoalkali antimonides, 85–86
MOSFET (metal–oxide–semiconductor field-effect transistors), 237
  GaAs plasma oxidation for, 277
MTF, *see* Mechanical transfer function; Modulation transfer functions
Multialkali antimonides
  mean free path for electron–electron scattering in, 112
  optical properties, 107
Multialkali photocathodes
  band-bending at surface of, 100

cesium role in, 97–99
sensitivity, 92

## N

NEA materials, *see* Negative electron affinity materials
Negative electron affinity
  requirements for, 114–115
  on semiconductor surfaces, 114
  silicon activation to, 121
Negative electron affinity cathodes, quantum efficiency, 157–158
Negative electron affinity GaP dynodes, in photomultiplier tubes, 135
Negative electron affinity materials, 112–140
  absorption coefficient for, 70
  activation techniques for, 128–129
  carrier transport and, 123–125
  cold cathodes and, 134–135
  dark current, 135
  diffusion length and doping for, 125–126
  electron equilibrium in, 65
  escape depth vs. diffusion depth for, 133
  photoemission for, 62
  in photoemitters beyond 1.1 $\mu$m, 136–140
  secondary emission, 133–134
  stability, 135–136
  surface escape probability for, 126–128
Negative electron affinity photocathodes, fabrication, 129–133
Niobium, nitriding rate for, 288
Nitriding, glow-discharge, 286–289
Nitrogen gas, role of in nitriding process, 286

## O

Opaque film, true quantum yield, 63
Optical absorption
  joint density of states in, 57
  absorption coefficient in, 55
  in photoemission, 55–59
Optical properties, of alkali–antimonide photocathodes, 101–112
Overcoatings
  aluminum oxide as, 25

# SUBJECT INDEX

evaporated aluminum mirrors with, 25–32
silicon monoxide for, 25–26
Oxide films, solution growth technique for, 213–214

## P

Pair production
　in electron transport, 60–61
　mean free path for, 60, 110
Pair-production collision, minimum energy loss in, 61
Parallel-plate-type reactor, 267–268
Perkin-Elmer spectrophotometer, 7
Phosphorus nitride films, plasma chemical vapor deposition for, 270–271
Photocathode photoemission, 90–94, see also Photoemission
　crystal structure in, 93–94
Photocathodes
　alkali antimonide, see Alkali antimonides
　high-sensitivity, 99
　multialkali, see Multialkali photocathodes
　fibromultiplier use and, 141
　silver–oxygen–cesium, see Silver–oxygen–cesium photocathodes
　thermionic emission from, 100
Photocathode spectral response, incident light and, 92–93
Photoelectrochemical solar cells, chemical deposition process for, 226
Photoelectrons, bulk generation, 55
Photoemission
　electron–electron scattering in, 60–62
　electron escape from surface in, 66–70
　electron transport in, 59–66
　mechanism of, 54–70
　negative electron affinity materials in, 62
　optical absorption in, 55–59
　as "surface" phenomenon, 54
Photoemission quantum efficiency, 64, 69, 112
Photoemissive materials, 53–158
　Ag–O–Cs(S–1) photocathode, 70–84
　alkali antimonides, 84–112
　applications, 140–157
　for image tubes, 145–157
　photoemission mechanisms and, 54–70
　in photomultipliers, 140–144

quantum efficiency improvement in, 112
resistance peak in, 95
spectral response, 80
Photoemissive properties, of alkali antimonides, 90–94
Photoemitters
　alkali antimonide type, 59
　quantum efficiency, 64, 69, 112
　transferred-electron type, 136–137
Photogenerated carriers, transport of, 123–125
Photoluminescence, in silicon–carbon–hydrogen films, 274–275
Photomultipliers
　dynode materials for, 141–144
　intended use for, 141
　photoemissive materials in, 140–144
Photomultiplier tubes
　with glass-bonded GaAs cathode, 132
　negative electron affinity–GaP dynodes in, 135
Photodetectors, chemical deposition process for, 228–230
Plasma carburizing, 283–285
Plasma chemical transport, film deposition by, 243
Plasma chemical vapor deposition, see also Chemical solution deposition
　for aluminum nitride films, 269–270
　of aluminum oxide films, 264–265
　for boron nitride films, 271–272
　of boron on tungsten, 257
　carbonaceous films in, 251–252
　carbon coatings in, 246
　deposition rate in, 255
　electron energy loss analysis in, 257
　film preparation and properties in, 246–275
　glow-discharge nitriding and, 286–289
　ion-beam film deposition in, 246–247
　ionization degree in, 256
　on oxide films, 258–265
　for phosphorus nitride films, 270–271
　plasma carburizing and, 283–285
　plasma oxidation and, 275–283
　of silicon carbide films, 272–275
　on silicon dioxide films, 258–265
　for silicon nitride films, 265–272
　techniques and systems in, 238–245
　TEOS vapors in, 260–261

for titanium carbide films, 275
types of, 239–241
Plasma deposition reactor potentials, 241
Plasma deposition system, Roster–Engle design, 242
Plasma etching, 237
Plasma-grown gallium arsenide films, 277–282
 arsenic deficiency in, 280
 composition and electrical characteristics, 279–280
Plasma-grown silicon oxide films, 282–283
Plasma oxidation
 of GaAs devices, 277–282
 in magnetic field, 278
 in oxide-layer production on GaAs, 237
 rf oxidation and, 282–283
 for silicon dioxide films, 282
 for thin insulating films, 275–277
Plasmon scattering, in free-electron metal, 62–63
Platinum, reflectance, 15
Platinum mirror coatings, 17
Polarizer
 four-mirror, 33–34
 reflection type, 33–34
 three-mirror, 33
Polycrystalline silica, plasma-chemical transport deposition of, 243
Positive electron affinity, vs. negative electron affinity, in band-gap energy, 113
Potassium antimonide
 crystal structure, 88
 optical absorption, 103
Potassium cesium antimonide, superficial oxidation, 92
Potential, distribution of in space, in tunnel, and flat-bed reactors, 240
p-type material, in electron affinity reduction, 70
Pulsed-plasma deposition system, 244
Pyrolytic chemical vapor deposition, 238, *see also* Spray pyrolysis
 for silicon carbide films, 272

## Q

Quantum efficiency
 of Ag–O–Cs photocathodes, 84

 of negative electron affinity cathodes, 157–158
 in photoemission, 64, 69, 112
Quantum yield, for copper sample, 63–64
Quartz crystal oscillator, 73

## R

Reactive evaporation process, 26
Red phosphorus, chemical transport of, 245
Reflectance
 angle of incidence in, 18–23
 defined, 3
 optical constants and, 19
 parallel vs. average components of, 23
Reflectance-enhancing surface films, 22–42
Reflectance measurements
 instruments for, 4–5
 for mirror coatings, 3–8
Reflectance standard, 4
Reflecting coatings, deposition of, 8–16
Reflection mode secondary emission, 133
Reflection-type mirrors, metal–dielectric mirrors as, 42–45
Reflection-type polarizers, for 10.6-$\mu$m $CO_2$ laser emission line, 32–34
Reflectivity
 defined, 3
 in optical absorption, 55
Reflectometer
 in reflectance measurements, 4–5
 Strong-type, 5–6
Refraction index, 18
Rhodium, nonabsorbing surface films and, 24–25
Rhodium mirror coatings
 optical constants for, 19
 reflectance, 15–19, 39–42
Roster–Engle plasma deposition system, 242
Rubidium antimonide
 optical absorption coefficient, 105
 structure, 88

## S

Santa Barbara Research Center, 201

# SUBJECT INDEX

Sapphire, epitaxial layer growth on, 130
Secondary-electron emission
  in alkali antimonide photocathodes, 101
  for negative electron affinity materials, 132–134
Signal generating tubes, 152–157
Silicon carbide, optical band gap for, 272
Silicon carbide films, pyrolytic decomposition, 272
Silicon–carbon–hydrogen films, photoluminescence in, 274
Silicon dioxide coatings, 225
Silicon dioxide films, in thermally activated chemical vapor deposition, 258–268
Silicon monoxide, evaporation, 25–26
Silicon monoxide coatings
  adsorptance of in UV region, 26
  deposition, 28
  IR reflectance data for, 29–30
  for metal–dielectric mirrors, 44–45
  oxidation state increase in, 28
  reflectance vs. wavelength data for, 32
Silicon monoxide films
  as overcoatings, 25–26
  reactive evaporation, 26
Silicon nitride deposition system, 243
Silicon nitride films
  hydrogen content, 269
  nitrogen as reducing agent in, 269
  parallel-plate-type reactor and, 267–268
  as passivating layers, 266
  plasma chemical vapor deposition for, 265–272
  pyrolytic dissociation and, 265–266
Silicon oxide coatings, in silver film delamination, 38
Silicon oxides, as protective layer for mirror coatings, 2, see also Silicon monoxide
Silicon slice, transmittance curves for, 253
Silver
  evaporation, 15–16
  incidence angle, 20
  nonabsorbing surface films and, 23
  photoemission from, 81
  reflectance, 20–23
  rhodium and, 22
Silver films
  absorption behavior, 71
  evaporated, 24
  growth, 71
  hydrogen sulfide and, 37
  Maxwell–Garnett absorption for, 71–72
  for silver–oxygen–cesium photocathode, 71
Silver mirror coatings, optical constants and reflectance, 17–19
Silver mirrors
  adherent, 36–39
  IR reflectance and polarization characteristics for, 30–31
Silver oxide
  formation, 73
  as insulator, 73
  silver layer deposition on, 73–74
Silver oxide film, cesium-vapor treatment of, 74
Silver–oxygen cathodes, high-sensitivity, 72
Silver–oxygen–cesium, Maxwell–Garnett absorption for, 79
Silver–oxygen–cesium photocathode, 70–84
  discovery, 54, 70
  electrical properties, 76
  light absorption, 78
  optical properties, 78–79
  photoemission from, 80–84
  preparation, 71–75
  properties, 76–84
  quantum efficiency, 84
  thermionic emission, 76–77
Sodium antimonide
  absorption coefficient for, 104
  band gap and electron affinity for, 96
  crystal structure, 88
Sodium–potassium antimonide structure, 89
Sodium selenosulfate, synthesis, 203
Solar cells, photoelectrochemical, 226
Solar films, spray-deposited, 224–225
Solution-grown chalcogenide films, see also Solution-grown films
  optical properties, 218–219
  photoconducting characteristics, 230
Solution-grown films, see also Solution growth process
  electron transport parameters, 222
  epitaxial growth, 214
  optical and electrical properties, 218–222
  photoconductivity, 223
  scanning electron microscope studies, 216
  structure, 214–217

transmission electron microscopy, 214
transport properties, 217–223
Solution growth chemistry, 202–206
  hydrogen sulfide in, 202
Solution growth process
  complexing agents in, 203, 208
  film growth and deposition parameters in, 206–211
  ions and, 203, 208
  for multicomponent films, 212–213
  in organic film deposition, 201–211
  oxide films in, 213
  ph value in, 209
Spectrophotometer, schematic diagram, 7
Spinel, epitaxial layer growth on, 130
Spray pyrolysis, *see also* Spray pyrolytic process
  atomization process in, 171
  block diagram, 170
  cadmium sulfide in, 169, 172–173, 176
  carbide and nitride films in, 177
  chemical aspects, 174–177
  defined, 168–169
  electrical and optical properties imparted by, 192–201
  growth kinetics in, 172–174
  oxide films in, 176–177
  spray setup for, 169–171
  sulfides and selenides for, 174–176
  for thin-film solar cells, 224–225
Spray pyrolytic process, 169–177
  characteristics, 178–181
  copyrolysis in, 183–187
  film composition in, 179–181
  growth rate in, 178
  multicomponent doping and alloying effects in, 181–187
  physical parameters of oxide films in, 184–185
  spraying solutions used in, 186
  stannite, adamantine, and chalcopyrite compounds in, 182
  substrate effects in, 178–179
Steel
  glow-discharge nitriding of, 286–289
  plasma carburizing of, 283–285
Strong-type reflectometer, 5–6
Surface barrier, electron affinity and, 66

Surface escape probability
  band-bending and, 127
  from negative electron affinity materials, 106–128
Surface mirrors, vacuum systems for film deposition on, 2, *see also* Front surface mirrors; Mirror coatings

## T

Television camera tubes, 152–157
Tellurium ions, in spray pyrolysis, 203
TEOS (tetraethoxysilane) vapors, in chemical vapor deposition, 260–261
Thin Films
  growth of in spray pyrolytic process, 206
  plasma chemical vapor deposition, 237–287
Thin-film solar cells, spray-deposited
Thin silver films, *see also* Silver films
  light transmission of, 72
  optical properties, 71
Three-mirror polarizer, 33–34
Titanium, nitriding rate for, 288
Titanium carbide films
  plasma CVD for, 275
  properties, 193
  in spray pyrolytic process, 190
TM, *see* Transmission mode
Tokamaks, fusion, 237
Transferred-electron-type photoemitter, 136–137, *see also* Photoemission; Photoemitters
Transmission electron microscopy, 214
Transmission mode, secondary emission in, 103
Transmission mode photocathode quantum efficiency, calculation of, 124–125
Transmission mode silicon photocathode, 132
Transparent conducting coatings, 228
Transparent conducting oxides, properties, 200
Trialkali antimonides, 186–187
Trialkali cathodes, electrical properties, 97
Tunnel reactor potentials, 241

## U

Ultraviolet irradiation, silicon monoxide coating effects on, 27–28
Ultraviolet region, silicon monoxide coating absorptance in, 26
Umklapp process, in electron transfer, 60
Uncoated silver mirrors, IR reflectance and polarization data for, 330–331
Ural Polytechnic Institute, 201
UV, *see* Ultraviolet region

## V

Vacuum ultraviolet spectroscopy, coatings for, 2
Vapor deposition, *see* Chemical vapor deposition; Plasma chemical vapor deposition

## W

Water absorption, in evaporated dielectric films, 46–49

## Y

Yttria-coated aluminum mirrors, reflectance, 35

## Z

Zinc oxide films, spray pyrolysis, 190
Zinc selenide films, properties, 194
Zinc sulfide coatings, 48
Zinc sulfide films, electron transparent parameters, 197